Multiple Attribute Decision Making

METHODS AND APPLICATIONS

Multiple Attribute Decision Making

METHODS AND APPLICATIONS

Gwo-Hshiung Tzeng
Jih-Jeng Huang

CRC Press
Taylor & Francis Group
Boca Raton London New York

CRC Press is an imprint of the
Taylor & Francis Group, an **informa** business
A CHAPMAN & HALL BOOK

First published in paperback 2024

First published 2011
2385 NW Executive Center Drive, Suite 320, Boca Raton FL 33431

and by CRC Press
4 Park Square, Milton Park, Abingdon, Oxon, OX14 4RN

CRC Press is an imprint of Taylor & Francis Group, LLC

Publisher's Note
The publisher has gone to great lengths to ensure the quality of this reprint but points out that some imperfections in the original copies may be apparent.

Library of Congress Cataloging-in-Publication Data

Tzeng, G. H. (Gwo Hshiung)
Multiple attribute decision making : methods and appliations / Gwo-Hshiung Tzeng, Jih-Jeng Huang.
p. cm.
Includes bibliographical references and index.
ISBN 978-1-4398-6157-8 (hardcover : alk. paper)
1. Multiple criteria decision making. 2. Fuzzy sets. I. Huang, Jih-Jeng. II. Title.

T57.95.T94 2011
658.4'03--dc23 2011022162

ISBN: 978-1-4398-6157-8 (hbk)
ISBN: 978-1-032-92139-6 (pbk)
ISBN: 978-0-429-11070-2 (ebk)

DOI: 10.1201/b11032

Visit the Taylor & Francis Web site at
http://www.taylorandfrancis.com

and the CRC Press Web site at
http://www.crcpress.com

Contents

Part II Applications of MADM

Preface

As Herbert Simon pointed out, "most people are only partly rational, and are in fact emotional or irrational in the remaining part of their actions." These so-called irrational behaviors sometimes result from the restrictions of humans to evaluate trade-off alternatives when there are more than three criteria. However, the pursuit of a method to make an ideal decision has never been given up by scholars and practitioners. This purpose was the motivation behind the emergence of multiple criteria decision making (MCDM).

MCDM is a discipline aimed at supporting decision makers who are faced with numerous and conflicting alternatives to make an optimal decision. To achieve this purpose, two critical questions should be unlocked: preference structure and weights. Therefore, for the past 50 years, scholars have proposed various functions to try to represent the true preference structure of a decision maker and the correct weights of criteria, and these efforts will certainly be ongoing for the next 50 years.

This book is divided into two parts: methodologies and applications. In the methodological part, we focus on explaining the theory of each method. Then, we give a numerical example which can be calculated without a computer so that readers can truly understand the procedures of MCDM methods. Another central concern in this paper is the integration of the theory and practice of MCDM. Therefore, in the applications section we present various methods used in dealing with realistic MCDM problems.

We believe that the book can be of value to the following groups with respect to their own objectives:

- Undergraduate and graduate students who wish to extend their knowledge of the methods of MCDM or publish papers in the journals of the OR/MS field.
- Practitioners who seek to make an ideal decision by using MCDM methods.

Finally, we hope all our readers are satisfied with this book and reap great rewards from it.

Gwo-Hshiung Tzeng and Jih-Jeng Huang

Authors

Gwo-Hshiung Tzeng was born in 1943 in Taiwan. In 1967 he received a Bachelor of Business Management from the Tatung Institute of Technology (now Tatung University), Taiwan; in 1971, he received a Master of Urban Planning from Chung Hsing University (now Taipei University), Taiwan; and in 1977, he received a PhD in Management Science from Osaka University, Osaka, Japan.

He was an associate professor at Chiao Tung University, Taiwan, from 1977 to 1981, a research associate at Argonne National Laboratory from July 1981 to January 1982, a visiting professor in the Department of Civil Engineering at the University of Maryland, College Park, from August 1989 to August 1990, a visiting professor in the Department of Engineering-Economic Systems, Energy Modeling Forum, at Stanford University, from August 1997 to August 1998, a professor at Chaio Tung University from 1981 to 2003, and a chair professor at National Chiao Tung University. His current research interests include statistics, multivariate analysis, network routing and scheduling, multiple criteria decision making, fuzzy theory, and hierarchical structure analysis for application to technology management, energy, the environment, transportation systems, transportation investment, logistics, location, urban planning, tourism, technology management, electronic commerce, global supply chains, etc. In March 2009, his paper "Compromise solution by MCDM methods: A comparative analysis of VIKOR and TOPSIS," as published in the *European Journal of Operational Research* (July 16, 156(2), 445–455, 2004) was identified by Thomson Reuters' *Essential Science Indicators*^{SM} as one of the most cited papers in the field of economics (ESI Highly Cited Paper).

He has received the MCDM Edgeworth-Pareto Award from the International Society on Multiple Criteria Decision Making (June 2009), the world Pinnacle of Achievement Award 2005, the national distinguished chair professor and award (highest honor offered) from the Ministry of Education Affairs of Taiwan, distinguished research award three times, and distinguished research fellow (highest honor offered) twice from the National Science Council of Taiwan. He has been an IEEE Fellow since September 30, 2002. He organized a Taiwan affiliate chapter of the International Association of Energy Economics in 1984 and was the chairman of the Tenth International Conference on Multiple Criteria Decision Making, July 19–24, 1992, in Taipei; the cochairman of the 36th International Conference on Computers and Industrial Engineering, June 20–23, 2006, Taipei, Taiwan; and the chairman of the International Summer School on Multiple Criteria Decision Making 2006, July 2–14, Kainan University, Taiwan. He is a member of IEEE, IAEE, ISMCDM, World Transport, the Operations Research Society of Japan, the Society of Instrument and Control Engineers Society of Japan, the City Planning Institute of Japan, the Behaviormetric Society of Japan, and the Japan Society for Fuzzy Theory and Systems, and participates in many societies in Taiwan. He is editor-in-chief of the *International Journal of Information Systems for Logistics and Management*, and so on.

Jih-Jeng Huang was born in 1977 in Taiwan. He is currently assistant professor of Computer Science and Information Management at Soochow University (Taiwan) and teaches research method, multivariate analysis, capital asset and pricing models, etc. He received his PhD in Information Management from the National Taiwan University. His current research interests include multiple criteria decision making, knowledge management, behavioral economics and finance, and data analysis. He has published on these interests widely in journals and conference proceedings.

1 Introduction

1.1 PROFILE OF MULTIPLE CRITERION DECISION MAKING

Decision-making processes involve a series of steps: identifying the problems, constructing the preferences, evaluating the alternatives, and determining the best alternatives (Simon 1977; Keendy and Raiffa 1993; Kleindorfer, Kunreuther, and Schoemaker 1993). Generally speaking, three kinds of formal analysis can be employed to solve decision-making problems (Bell, Raiffa, and Tversky 1988; Kleindorfer et al. 1993):

Descriptive analysis is concerned with the problems that decision makers (DM) actually solve.
Prescriptive analysis considers the methods that DM ought to use to improve their decisions.
Normative analysis focuses on the problems that DM should ideally address.

In this book, we limit our topics to normative analysis and prescriptive analysis, since descriptive analysis (or so-called behavior decision research) is especially addressed in the fields of psychology, marketing, and consumer research (Kahneman and Tversky 2000). Meanwhile, normative analysis and prescriptive analysis are dealt with in the fields of decision science, economics, and operations research (OR).

Decision making is extremely intuitive when considering single criterion problems, since we only need to choose the alternative with the highest preference rating. However, when DM evaluate alternatives with multiple criteria, many problems, such as weights of criteria, preference dependence, and conflicts among criteria, seem to complicate the problems and need to be overcome by more sophisticated methods.

In order to deal with multiple criteria decision-making (MCDM) problems, the first step is to figure out how many attributes or criteria exist in the problem and how to grasp the way of the problems (i.e., identifying the problems). Next, we need to collect the appropriate data or information in which the preferences of DM can be correctly reflected upon and considered (i.e., constructing the preferences). Further work builds a set of possible alternatives or strategies in order to guarantee that the goal will be reached (i.e., evaluating the alternatives). Through these efforts, the next step is to select an appropriate method to help us to evaluate and outrank or improve the possible alternatives or strategies (i.e., finding and determining the best alternative).

To facilitate systematic research in the field of MCDM, Hwang and Yoon (1981) suggested that MCDM problems can be classified into two main categories: multiple attribute decision making (MADM) and multiple objective decision making (MODM), based on the different purposes and different data types. The former is applied in the evaluation facet, which is usually associated with a limited number of predetermined alternatives and discrete preference ratings. The latter is especially suitable for the design/planning facet, which aims to achieve the optimal or aspired goals by considering the various interactions within the given constraints. However, conventional MCDM only considers the crisp decision problems and lacks a general paradigm for specific real-world problems, such as group decisions and uncertain preferences.

Most MCDM problems in the real world, therefore, should naturally be regarded as fuzzy MCDM problems (Zadeh 1965; Bellman and Zadeh 1970), which consist of goals, aspects (or dimensions), attributes (or criteria), and possible alternatives (or strategies). More specifically, we can classify MCDM problems in the fuzzy environment into two categories: fuzzy multiple attribute decision making (FMADM) and fuzzy multiple objective decision making (FMODM), based on the concepts of MADM and MODM.

The profile of MCDM is illustrated in Figure 1.1.

1.2 HISTORICAL DEVELOPMENT OF MULTIPLE ATTRIBUTE DECISION MAKING

The historical origins of MADM can be traced back to correspondence between Nicolas Bernoulli (1687–1759) and Pierre Rémond de Montmort (1678–1719), discussing the St. Petersburg paradox. The St. Petersburg game denotes the problem:

> "A game is played by flipping a fair coin until it comes up tails, and the total number of flips, n, determines the prize, which equals $\$2 \times n$. If the coin comes up heads the first time, it is flipped again, and so on. The problem arises: how much are you willing to pay for this game? (Bernstein 1996)"

According to the expected value theory, it can be calculated that $EV = \sum_{n=1}^{\infty}(1/2)^n \times 2^n$ and the expected value will go to infinity. However, this result obviously goes against human behavior since no one is willing to pay more than \$1000 for this game. The answer to the St. Petersburg paradox was unavailable until Daniel Bernoulli (1700–1782) published his influential research on utility theory in 1738. We ignore the concrete discussions describing the solution of the St. Petersburg paradox in detail, but focus on the conclusion that humans make decisions based not on the expected value but the utility value. The implication of the utility value is that humans choose the alternative with the highest utility value when confronting the MADM problems.

In 1947, von Neumann and Morgenstern published their famous book, *Theory of Games and Economic Behavior*, to conceive a mathematical theory of economic and social organization in detail, based on game theory. There is no doubt that the great work of von Neumann and Morgenstern indeed opens the door to MADM. Roughly speaking, the methods for dealing with MADM problems can be mainly divided into multiple attribute utility theory (MAUT) and outranking methods (especially refer to ELECTRE [Benayoun, Roy, and Sussman 1966; Roy 1968] and PROMETHEE [Brans, Mareschal, and Vincke 1984] methods).

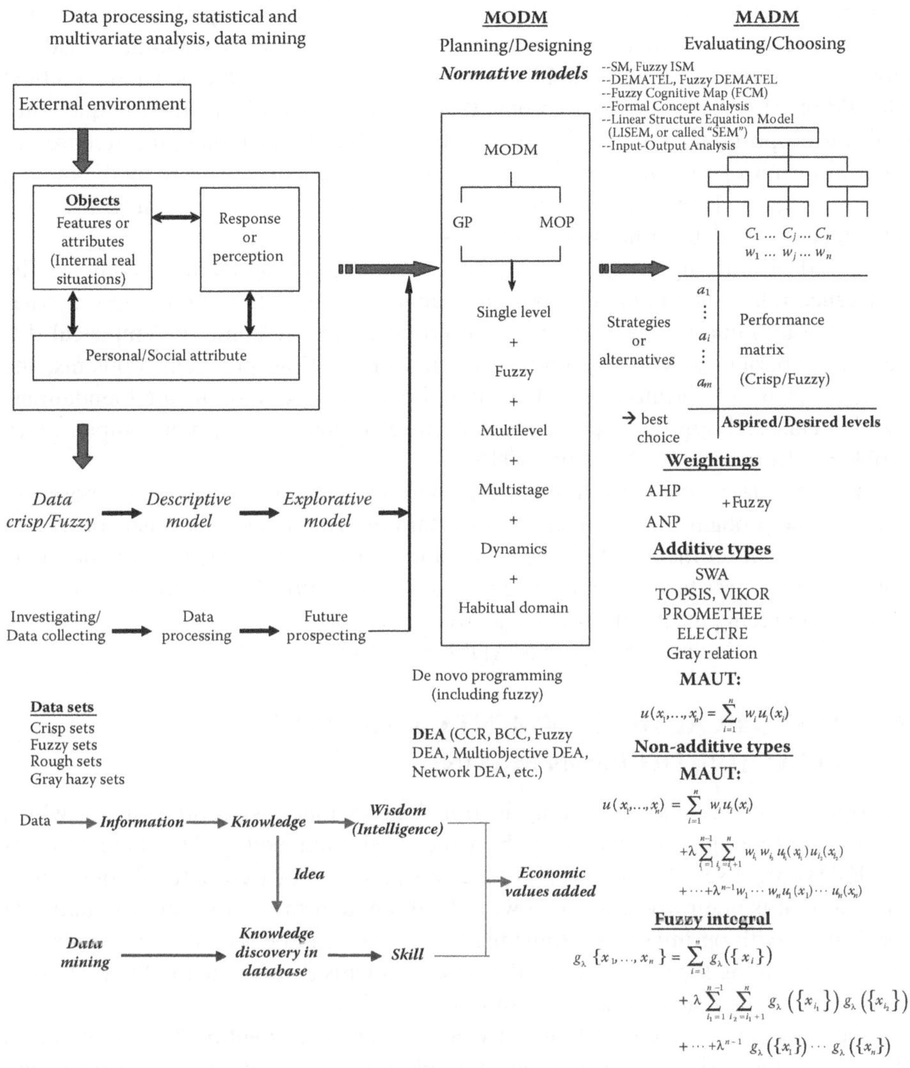

FIGURE 1.1 Profile of MCDM.

On the basis of Bernoulli's utility theory, MAUT determines the DM' preferences, which can usually be represented as a hierarchical structure, by using an *appropriate* utility function. By evaluating the utility function, a decision maker can easily determine the best alternative with the highest utility value. Although many papers have been proposed in determining the appropriate utility function of MAUT (Fishburn 1970), the main criticism of MAUT concentrates on the unrealistic assumption of preferential independence (Grabisch 1995; Hillier 2000).

Preferential independence can be described as follows: the preference outcome of one criterion over another criterion is not influenced by the remaining criteria. However, it should be highlighted that the criteria are usually interactive in practical MCDM problems. In order to overcome that non-additive problem, the Choquet integral was proposed (Choquet 1953; Sugeno 1974). The Choquet integral can represent a certain kind of interaction among criteria using the concept of redundancy and support/synergy. However, another critical problem of the Choquet integral arises: how to correctly determine fuzzy measures?

Instead of building complex utility functions, outranking methods compare the preference relations among alternatives to acquire information on the best alternative. Although outranking methods were proposed to overcome the empirical difficulties experienced with the utility function in handling practical problems, the main criticisms of outranking methods have been the lack of axiomatic foundations, such as classical aggregate problems, structural problems, and non-compensatory problems (Bouyssou and Vansnick 1986).

In 1965, fuzzy sets (Zadeh 1965; Bellman and Zadeh 1970) were proposed to confront the problems of linguistic or uncertain information and be a generalization of conventional set theory. With the successful applications in the field of automatic control, fuzzy sets have recently been incorporated into MADM for dealing with MADM problems in situations of subjective uncertainty.

For the holistic development of MADM, refer to Figure 1.2.

1.3 HISTORICAL DEVELOPMENT OF MULTIPLE OBJECTIVE DECISION MAKING

Multiple objective decision making is aimed at optimal design problems in which several (conflicting) objectives are to be achieved simultaneously. The characteristics of MODM are a set of (conflicting) objectives and a set of well-defined constraints. Therefore, it is naturally associated with the method of mathematical programming for dealing with optimization problems. However, it can be seen that two main difficulties involving the trade-off and the scale problems complicate the MODM problems through the mathematical programming model.

The trade-off problem is that since a final optimal solution is usually given through mathematical programming, multiple objectives have to transform it into a weighted single objective. Therefore, a process of obtaining trade-off information between the considered objectives should first be identified. Note that if the trade-off information is unavailable, Pareto solutions should be derived. The scaling problem, on the other hand, is that as the number of dimensions increases beyond the capacity, it suffers from the problem of the curse of dimensionality, i.e., the computational cost increases tremendously. Recently, many evolution algorithms, such as genetic algorithms (Holland 1975), genetic programming (Koza 1992), and evolution strategy (Rechenberg 1973) have been suggested to handle this problem.

Since Kuhn and Tucker (1951) published multiple objectives using the vector optimization concept and Yu (1973) proposed the compromise solution method to cope with MODM problems, considerable work has been done on various applications, such as transportation investment and planning, econometric and development

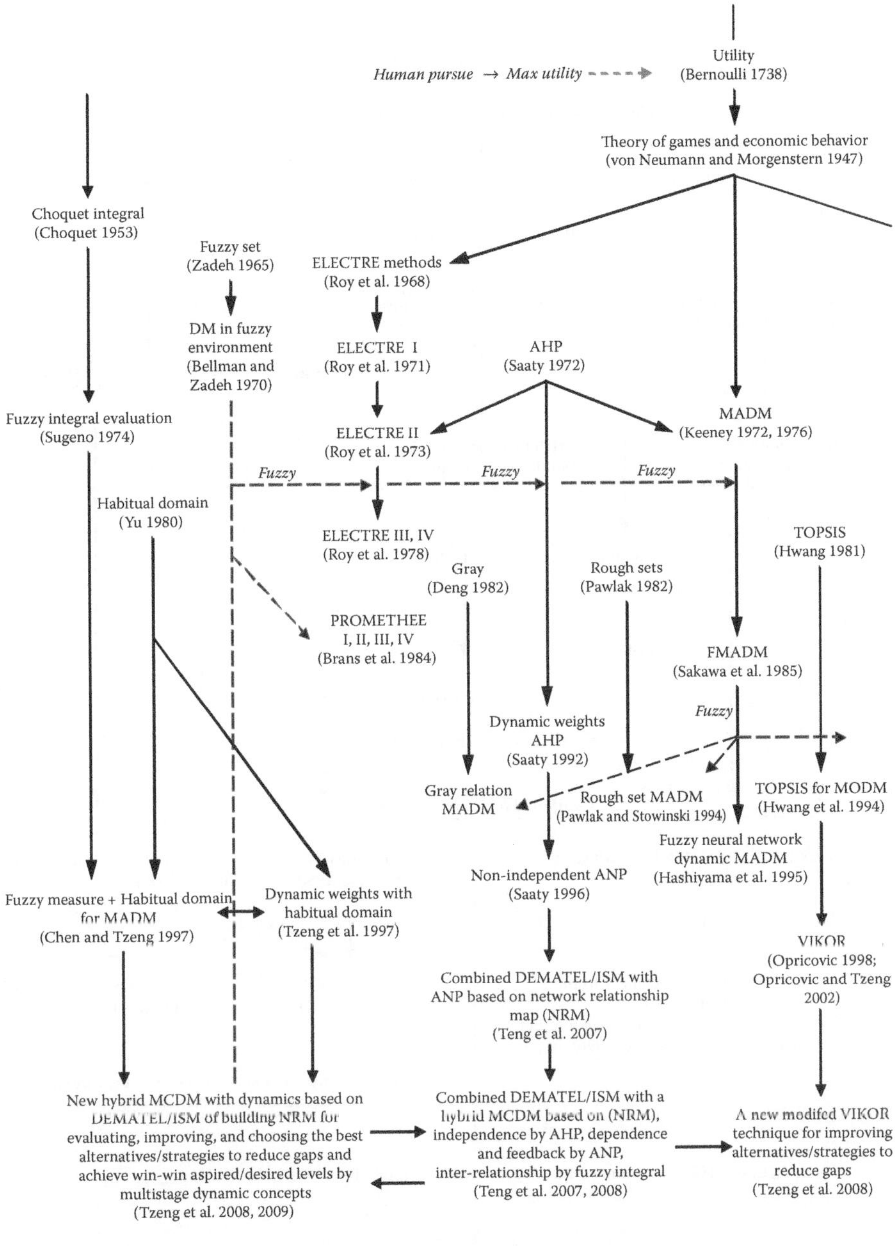

FIGURE 1.2 Development of MADM.

planning, financial planning, conducting business and the selection of investment portfolios, land-use planning, water resource management, public policy and environmental issues, and so on. The theatrical work is extended from simple multiple objective programming to multilevel multiobjective programming and multistage multiobjective programming for confronting more complicated real-world problems.

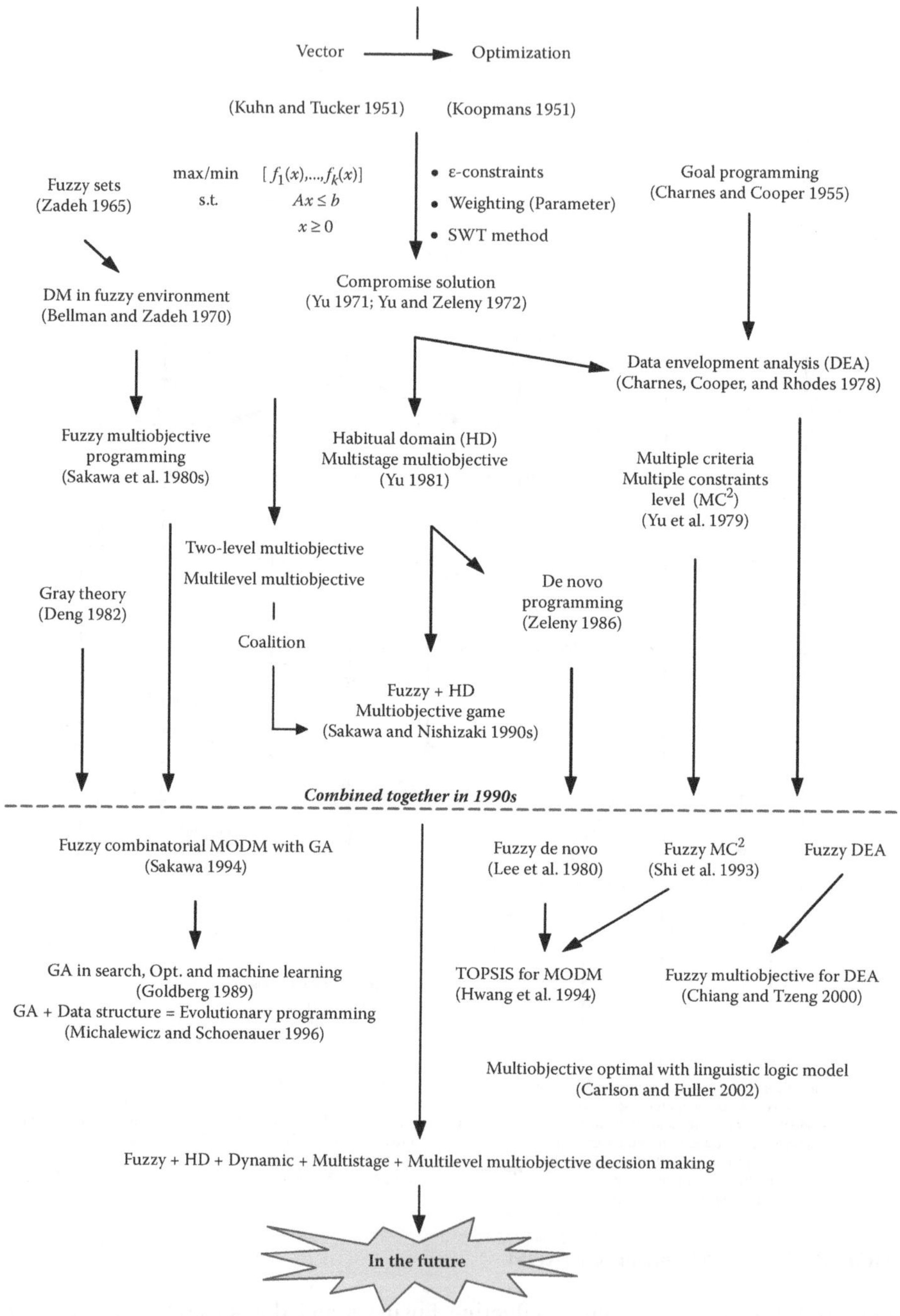

FIGURE 1.3 Development of MODM.

On the other hand, the conventional MODM seems to ignore the problem of subjective uncertainty. It can be seen that since the objectives and constraints may involve linguistic and fuzzy variables, fuzzy numbers should be incorporated into MODM for dealing with more extensive problems. After Bellman and Zadeh (1970) proposed the concepts of decision making under fuzzy environments, many distinguished works guided the study of fuzzy multiple objective linear programming (FMOLP), such as Hwang and Yoon (1981), Zimmermann (1978), Sakawa (1983, 1984a, 1984b), and Lee and Li (1993).

FMOLP formulates the objectives and the constraints as fuzzy sets, with respect to their individual linear membership functions. The decision set is defined by the intersection of all fuzzy sets and the relevant hard constraints. A crisp solution is generated by selecting the optimal solution, such that it has the highest degree of membership in the decision set. For further discussions, readers can refer to Zimmermann (1978), Werners (1987), and Martinson (1993).

For the historical development of MODM, Please refer to Figure 1.3.

1.4 INTRODUCTION TO FUZZY SETS

In this section, we do not introduce all topics about fuzzy sets in detail but concentrate on the basic concepts of fuzzy sets and the arithmetic operations of fuzzy numbers for the purpose of this book.

1.4.1 BASIC CONCEPTS

In contrast to classical set theory for coping with Boolean logic problems, fuzzy sets were proposed to represent the degree of elements belonging to the specific sets. Instead of using the characteristic function as a mapping function, a fuzzy subset $\tilde{A}$ of a universal set X can be defined by its membership function $\mu_{\tilde{A}}(x)$ as

$$\tilde{A} = \left\{ \left(x, \mu_{\tilde{A}}(x) \right) \mid x \in X \right\}, \tag{1.1}$$

where $x \in X$ denotes the elements belonging to the universal set, and

$$\mu_{\tilde{A}}(x) : X \to [0,1]. \tag{1.2}$$

Given a discrete finite set $X = \{x_1, x_2, \ldots, x_n\}$, a fuzzy subset $\tilde{A}$ of X can also be represented as

$$\tilde{A} = \sum_{i=1}^{n} \mu_{\tilde{A}}(x_i)/x_i. \tag{1.3}$$

For a continuous case, a fuzzy set $\tilde{A}$ of X can be represented as

$$\tilde{A} = \int_X \mu_{\tilde{A}}(x)/x. \tag{1.4}$$

Next, we present some definitions that will be used in the presented FMADM models as follows.

Definition 1.1

Let a fuzzy subset $\tilde{A}$ of a set X be considered; the support of $\tilde{A}$ is a crisp set of X defined by

$$\text{supp}(\tilde{A}) = \{x \in X \mid \mu_{\tilde{A}}(x) > 0\}. \tag{1.5}$$

Definition 1.2

The α-cut of a fuzzy subset $\tilde{A}$ of X can be defined by

$$\tilde{A}(\alpha) = \{x \in X \mid \mu_{\tilde{A}}(x) \geq \alpha\}, \quad \forall \alpha \in [0,1]. \tag{1.6}$$

Definition 1.3

Let a fuzzy subset $\tilde{A}$ of a set X be considered; the high of $\tilde{A}$ is the least upper bound (sup) of $\mu_{\tilde{A}}(x)$ and is defined by

$$h(A) = \sup_{x \in X} \mu_{\tilde{A}}(x). \tag{1.7}$$

Definition 1.4

A fuzzy subset $\tilde{A}$ of a set X is said to be normal if and only if its height is unity and called subnormal if and only if its height is not unity.

Fuzzy sets were originally proposed to deal with problems of subjective uncertainty. Subjective uncertainty results from using linguistic variables to represent the problem or the event. Note that a linguistic variable is a variable that is expressed by verbal words or sentences in a natural or artificial language. For example, linguistic variables with triangular fuzzy numbers may take on effect-values such as "very high (very good)," "high (good)," "fair," "low (bad)," and "very low (very bad)," as shown in Figure 1.4, to indicate the membership functions of the expression values.

$\mu_{\tilde{A}}(x)$

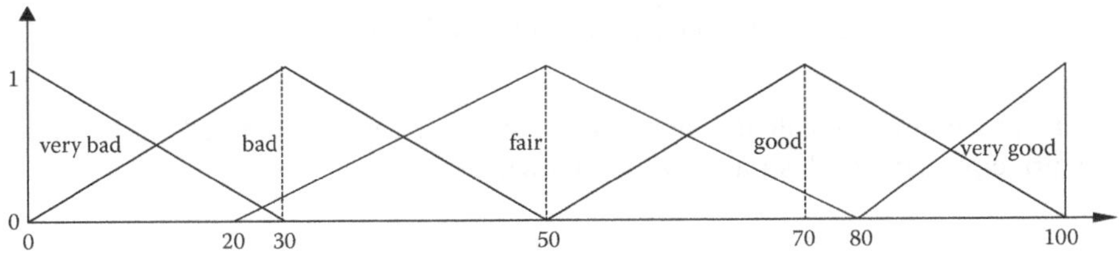

FIGURE 1.4 Membership function of the five levels of linguistic variables.

The adoption of linguistic variables has recently become widespread and is used to assess the linguistic ratings given by the evaluators. Furthermore, linguistic variables are also employed as a way to measure the achievement of the performance value for each criterion. Since the linguistic variables can be defined by the corresponding membership function and the fuzzy interval, we can naturally manipulate the fuzzy numbers to deal with the FMADM problems.

1.4.2 Fuzzy Arithmetic Operations

The fuzzy arithmetic operations involve adding, subtracting, multiplying, and dividing fuzzy numbers. Generally, these fuzzy arithmetic operations are derived based on the extension principle and α-cut arithmetic. For more detailed discussions about fuzzy arithmetic operations, refer to Dubois et al. (1993, 2000), Dubois and Prade (1987, 1988), Kaufmann and Gupta (1985, 1988), and Mares (1994). In this section, we briefly introduce the fuzzy arithmetic operations according to the extension principle and α-cut arithmetic, respectively, as follows.

1.4.3 Extension Principle

Let $\tilde{m}$ and $\tilde{n}$ be two fuzzy numbers and z denote the specific event. Then, the membership functions of the four basic arithmetic operations for $\tilde{m}$ and $\tilde{n}$ can be defined by

$$\mu_{\tilde{m}+\tilde{n}}(z) = \sup_{x,y}\left\{\min\left(\tilde{m}(x), \tilde{n}(y)\right) \mid x + y = z\right\}; \tag{1.8}$$

$$\mu_{\tilde{m}-\tilde{n}}(z) = \sup_{x,y}\left\{\min\left(\tilde{m}(x), \tilde{n}(y)\right) \mid x - y = z\right\}; \tag{1.9}$$

$$\mu_{\tilde{m}\times\tilde{n}}(z) = \sup_{x,y}\left\{\min\left(\tilde{m}(x), \tilde{n}(y)\right) \mid x \times y = z\right\}; \tag{1.10}$$

$$\mu_{\tilde{m}\div\tilde{n}}(z) = \sup_{x,y}\left\{\min\left(\tilde{m}(x),\tilde{n}(y)\right) \mid x \div y = z\right\}. \tag{1.11}$$

The procedures for calculating two fuzzy numbers, $\tilde{m}$ and $\tilde{n}$, based on the extension principle can be illustrated by the following example:

x	1	3	5	7	9
y	1	3	5	7	9
$\mu_{\tilde{m}}(x)$	0.2	0.4	0.6	1.0	0.8
$\mu_{\tilde{n}}(y)$	0.1	0.3	1.0	0.7	0.5

$$\mu_{\tilde{m}+\tilde{n}}(10) = \sup\{0.2, 0.4, 0.6, 0.3, 0.1\} = 0.6;$$

$$\mu_{\tilde{m}-\tilde{n}}(-2) = \sup\{0.2, 0.4, 0.6, 0.5\} = 0.6;$$

$$\mu_{\tilde{m}\times\tilde{n}}(9) = \sup\{0.2, 0.3, 0.1\} = 0.3;$$

$$\mu_{\tilde{m}\div\tilde{n}}(3) = \sup\{0.1, 0.3\} = 0.3.$$

Next, we provide another method to derive the fuzzy arithmetic operations based on the concept of α-cut.

1.4.4 α-Cut Arithmetic

Let $\tilde{m} = [m^l, m^m, m^u]$ and $\tilde{n} = [n^l, n^m, n^u]$ be two fuzzy numbers in which the superscripts l, m, and u denote the infimum, the mode, and the supremum, respectively. The standard fuzzy arithmetic operations can be defined using the concepts of α-cut as follows:

$$\tilde{m}(\alpha) + \tilde{n}(\alpha) = \left[m^l(\alpha) + n^l(\alpha), m^u(\alpha) + n^u(\alpha)\right]; \tag{1.12}$$

$$\tilde{m}(\alpha) - \tilde{n}(\alpha) = \left[m^l(\alpha) - n^u(\alpha), m^u(\alpha) - n^l(\alpha)\right]; \tag{1.13}$$

$$\tilde{m}(\alpha) \div \tilde{n}(\alpha) \approx \left[m^l(\alpha), m^u(\alpha)\right] \times \left[1/n^u(\alpha), 1/n^l(\alpha)\right]; \tag{1.14}$$

$$\tilde{m}(\alpha) \times \tilde{n}(\alpha) \approx [M, N], \tag{1.15}$$

where (α) denotes the α-cut operation, ≈ is the approximation operation, and

$$M = \min\left\{m^l(\alpha)n^l(\alpha), m^l(\alpha)n^u(\alpha), m^u(\alpha)n^l(\alpha), m^u(\alpha)n^u(\alpha)\right\};$$

$$N = \max\left\{m^l(\alpha)n^l(\alpha), m^l(\alpha)n^u(\alpha), m^u(\alpha)n^l(\alpha), m^u(\alpha)n^u(\alpha)\right\}.$$

An example is also given to illustrate the computation of fuzzy numbers. Let there be two fuzzy numbers $\tilde{m}$= [3, 5, 8] and $\tilde{n}$ = [2, 4, 6]. Then

$$\tilde{m}(\alpha)+\tilde{n}(\alpha)=\left[(3+2\alpha)+(2+2\alpha),(8-3\alpha)+(6-2\alpha)\right];$$

$$\tilde{m}(\alpha)-\tilde{n}(\alpha)=\left[(3+2\alpha)-(2+2\alpha),(8-3\alpha)-(6-2\alpha)\right];$$

$$\tilde{m}(\alpha)\div\tilde{n}(\alpha)\approx\left[\frac{(3+2\alpha)}{(6-2\alpha)},\frac{(8-3\alpha)}{(2+2\alpha)}\right];$$

$$\tilde{m}(\alpha)\times\tilde{n}(\alpha)\approx\left[(3+2\alpha)(2+2\alpha),(8-3\alpha)(6-2\alpha)\right].$$

1.4.5 Ranking Fuzzy Numbers

It is clear that since the fuzzy arithmetic operations are based on the α-cut arithmetic result in the fuzzy interval, it is not always obvious how to determine the optimal alternative and this involves the problem of ranking fuzzy numbers or defuzzification. In previous works, the procedure of defuzzification has been proposed to locate the best non-fuzzy performance (BNP) value. Defuzzified fuzzy ranking methods generally can be divided into the following four categories: preference relation methods, a fuzzy mean and spread method, fuzzy scoring methods, and linguistic methods (Chen and Hwang 1992).

Even though more than 30 defuzzified methods have been proposed in the past 20 years, only the center of area (CoA) is described in this book for its simplicity and usefulness. Consider the preference ratings of an alternative with n attributes is represented using the fuzzy number, the BNP values of the alternative using the CoA can be formulated as:

$$S=\frac{\sum_{i=1}^{n}x_i\mu(x_i)}{\sum_{i=1}^{n}\mu(x_i)}, \tag{1.16}$$

where x_i denotes the preference ratings of the ith attribute and $\mu(x_i)$ is the corresponding membership function.

1.5 OUTLINE OF THIS BOOK

The organization of this book is according to the processes of decision making as follows: identifying the problems, building network relationship maps (NRM), constructing the preferences, evaluating and improving the alternatives to reduce the gaps, and then finding and determining the best alternative for achieving the aspired/desired levels. However, it can be realized that identifying the problems, constructing the preferences, and improving/finding and determining the best alternative are very intuitive for DM. That is, the contents of this book focus on the process of evaluating the alternatives and can be divided into two main steps: determining the relative weights of criteria and aggregating the ratings of alternatives. It is a non-trivial and tough problem to derive the relative weights of each criterion, since criteria are usually interactive and dependent on each other. On the other hand, the different aggregated operators indicate the different DM' preferences and should be appropriately determined to obtain the correct overall ratings of each alternative.

In Chapters 2 and 3, the analytic hierarchy process (AHP) and the analytic network process (ANP) are presented to determine the relative importance of criteria. AHP is used to derive the relative weights of criteria using the pairwise comparison between criteria in a hierarchical system. By releasing the restriction of the hierarchical structure, ANP was proposed to derive the relative weights of criteria in a network structure.

In Chapters 4 through 10, several FMADM methods are introduced as follows. In Chapter 4, the fuzzy simple additive weighting (FSAW) method is extended to the simple additive weighting (SAW) method by considering fuzzy numbers. The SAW method is the best known and the most adopted MADM model for only considering the weights of criteria and the additive form.

Chapter 5 presents the fuzzy TOPSIS method to evaluate the alternative by considering the bi-objectives, including the positive and the negative ideal points under the situation of subjective uncertainty. The concept of TOPSIS is that the best alternative should minimize the distance from the positive ideal point (PIS) and maximize the distance from the negative ideal point (NIS).

Outranking methods, ELECTRE and PROMETHEE, are described in Chapters 6 (ELECTREs I, II, III, IV) and Chapter 7 (PROMETHEEs I, II, III, IV). In contrast to the utility-based methods, outranking methods use the preferred relations among alternatives to determine the best alternative. These methods are widely employed in Europe, especially in France, to deal with specific MADM problems.

The fuzzy integral technique was proposed to consider the problem of preference dependence in MADM problems and will be discussed in Chapter 8. Preference independence is the foundation of MAUT. However, it can be seen that the situation of preference dependence among criteria which usually happens is not considered in the conventional MADM models.

Several applications of the above models for dealing with practical MADM problems are given in Chapter 9. In the Appendix, three structural models, interpretive structural modeling (ISM), decision making trial and evaluation laboratory (DEMATEL), and fuzzy cognition maps (FCM), are introduced to construct a hierarchical/network system. Since the importance weights generated by using the AHP/ANP method are based on the specific hierarchical/network system, the crucial issue of what the appropriated structure is should be discussed first.

Part I

Concepts and Theory of MADM

2 Analytic Hierarchy Process

Since Bernoulli (1738) proposed the concept of utility function to reflect human persuit, such as maximum satisfaction, and von Neumann and Morgenstern (1947) presented the theory of game and economic behavior model, which expanded the studies on human economic behavior for multiple attribute decision making (MADM) problems, an increasing amount of literature has been engaged in this field. Roughly speaking, the procedures of MADM can be summarized in five main steps as follows (Dubois and Prade 1980):

Step 1: Define the nature of the problem;
Step 2: Construct a hierarchy system for its evaluation (Figure 2.1);
Step 3: Select the appropriate evaluation model;
Step 4: Obtain the relative weights and performance score of each attribute with respect to each alternative;
Step 5: Determine the best alternative according to the synthetic utility values, which are the aggregation value of relative weights, and performance scores corresponding to alternatives.

If the overall scores of the alternatives are fuzzy, we can add Step 6 to rank the alternatives for choosing the best one.

Step 6: Outrank the alternatives referring to their synthetic fuzzy utility values from Step 5.

It should be highlighted that Keeney and Raiffa (1976) suggest that five principles must be followed when criteria are being formulated: (1) completeness, (2) operationality, (3) decomposability, (4) non-redundancy, and (5) minimum size.

On the basis of dealing with MADM problems, the analytic hierarchy process (AHP) was proposed to derive the relative weights according to the appropriate hierarchical system. In this chapter, four methods, including the eigenvalue method, the geometric mean method, the linear programming method, and the lambda-max method, are proposed to derive the weights using the AHP. Among these methods, only the eigenvalue method is employed to deal with crisp numbers and the other methods are adopted to handle the AHP under fuzzy numbers.

2.1 EIGENVALUE METHOD

AHP was proposed by Saaty (1977, 1980) to model subjective decision-making processes based on multiple attributes in a hierarchical system. From that moment on, it has been widely used in corporate planning, portfolio selection, and benefit/cost

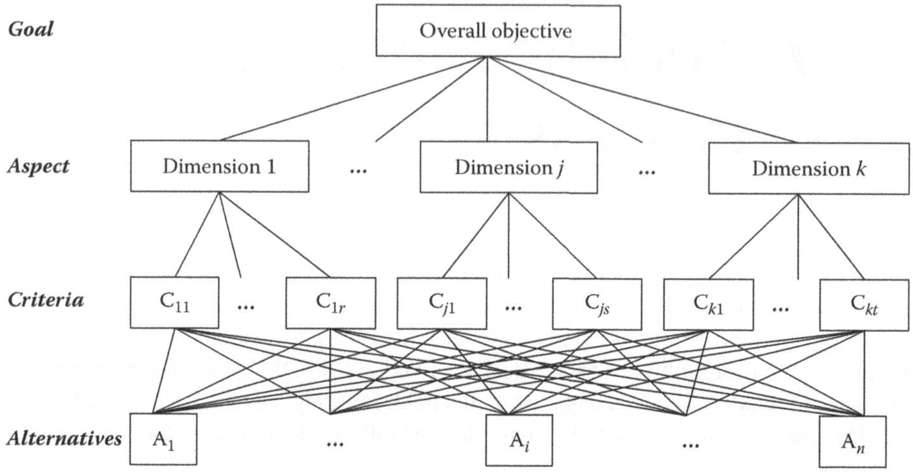

FIGURE 2.1 Hierarchical system for MADM.

analysis by government agencies for resource allocation purposes. It should be highlighted that all decision problems are considered as a hierarchical structure in the AHP. The first level indicates the goal for the specific decision problem. In the second level, the goal is decomposed of several criteria and the lower levels can follow this principal to divide into other sub-criteria. Therefore, the general form of the AHP can be depicted as shown in Figure 2.2.

The four main steps of the AHP can be summarized as follows:

Step 1: Set up the hierarchical system by decomposing the problem into a hierarchy of interrelated elements;

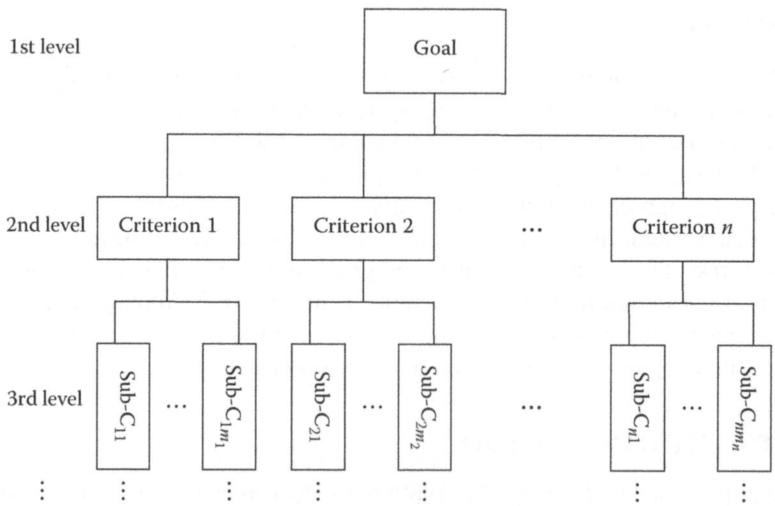

FIGURE 2.2 The hierarchical structure of the AHP.

Step 2: Compare the comparative weight between the attributes of the decision elements to form the reciprocal matrix;

Step 3: Synthesize the individual subjective judgment and estimate the relative weight;

Step 4: Aggregate the relative weights of the elements to determine the best alternatives/strategies.

If we wish to compare a set of n attributes pairwise according to their relative importance weights, where the attributes are denoted by $a_1, a_2, \ldots, a_n$ and the weights are denoted by $w_1, w_2, \ldots, w_n$, then the pairwise comparisons can be represented by questionnaires with subjective perception as:

$$\boldsymbol{A} = \begin{bmatrix} a_{11} & \cdots & a_{1j} & \cdots & a_{1n} \\ \vdots & & \vdots & & \vdots \\ a_{i1} & \cdots & a_{ij} & \cdots & a_{in} \\ \vdots & & \vdots & & \vdots \\ a_{n1} & \cdots & a_{nj} & \cdots & a_{nn} \end{bmatrix}, \tag{2.1}$$

where $a_{ij} = 1/a_{ji}$ (positive reciprocal) and $a_{ij} = a_{ik}/a_{jk}$. Note that in realistic situations, w_i/w_j is usually unknown. Therefore, the problem for the AHP is to find a_{ij} such that $a_{ij} \cong w_i/w_j$.

Let a weight matrix be represented as:

$$\boldsymbol{W} = \begin{matrix} & w_1 & \cdots & w_j & \cdots & w_n \\ \begin{matrix} w_1 \\ \vdots \\ w_i \\ \vdots \\ w_n \end{matrix} & \left[\begin{matrix} w_1/w_1 \\ \vdots \\ w_i/w_1 \\ \vdots \\ w_n/w_1 \end{matrix}\right. & \begin{matrix} \cdots \\ \\ \cdots \\ \\ \cdots \end{matrix} & \begin{matrix} w_1/w_j \\ \vdots \\ w_i/w_j \\ \vdots \\ w_n/w_j \end{matrix} & \begin{matrix} \cdots \\ \\ \cdots \\ \\ \cdots \end{matrix} & \left.\begin{matrix} w_1/w_n \\ \vdots \\ w_i/w_n \\ \vdots \\ w_n/w_n \end{matrix}\right] \end{matrix}.$$

By multiplying $\boldsymbol{W}$ by $\boldsymbol{w}$ yield

$$\boldsymbol{W} \times \boldsymbol{w} = \begin{matrix} & w_1 & \cdots & w_j & \cdots & w_n \\ \begin{matrix} w_1 \\ \vdots \\ w_i \\ \vdots \\ w_n \end{matrix} & \left[\begin{matrix} w_1/w_1 \\ \vdots \\ w_i/w_1 \\ \vdots \\ w_n/w_1 \end{matrix}\right. & \begin{matrix} \cdots \\ \\ \cdots \\ \\ \cdots \end{matrix} & \begin{matrix} w_1/w_j \\ \vdots \\ w_i/w_j \\ \vdots \\ w_n/w_j \end{matrix} & \begin{matrix} \cdots \\ \\ \cdots \\ \\ \cdots \end{matrix} & \left.\begin{matrix} w_1/w_n \\ \vdots \\ w_i/w_n \\ \vdots \\ w_n/w_n \end{matrix}\right] \end{matrix} \begin{bmatrix} w_1 \\ \vdots \\ w_j \\ \vdots \\ w_n \end{bmatrix} = n \begin{bmatrix} w_1 \\ \vdots \\ w_j \\ \vdots \\ w_n \end{bmatrix} \tag{2.2}$$

TABLE 2.1
Ratio Scale in the AHP

Intensity	1	3	5	7	9	2, 4, 6, 8
Linguistic	Equal	Moderate	Strong	Demonstrated	Extreme	Intermediate value

or

$$(\boldsymbol{W}-n\boldsymbol{I})\boldsymbol{w}=\boldsymbol{0}. \tag{2.3}$$

Table 2.1 represents the ratio scale that is employed to compare the importance weight between criteria according to the linguistic meaning from 1 to 9 to denote equal importance to extreme importance.

Since solving Equation 2.3 is the eigenvalue problem, we can derive the comparative weights by finding the eigenvector $\boldsymbol{w}$ with respective λ_{max} that satisfies $\boldsymbol{Aw} = \lambda_{max}\boldsymbol{w}$, where λ_{max} is the largest eigenvalue of the matrix $\boldsymbol{A}$, i.e., find the eigenvector $\boldsymbol{w}$ with respective λ_{max} for $(\boldsymbol{A} - \lambda_{max}\boldsymbol{I})\boldsymbol{w} = \boldsymbol{0}$.

Furthermore, in order to ensure the consistency of the subjective perception and the accuracy of the comparative weights, two indices, including the consistency index (*C.I.*) and the consistency ratio (*C.R.*), are suggested. The equation of the *C.I.* can be expressed as:

$$C.I. = (\lambda_{max} - n)/(n-1), \tag{2.4}$$

where λ_{max} is the largest eigenvalue, and n denotes the numbers of the attributes. Saaty (1980) suggested that the value of the *C.I.* should not exceed 0.1 for a confident result. On the other hand, the *C.R.* can be calculated as:

$$C.R. = \frac{C.I.}{R.I.},$$

where *R.I.* refers to a random consistency index, which is derived from a large sample of randomly generated reciprocal matrices using the scale 1/9, 1/8,…,1,…,8, 9. The *R.I.* with respect to different size matrices is shown in Table 2.2.

TABLE 2.2
The *R.I.* for Different Size Matrices

Number of elements	3	4	5	6	7	8	9	10	11	12	13
R.I.	0.52	0.89	1.11	1.25	1.35	1.40	1.45	1.49	1.51	1.54	1.56

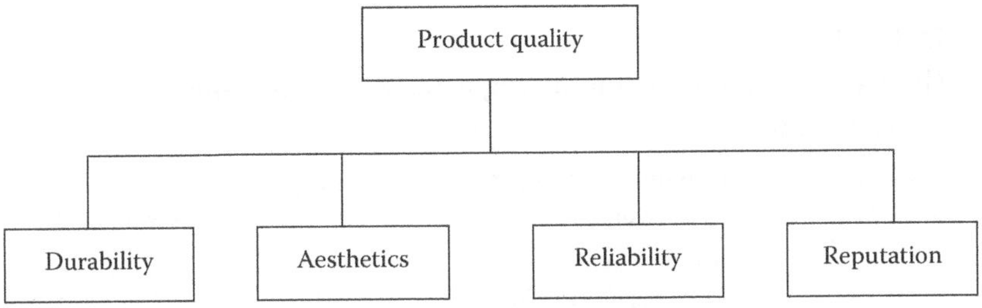

FIGURE 2.3 The hierarchy structure of the product quality.

The $C.R.$ should be under 0.1 for a reliable result and 0.2 is the maximum tolerated level. Next, we provide a numerical example to demonstrate the procedure of the AHP in detail.

Example 2.1

Consider that the product quality of a company can be evaluated using four criteria: durability, aesthetics, reliability, and reputation. The decision maker wants to determine the weights of the criteria using the AHP so that he/she can allocate the appropriate budgets to obtain the optimal product quality. The hierarchical structure adopted in this example to deal with the problem of the product quality can be depicted as shown in Figure 2.3.

The pairwise comparison of each criterion can be described as shown in Table 2.3. Using the eigenvalue method, we can derive the largest eigenvalue $\lambda_{max} = 4.1701$ and the eigenvector $\boldsymbol{r}' = [0.3145, 0.1168, 0.9398, 0.0655]$. By normalizing the eigenvector, we can obtain the weights vector $\boldsymbol{w}' = [0.2189, 0.0813, 0.6542, 0.0456]$. In addition, from the values of $C.I. = 0.0567$ and $C.R. = 0.0637$, we can conclude that the consistency of the subjective perception is satisfied.

However, in realistic problems, the perception of a decision maker is usually vague, fuzzy, or linguistic. For example, the linguistic expression that one criterion is "much more" important than another may be expressed by the ratios 7/1 or 9/1. Therefore, the fuzzy pair comparisons among criteria are more suitable for this situation. In the following sections, we present three kinds of methods to calculate the fuzzy AHP as follows.

TABLE 2.3
The Pairwise Comparison of Each Criterion

Quality Criteria	Durability	Aesthetics	Reliability	Reputation
Durability	1	3	1/5	7
Aesthetics	1/3	1	1/7	2
Reliability	5	7	1	9
Reputation	1/7	1/2	1/9	1

TABLE 2.4
The Pairwise Comparison of Linguistic Variables Using Fuzzy Numbers

Intensity of Fuzzy Scale	Definition of Linguistic Variables	Fuzzy Number	User-defined
$\tilde{1}$	Similar importance (SI)	(L,M,U)	= (__,1,__)
$\tilde{3}$	Moderate importance (MI)	(L,M,U)	= (__,3,__)
$\tilde{5}$	Intense importance (II)	(L,M,U)	= (__,5,__)
$\tilde{7}$	Demonstrated importance (DI)	(L,M,U)	= (__,7,__)
$\tilde{9}$	Extreme importance (EI)	(L,M,U)	= (__,9,__)
$\tilde{2},\tilde{4},\tilde{6},\tilde{8}$	Intermediate values	(L,M,U)	= (__,__,__)

2.2 GEOMETRIC MEAN METHOD

The geometric mean method was first employed by Buckley (1985) to extend the AHP to consider the situation of using linguistic variables (Zadeh 1965). The degrees of the pairwise comparison of linguistic variables can be expressed using the fuzzy numbers as shown in Table 2.4.

The corresponding membership function can be depicted as shown in Figure 2.4.

Next, from the information of the pairwise comparison, we can form the fuzzy positive reciprocal matrix as follows:

$$\tilde{A} = \begin{bmatrix} \tilde{a}_{11} & \cdots & \tilde{a}_{1j} & \cdots & \tilde{a}_{in} \\ \vdots & & \vdots & & \vdots \\ \tilde{a}_{i1} & \cdots & \tilde{a}_{ij} & \cdots & \tilde{a}_{in} \\ \vdots & & \vdots & & \vdots \\ \tilde{a}_{n1} & \cdots & \tilde{a}_{nj} & \cdots & \tilde{a}_{nn} \end{bmatrix}, \tag{2.5}$$

where $\tilde{a}_{ij} \odot \tilde{a}_{ji} \approx 1$ and $\tilde{a}_{ij} \cong w_i / w_j$.

Then, the geometric mean method for finding the final fuzzy weights of each criterion can be formulated as follows:

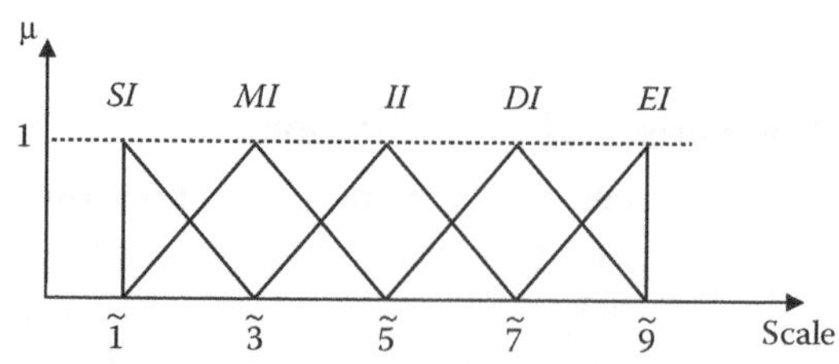

FIGURE 2.4 The membership function of linguistic variables.

$$\tilde{w}_i = \tilde{r}_i \odot \left(\tilde{r}_1 \oplus \tilde{r}_2 \oplus \cdots \oplus \tilde{r}_n\right)^{-1}, \tag{2.6}$$

where

$$\tilde{r}_i = \left(\tilde{a}_{i1} \odot \tilde{a}_{i2} \odot \cdots \odot \tilde{a}_{in}\right)^{1/n}. \tag{2.7}$$

Example 2.2

Reconsider the problem of Example 2.1 for the linguistic variable situation. Assume that we can represent the pairwise comparison of each criterion using the fuzzy numbers as shown in Table 2.5.

By using Equation 2.6, we can obtain the fuzzy weights of each criterion as:

$$\tilde{w}_1 = [0.1623, 0.2203, 0.3007];$$

$$\tilde{w}_2 = [0.0574, 0.0855, 0.1264];$$

$$\tilde{w}_3 = [0.5080, 0.6483, 0.8152];$$

$$\tilde{w}_4 = [0.0356, 0.0459, 0.0679],$$

where

$$\tilde{r}_1 = [1.1892, 1.4316, 1.6818];$$

$$\tilde{r}_2 = [0.4204, 0.5555, 0.7071];$$

$$\tilde{r}_3 = [3.7224, 4.2129, 4.5590];$$

$$\tilde{r}_4 = [0.2608, 0.2985, 0.3799].$$

The fuzzy weights of each criterion can also be defuzzified by center of area (CoA) in order to obtain a crisp solution as:

$$w_1 = 0.2278; w_2 = 0.0898; w_3 = 0.6572; w_4 = 0.0498.$$

TABLE 2.5
The Pairwise Comparison of the Fuzzy Language

Quality Criteria	Durability	Aesthetics	Reliability	Reputation
Durability	[1,1,1]	[2,3,4]	[1/6,1/5,1/4]	[6,7,8]
Aesthetics	[1/4,1/3,1/2]	[1,1,1]	[1/8,1/7,1/6]	[1,2,3]
Reliability	[4,5,6]	[6,7,8]	[1,1,1]	[8,9,9]
Reputation	[1/8,1/7,1/6]	[1/3,1/2,1]	[1/9,1/9,1/8]	[1,1,1]

Although the geometric mean method makes it very easy to extend the AHP for considering the fuzzy situation, the main shortcoming of this method is the problem of the irrational fuzzy interval. There are two clear reasons for this irrational interval. First, the multiplication of fuzzy numbers will increase the fuzzy interval. Second, the geometric mean method does not consider the condition, such that the sum of the weights equals 1. In order to overcome the problem above, many mathematical programming models have been proposed to derive the weights in the FAHP (Fuzzy AHP).

2.3 LINEAR PROGRAMMING METHOD

In this section, we only describe one of many mathematical programming methods, which was proposed by Mikhailov (2000, 2003) to derive the weights of the FAHP. Other mathematical programming methods have similar concepts to the method above. The linear programming method for deriving the weights of the FAHP can be described as follows.

Given a fuzzy positive reciprocal matrix $\tilde{\boldsymbol{A}} = [\tilde{a}_{ij}]_{n \times n}$, the fuzzy pairwise comparison judgments can be described using the following interval judgment:

$$l_{ij} \tilde{\leq} \frac{w_i}{w_j} \tilde{\leq} u_{ij}, \quad i = 1,2,\ldots,n-1;\ j = 1,2,\ldots,n;\ i < j. \tag{2.8}$$

With the specific level α-cut, the judgment degree of uncertainty can be represented as:

$$l_{ij}(\alpha) \tilde{\leq} \frac{w_i}{w_j} \tilde{\leq} u_{ij}(\alpha), \quad \alpha \in [0,1];\ i = 1,2,\ldots,n-1;\ j = 1,2,\ldots,n;\ i < j. \tag{2.9}$$

By multiplying w_j into Equation 2.9, we can represent the inequalities above as a set of single-sided fuzzy constraints:

$$\begin{aligned} w_i - w_j u_{ij}(\alpha) &\tilde{\leq} 0; \\ -w_i + w_j l_{ij}(\alpha) &\tilde{\leq} 0, \end{aligned} \tag{2.10}$$

or the matrix form:

$$\boldsymbol{R}w \tilde{\leq} 0, \tag{2.11}$$

where the matrix $\boldsymbol{R} \in \Re^{2m \times n}$.

In order to measure the consistent satisfaction of the interval judgment, the linear membership function is employed as follows:

$$\mu_k(\boldsymbol{R}_k\boldsymbol{w}) = \begin{cases} 1 - \dfrac{\boldsymbol{R}_k\boldsymbol{w}}{d_k}, & \boldsymbol{R}_k\boldsymbol{w} \le d_k; \\ 0, & \boldsymbol{R}_k\boldsymbol{w} > d_k, \end{cases} \tag{2.12}$$

where d_k is a tolerance parameter, denoting the admissible interval of approximate satisfaction of the crisp inequality $\boldsymbol{R}_k\boldsymbol{w} \le 0$.

The optimal weights are a crisp vector and can be represented as follows:

$$\mu_D(\boldsymbol{w}) = \max_{w}\left\{\min_{k=1,2,\ldots,m}\left[\mu_1(\boldsymbol{R}_1\boldsymbol{w}),\ldots,\mu_m(\boldsymbol{R}_m\boldsymbol{w})\right] \mid w_1 + w_2 + \cdots + w_n = 1\right\}. \tag{2.13}$$

Now, we can use the max-min operator to derive the optimal solution by solving the following linear programming model:

$$\max \lambda \tag{2.14}$$

$$\text{s.t. } \lambda \le 1 - \frac{\boldsymbol{R}_k\boldsymbol{w}}{d_k},$$

$$\sum_{i=1}^{n} w_i = 1, \quad w_i > 0,\ i = 1,2,\ldots,n;\ k = 1,2,\ldots,2m,$$

where λ denotes the degree of satisfaction and can be represented as the *C.I.* For $\lambda^* \ge 1$, the decision maker's judgments are totally consistent. On the other hand, if $\lambda^* \le 0$, the judgments of the decision maker's comparison are totally inconsistent. Furthermore, $0 < \lambda^* < 1$ indicates the degree of inconsistent judgments of the decision maker.

Example 2.3

In order to modify the problem of the geometric mean method, we can reconsider the problem of Example 2.2 using the linear programming model above. With the specific α-cut = 1, 0.2, 0.5, 0.8, and 1, we can obtain the particular weightvector under the specific degree of uncertainty and the degree of satisfaction as shown in Table 2.6.

In this section, only one kind of mathematical programming method is used to derive the FAHP. However, other variations can be easily modeled according to the purpose of the AHP as follows.

For the AHP, a near consistent matrix $\boldsymbol{A}$ with a small reciprocal multiplicative perturbation of a consistent matrix is given by Saaty (2003):

$$\boldsymbol{A} = \boldsymbol{W} \bullet \boldsymbol{E}, \tag{2.15}$$

TABLE 2.6
Various Weightvectors with the Specific α-cut

α-cut	w_1	w_2	w_3	w_4	λ*
0	0.2281	0.1579	0.5614	0.0523	0.9123
0.2	0.2312	0.1482	0.5680	0.0526	0.9052
0.5	0.2350	0.1361	0.5765	0.0524	0.8947
0.8	0.2381	0.1263	0.5836	0.0520	0.8845
1	0.2398	0.1207	0.5878	0.0517	0.8777

where • denotes the Hadamard product, $\boldsymbol{W} = [w_{ij}]_{n\times n}$ is the matrix of weight ratios, and $\boldsymbol{E} \equiv [\varepsilon_{ij}]_{n\times n}$ is the perturbation matrix, where $\varepsilon_{ij} = \varepsilon_{ji}^{-1}$.

From $\boldsymbol{Aw} = \lambda_{\max} \boldsymbol{w}$, it can be seen that

$$\sum_{j=1}^{n} a_{ij} w_j - \lambda_{\max} w_i = 0, \tag{2.16}$$

and

$$\lambda_{\max} = \sum_{j=1}^{n} a_{ij}\, w_j \Big/ w_i = \sum_{j=1}^{n} \varepsilon_{ij} \left(\text{according to Equation 2.15}\right). \tag{2.17}$$

On the other hand, the multiplicative perturbation can be transformed to an additive perturbation of a consistent matrix such that

$$\sum_{j=1}^{n} \frac{w_i}{w_j} \varepsilon_{ij} = \sum_{j=1}^{n} \frac{w_i}{w_j} + v_{ij}, \tag{2.18}$$

where v_{ij} is the additive perturbation.

Since $\sum_{j=1}^{n} a_{ij}\, w_j / w_i = \sum_{j=1}^{n} \varepsilon_{ij}$, we can rewrite Equation 2.18 as

$$\sum_{j=1}^{n} \left(\frac{w_i}{w_j} a_{ij} \frac{w_j}{w_i} \right) = \sum_{j=1}^{n} \left(\frac{w_i}{w_j} \varepsilon_{ij} \right) = \sum_{j=1}^{n} \left(\frac{w_i}{w_j} + v_{ij} \right), \tag{2.19}$$

and

$$\sum_{j=1}^{n} v_{ij} = \sum_{j=1}^{n} \left(a_{ij} - \frac{w_i}{w_j} \right). \tag{2.20}$$

On the basis of Equations 2.17 through 2.20, it can be seen that $\lambda_{max} = n$ if and only if all $\varepsilon_{ij} = 1$ or $v_{ij} = 0$, which is equivalent to having all $a_{ij} = w_i/w_j$, indicating a consistent situation. Therefore, the AHP can be transformed to minimize the objective, Equation 2.20, such that the sum of weights equals 1. Similarly, the FAHP can also be derived with the same concepts above.

Although the linear programming method above can provide the appropriate weight by considering the sum of the weights equals 1, it cannot show the fuzzy interval of the weight. Since the fuzzy interval of the weight may provide some information for the decision maker to understand the variant degree of the uncertainty, we propose the fuzzy lambda-max method for providing a sound FAHP.

2.4 FUZZY LAMBDA-MAX METHOD

The fuzzy lambda-max method was proposed by Csutora and Buckley (2001) to modify the conventional FAHP. The main advantage of this method is that it soundly provides the rational fuzzy interval of weights and considers the weighting condition, such that the sum of the weights equals 1. The concepts of the fuzzy lambda-max method can be described as follows.

Let $1^T = (1,1,\ldots,1)$ be a vector of length m of all ones and $\mathbf{\Lambda}$ be any positive reciprocal matrix where sum(l) is the sum of all the elements in $\mathbf{\Lambda}^l$, $l = 1,2,\ldots.\infty$.
Define

$$z = \lim_{l\to\infty}\left(\frac{\mathbf{\Lambda}^l \cdot 1}{\text{sum}(l)}\right), \tag{2.21}$$

then if

$$w = \left(\sum_{i}^{m} z_i\right)^{-1} z, \tag{2.22}$$

where w is the unique, positive, and normalized eigenvector of $\mathbf{\Lambda}$ corresponding to λ_{max}. Next, we can calculate the fuzzy eigenvector by fuzzifying Equations 2.21 and 2.22 as follows.

Let $\tilde{\mathbf{\Lambda}}$ be a fuzzy positive and reciprocal matrix and choose the specific value $\alpha \in [0,1]$. Let $\Gamma(\alpha) = \Pi\{\tilde{a}_{ij}[\alpha] \mid 1 \le i < j \le n\}$ and $v \in \Gamma(\alpha)$ where $v = (a_{12},\ldots, a_{1n}, a_{23},\ldots, a_{n-1,n})$. Then, we can define the positive and reciprocal matrix $\mathbf{\Lambda} = [e_{ij}]$ as follows: (1) $e_{ij} = a_{ij}$ if $1 \le i < j \le n$; (2) $e_{ii} = 1$, $1 \le i \le n$; and (3) $e_{ji} = a_{ij}^{-1}$ if $1 \le i < j \le n$.
Let

$$z = \lim_{l\to\infty}\left(\frac{\mathbf{\Lambda}^l \cdot 1}{\text{sum}(l)}\right), \tag{2.23}$$

and define $\boldsymbol{w}_v = (\sum_i^m z_i)^{-1}\boldsymbol{z}$, where $\boldsymbol{w}_v^{\mathrm{T}} = (w_{v1},\ldots,w_{vn})$. We have described a continuous mapping $\Phi_i(v) = w_{vi}$, $1 \le i \le n$, for each α in [0,1]. Then, the fuzzy eigenvector can be obtained as

$$\tilde{w}_i = \left[w_i^l, w_i^c, w_i^u\right], \quad \forall 1 \le i \le n. \tag{2.24}$$

where

$$w_i^l = \min\left\{ w_{vi} \,\middle|\, \sum_{i=1}^{m} w_{vi} = 1,\, v \in \Gamma(\alpha)\right\}, \tag{2.25}$$

$$w_i^c = \max/\min\left\{ w_{vi} \,\middle|\, \sum_{i=1}^{m} w_{vi} = 1,\, v \in \Gamma(1)\right\}, \tag{2.26}$$

and

$$w_i^u = \max\left\{ w_{vi} \,\middle|\, \sum_{i=1}^{m} w_{vi} = 1, v \in \Gamma(\alpha)\right\}, \quad \forall 1 \le i \le n. \tag{2.27}$$

TABLE 2.7
Various Fuzzy Weights Vectors Using the Fuzzy Lambda-max Method

α-cut	$\tilde{w}_1$	$\tilde{w}_2$	$\tilde{w}_3$	$\tilde{w}_4$
0	[0.1740, 0.2203, 0.2764]	[0.5971, 0.0855, 0.1203]	[0.5735, 0.6483, 0.7091]	[0.0370, 0.0459, 0.0665]
0.2	[0.1830, 0.2203, 0.2645]	[0.0649, 0.0855, 0.1128]	[0.5882, 0.6483, 0.6963]	[0.0388, 0.0459, 0.0621]
0.5	[0.1967, 0.2203, 0.2474]	[0.0725, 0.0855, 0.1022]	[0.6104, 0.6483, 0.6778]	[0.0414, 0.0459, 0.0559]
0.8	[0.2107, 0.2203, 0.2309]	[0.0802, 0.0855, 0.0920]	[0.6329, 0.6483, 0.6600]	[0.0441, 0.0459, 0.0499]
1	[0.2203, 0.2203, 0.2203]	[0.0855, 0.0855, 0.0855]	[0.6483, 0.6483, 0.6483]	[0.0459, 0.0459, 0.0459]

Example 2.4

Using the fuzzy lambda-max method to reconsider the problem of Example 2.2 with the specific α-cut = 0, 0.2, 0.5, 0.8, and 1, we can obtain the particular fuzzy weights vector under the specific degree of uncertainty as shown in Table 2.7.

On the basis of Table 2.5, it can be seen that the lambda-max method can provide flexible and rational results. The reason is that, clearly, first, the lambda-max method can provide the fuzzy interval of the weights. Second, just like the linear programming method, the lambda-max method considers the weighting condition such that the sum of the weights is equal to one.

Although the fuzzy lambda-max method can reflect the uncertain degree of weights by deriving the fuzzy weights vector, it may not intuitively determine the best alternative because the overall ratings of alternatives are fuzzy. In this situation, the additional procedure of outranking fuzzy numbers is needed. On the other hand, the linear programming methods can derive the crisp weights vector and determine the alternative intuitively. A decision maker can choose the appropriate method of FAHP depending on his/her purpose.

3 Analytic Network Process and Fuzzy Analytic Network Process

With the successful applications of the analytic hierarchy process (AHP) (Saaty 1977, 1980; Saaty and Vargas 1998) in multiple criteria decision making (MCDM), the analytic network process (ANP) was proposed by Saaty (1996) for extending the AHP to release the restrictions of the hierarchical structure, which indicates that the criteria are independent from each other. By raising the supermatrix into the limiting powers, the global priority vectors can be obtained with the specific network structure for determining dependence and feedback problems among criteria.

3.1 ANALYTIC NETWORK PROCESS

The first step of the ANP is to compare the criteria in the whole system to form the supermatrix. This is done through pairwise comparisons by asking "How much importance does a criterion have compared to another criterion, with respect to our interests or preferences?" The relative importance value can be determined using a scale from 1 to 9 for representing equal importance to extreme importance (Saaty 1980, 1996). The general form of the supermatrix can be described as follows:

$$
\begin{array}{cc}
 & \begin{array}{ccccccccccc} & C_1 & & & C_2 & & \cdots & & C_m & \\ e_{11} & \cdots & e_{1n_1} & e_{21} & \cdots & e_{2n_2} & \cdots & e_{m1} & \cdots & e_{mn_m} \end{array} \\
W = \begin{array}{cc} C_1 & \begin{array}{c} e_{11} \\ e_{12} \\ \vdots \\ e_{1n_1} \end{array} \\ C_2 & \begin{array}{c} e_{21} \\ e_{22} \\ \vdots \\ e_{2n_2} \end{array} \\ \vdots & \vdots \\ C_m & \begin{array}{c} e_{m1} \\ e_{m2} \\ \vdots \\ e_{mn_m} \end{array} \end{array} & \begin{bmatrix} \mathbf{W}_{11} & \mathbf{W}_{12} & \cdots & \mathbf{W}_{1m} \\ \mathbf{W}_{21} & \mathbf{W}_{22} & \cdots & \mathbf{W}_{2m} \\ \vdots & \vdots & \vdots & \vdots \\ \mathbf{W}_{m1} & \mathbf{W}_{m2} & \cdots & \mathbf{W}_{mm} \end{bmatrix}
\end{array}
$$

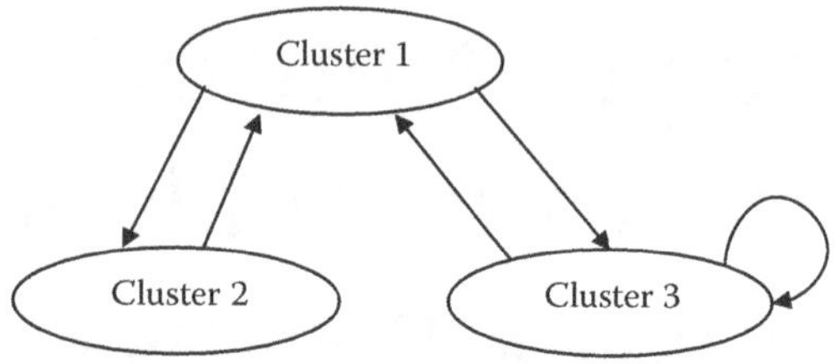

FIGURE 3.1 The network structure of Case 1.

where C_m denotes the mth cluster, e_{mn} denotes the nth element in the mth cluster, and $\mathbf{W}_{ij}$ is the principal eigenvector of the influence of the elements compared in the jth cluster to the ith cluster. In addition, if the jth cluster has no influence on the ith cluster, then $\mathbf{W}_{ij} = 0$. Therefore, the form of the supermatrix depends heavily on the variety of the structure.

Several structures were proposed by Saaty, including hierarchy, holarchy, suparchy, intarchy, etc., to demonstrate how the structure is affected by the supermatrix. Here, two simple cases, as shown in Figures 3.1 and 3.2, which both have three clusters, are used to demonstrate how to form the supermatrix based on the specific network structures.

Case 1. In Case 1, the supermatrix can be formed as the following matrix:

$$\mathbf{W} = \begin{array}{c} \\ C_1 \\ C_2 \\ C_3 \end{array} \begin{array}{c} \begin{array}{ccc} C_1 & C_2 & C_3 \end{array} \\ \begin{bmatrix} 0 & \mathbf{W}_{12} & \mathbf{W}_{13} \\ \mathbf{W}_{21} & 0 & 0 \\ \mathbf{W}_{31} & 0 & \mathbf{W}_{33} \end{bmatrix} \end{array}.$$

In Figure 3.2, the second case is more complex than the first case:

Case 2. Then, the supermatrix of Case 2 can be expressed as

$$\mathbf{W} = \begin{array}{c} \\ C_1 \\ C_2 \\ C_3 \end{array} \begin{array}{c} \begin{array}{ccc} C_1 & C_2 & C_3 \end{array} \\ \begin{bmatrix} \mathbf{W}_{11} & \mathbf{W}_{12} & \mathbf{W}_{13} \\ \mathbf{W}_{21} & \mathbf{W}_{22} & 0 \\ 0 & \mathbf{W}_{32} & 0 \end{bmatrix} \end{array}.$$

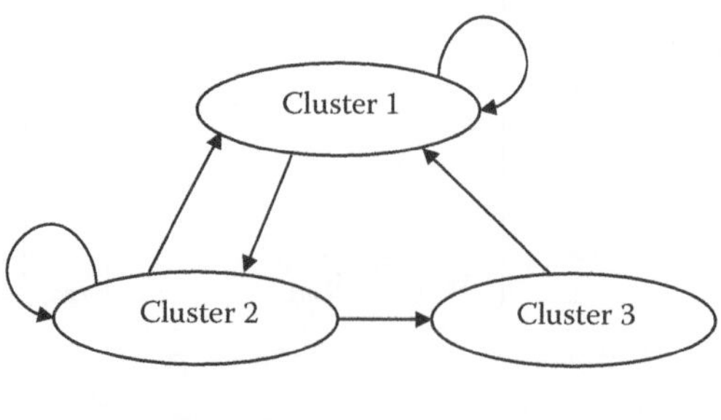

FIGURE 3.2 The network structure of Case 2.

After forming the supermatrix, the weighted supermatrix is derived by transforming all column sums to unity exactly. This step is very similar to the concept of a Markov chain for ensuring the sum of these probabilities of all states is equal to 1. Next, we raise the weighted supermatrix to limiting powers such as Equation 3.1 to get the global priority vectors or so-called weights:

$$\lim_{k\to\infty} \boldsymbol{W}^k. \tag{3.1}$$

In addition, if the supermatrix has the effect of cyclicity, the limiting supermatrix is not the only one. There are two or more limiting supermatrices in this situation and the Cesaro sum would be calculated to get the priority. The Cesaro sum is formulated as

$$\lim_{k\to\infty}\left(\frac{1}{N}\right)\sum_{r=1}^{N} W_r^k, \tag{3.2}$$

to calculate the average effect of the limiting supermatrix (i.e., the average priority weights) where $\boldsymbol{W}_r$ denotes the rth limiting supermatrix. Otherwise, the supermatrix would be raised to large powers to get the priority weights. Discussions of the mathematical processes of the ANP can refer to the literature (such as, Saaty 1996; Sekitani and Takahashi 2001) in more detail.

Example 3.1

In order to show the concrete procedures of the ANP, a simple example of system development is demonstrated to derive the priority of each criterion. As we know, the key to developing a successful system is the matching of human and technology factors. Assume the human factor can be measured by the criteria of business culture (C), end-user demand (E), and management (M). On the other hand, the technology factor can be measured by the criteria of employee ability (A), process (P), and resource (R). In addition, human and technology factors are interdependent, as shown in Figure 3.3.

The first step of the ANP is to compare the importance between each criterion. For example, the first matrix below is to ask the question "For the criterion of employee ability, how much more important is the human factor than the technology factor criteria." The other matrices can easily be formed with the same

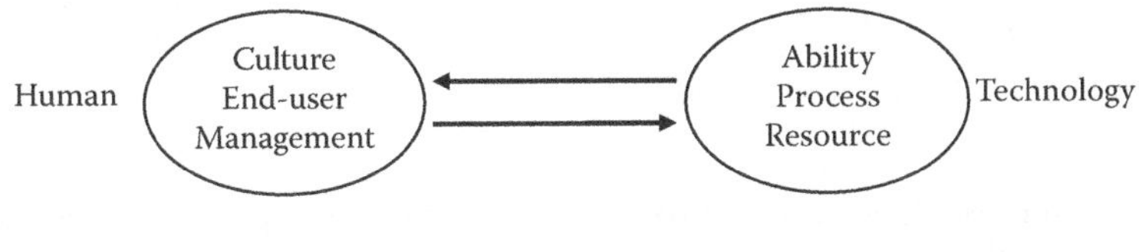

FIGURE 3.3 The network structure of the system development.

procedures. The next step is to calculate the influence (i.e., calculate the principal eigenvector) of the elements (criteria) in each component (matrix).

Ability	**Culture**	**End-user**	**Management**	**Eigenvector**	**Normalization**
Culture	1	3	4	0.634	0.634
End-user	1/3	1	1	0.192	0.192
Management	1/4	1	1	0.174	0.174
Process	**Culture**	**End-user**	**Management**	**Eigenvector**	**Normalization**
Culture	1	1	1/2	0.250	0.250
End-user	1	1	1/2	0.250	0.250
Management	2	2	1	0.500	0.500
Resource	**Culture**	**End-user**	**Management**	**Eigenvector**	**Normalization**
Culture	1	2	1	0.400	0.400
End-user	1/2	1	1/2	0.200	0.200
Management	1	2	1	0.400	0.400
Culture	**Ability**	**Process**	**Resource**	**Eigenvector**	**Normalization**
Ability	1	5	3	0.637	0.637
Process	1/5	1	1/3	0.105	0.105
Resource	1/3	3	1	0.258	0.258
End-user	**Ability**	**Process**	**Resource**	**Eigenvector**	**Normalization**
Ability	1	5	2	0.582	0.582
Process	1/5	1	1/3	0.109	0.109
Resource	1/2	3	1	0.309	0.309
Management	**Ability**	**Process**	**Resource**	**Eigenvector**	**Normalization**
Ability	1	1/5	1/3	0.136	0.136
Process	5	1	3	0.654	0.654
Resource	3	1/3	1	0.210	0.210

Now, we can form the supermatrix based on the above eigenvectors and the structure in Figure 3.3. Since the human factor can affect the technology factor, and vice versa, the supermatrix is formed as follows:

e_{11} ← C

		C	E	M	A	P	R
Cluster 1 Human	C	0	0	0	0.634	0.250	0.400
	E	0	0	0	0.192	0.250	0.200
	M	0	0	0	0.174	0.500	0.400
Cluster 2 Technology	A	0.637	0.582	0.136	0	0	0
	P	0.105	0.109	0.654	0	0	0
	R	0.258	0.309	0.210	0	0	0

(0.200) → $w_{1_2 2_3}$

(lower-right block) → W_{22}

Then, the weighted supermatrix is obtained by ensuring all columns add up to unity exactly.

$$
\begin{array}{c} \\ \mathbf{C} \\ \mathbf{E} \\ \mathbf{M} \\ \mathbf{A} \\ \mathbf{P} \\ \mathbf{R} \end{array}
\begin{array}{c}
\begin{array}{cccccc} \mathbf{C} & \mathbf{E} & \mathbf{M} & \mathbf{A} & \mathbf{P} & \mathbf{R} \end{array} \\
\begin{bmatrix}
0 & 0 & 0 & 0.634 & 0.250 & 0.400 \\
0 & 0 & 0 & 0.192 & 0.250 & 0.200 \\
0 & 0 & 0 & 0.174 & 0.500 & 0.400 \\
0.637 & 0.582 & 0.136 & 0 & 0 & 0 \\
0.105 & 0.109 & 0.654 & 0 & 0 & 0 \\
0.258 & 0.309 & 0.210 & 0 & 0 & 0
\end{bmatrix}
\end{array}.
$$

Finally, by calculating the limiting power of the weighted supermatrix, the limiting supermatrix can be obtained as follows:

$$
\begin{array}{c} \\ \mathbf{C} \\ \mathbf{E} \\ \mathbf{M} \\ \mathbf{A} \\ \mathbf{P} \\ \mathbf{R} \end{array}
\begin{array}{c}
\begin{array}{cccccc} \mathbf{C} & \mathbf{E} & \mathbf{M} & \mathbf{A} & \mathbf{P} & \mathbf{R} \end{array} \\
\begin{bmatrix}
0 & 0 & 0 & 0.464 & 0.464 & 0.464 \\
0 & 0 & 0 & 0.210 & 0.210 & 0.210 \\
0 & 0 & 0 & 0.324 & 0.324 & 0.324 \\
0.463 & 0.463 & 0.463 & 0 & 0 & 0 \\
0.284 & 0.284 & 0.284 & 0 & 0 & 0 \\
0.253 & 0.253 & 0.253 & 0 & 0 & 0
\end{bmatrix}
\end{array}
\quad (\text{when } k \text{ is even})
$$

and

$$
\begin{array}{c} \\ \mathbf{C} \\ \mathbf{E} \\ \mathbf{M} \\ \mathbf{A} \\ \mathbf{P} \\ \mathbf{R} \end{array}
\begin{array}{c}
\begin{array}{cccccc} \mathbf{C} & \mathbf{E} & \mathbf{M} & \mathbf{A} & \mathbf{P} & \mathbf{R} \end{array} \\
\begin{bmatrix}
0.464 & 0.464 & 0.464 & 0 & 0 & 0 \\
0.210 & 0.210 & 0.210 & 0 & 0 & 0 \\
0.324 & 0.324 & 0.324 & 0 & 0 & 0 \\
0 & 0 & 0 & 0.463 & 0.463 & 0.463 \\
0 & 0 & 0 & 0.284 & 0.284 & 0.284 \\
0 & 0 & 0 & 0.253 & 0.253 & 0.253
\end{bmatrix}
\end{array}
\quad (\text{when } k \text{ is odd}).
$$

As we see, the supermatrix has the effect of cyclicity, and in this situation the Cesaro sum (i.e., add the two matrices and divide by two) is used to obtain the final priorities as follows:

$$
\begin{array}{c} \\ \mathbf{C} \\ \mathbf{E} \\ \mathbf{M} \\ \mathbf{A} \\ \mathbf{P} \\ \mathbf{R} \end{array}
\begin{array}{c}
\begin{array}{cccccc} \mathbf{C} & \mathbf{E} & \mathbf{M} & \mathbf{A} & \mathbf{P} & \mathbf{R} \end{array} \\
\begin{bmatrix}
0.233 & 0.233 & 0.233 & 0.233 & 0.233 & 0.233 \\
0.105 & 0.105 & 0.105 & 0.105 & 0.105 & 0.105 \\
0.162 & 0.162 & 0.162 & 0.162 & 0.162 & 0.162 \\
0.231 & 0.231 & 0.231 & 0.231 & 0.231 & 0.231 \\
0.142 & 0.142 & 0.142 & 0.142 & 0.142 & 0.142 \\
0.127 & 0.127 & 0.127 & 0.127 & 0.127 & 0.127
\end{bmatrix}
\end{array}.
$$

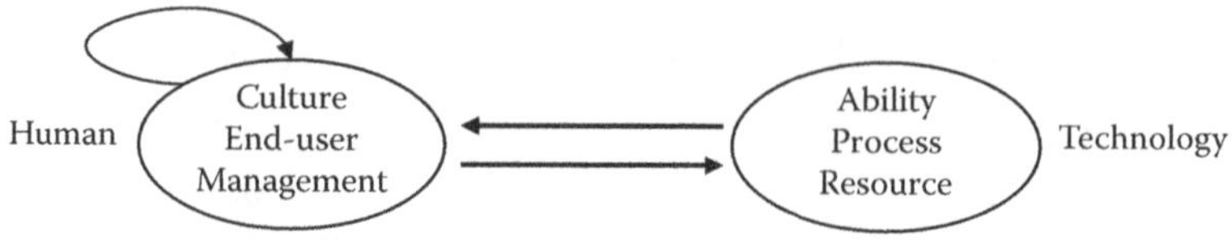

FIGURE 3.4 The network structure of system development with feedback effects.

In this example, the criterion of culture has the highest priority (0.233) in system development and the criterion of end-user has the least priority (0.105).

Example 3.2

In order to show the effect of the structure in the ANP, the other structure, which has the feedback effect on human factors, is considered as in Figure 3.4.

There are two methods to deal with the self-feedback effect. The first method is to simply place 1 in diagonal elements and the other method performs a pairwise comparison of the criteria on each criterion. In this example, we use the first method. With the same steps above, the unweighted supermatrix, the weighted supermatrix, and the limiting supermatrix can be obtained as follows, respectively:

$$\begin{array}{c} \\ \mathbf{C} \\ \mathbf{E} \\ \mathbf{M} \\ \mathbf{A} \\ \mathbf{P} \\ \mathbf{R} \end{array} \begin{array}{c} \begin{array}{cccccc} \mathbf{C} & \mathbf{E} & \mathbf{M} & \mathbf{A} & \mathbf{P} & \mathbf{R} \end{array} \\ \begin{bmatrix} 1 & 0 & 0 & 0.634 & 0.250 & 0.400 \\ 0 & 1 & 0 & 0.192 & 0.250 & 0.200 \\ 0 & 0 & 1 & 0.174 & 0.500 & 0.400 \\ 0.637 & 0.582 & 0.136 & 0 & 0 & 0 \\ 0.105 & 0.109 & 0.654 & 0 & 0 & 0 \\ 0.258 & 0.309 & 0.210 & 0 & 0 & 0 \end{bmatrix} \end{array},$$

$$\begin{array}{c} \\ \mathbf{C} \\ \mathbf{E} \\ \mathbf{M} \\ \mathbf{A} \\ \mathbf{P} \\ \mathbf{R} \end{array} \begin{array}{c} \begin{array}{cccccc} \mathbf{C} & \mathbf{E} & \mathbf{M} & \mathbf{A} & \mathbf{P} & \mathbf{R} \end{array} \\ \begin{bmatrix} 0.5 & 0 & 0 & 0.634 & 0.250 & 0.400 \\ 0 & 0.5 & 0 & 0.192 & 0.250 & 0.200 \\ 0 & 0 & 0.5 & 0.174 & 0.500 & 0.400 \\ 0.319 & 0.291 & 0.068 & 0 & 0 & 0 \\ 0.053 & 0.055 & 0.327 & 0 & 0 & 0 \\ 0.129 & 0.155 & 0.105 & 0 & 0 & 0 \end{bmatrix} \end{array},$$

$$\begin{array}{c} \\ \mathbf{C} \\ \mathbf{E} \\ \mathbf{M} \\ \mathbf{A} \\ \mathbf{P} \\ \mathbf{R} \end{array} \begin{array}{c} \begin{array}{cccccc} \mathbf{C} & \mathbf{E} & \mathbf{M} & \mathbf{A} & \mathbf{P} & \mathbf{R} \end{array} \\ \begin{bmatrix} 0.310 & 0.310 & 0.310 & 0.310 & 0.310 & 0.310 \\ 0.140 & 0.140 & 0.140 & 0.140 & 0.140 & 0.140 \\ 0.216 & 0.216 & 0.216 & 0.216 & 0.216 & 0.216 \\ 0.154 & 0.154 & 0.154 & 0.154 & 0.154 & 0.154 \\ 0.095 & 0.095 & 0.095 & 0.095 & 0.095 & 0.095 \\ 0.084 & 0.084 & 0.084 & 0.084 & 0.084 & 0.084 \end{bmatrix} \end{array}.$$

Since the effect of cyclicity does not exist in this example, the final priorities are directly obtained by limiting the power to converge. Although the criterion of culture also has the highest priority, the priority changes from 0.233 to 0.310. On the other hand, the lowest priority is resource (0.084) rather than end-user. Compared to the priorities of the two examples, the structures are the key to both the effects and the results. In addition, it should be highlighted that when we raise the weighted matrix to limiting power, the weighted matrix should always be the stochastic matrix.

The advantage of the ANP is that it is not only appropriate for both quantitative and qualitative data types, but it also can overcome the problem of interdependence and feedback between all features. Although the ANP has been widely applied to project selection (Lee and Kim 2000; Meade and Presley 2002), strategic decision (Karsak, Sozer, and Alptekin 2002; Sarkis 2003), and optimal scheduling (Momoh and Zhu 2003) recently, it seems to have been ignored in the problem of uncertainty. It is clear that due to the problem of incomplete information or human subjective uncertainty, it is hard even for experts to quantify the precise importance among criteria. Although the concepts of fuzzy sets have been incorporated in the AHP to consider the problem of uncertainty (Buckley 1985; van Laarhoven and Pedrycz 1983; Wagenknecht and Hartmann 1983), few papers extend the ANP to cope with uncertain human judgments.

Mikhailov and Singh (2003) proposed the fuzzy analytic network process (FANP) to extend the ANP fuzzy environments. Their method first derived crisp local weights from the fuzzy pairwise judgments by using the fuzzy preference programming (FPP) method (Mikhailov 2003), and then a crisp weighted supermatrix is formed and raised to a steady-state process to obtain global weights. In other words, their method can only derive fuzzy weights in the AHP rather than fuzzy weights in the ANP. In addition, scholars (Kahraman, Ertay, and Büyüközkan 2006; Büyüközkan 2004; Ertay et al. 2005; Mohanty et al. 2005) proposed another method to deal with the uncertain judgments in the ANP based on fuzzy arithmetic operations. However, their methods may result in the convergent and rational problems of fuzzy global weights, because of the use of standard fuzzy arithmetic operations to multiply and divide fuzzy numbers. In the next section, we propose another method to consider the ANP under fuzzy environments.

3.2 FUZZY ANALYTIC NETWORK PROCESS*

To incorporate the concept of uncertainty into the ANP, fuzzy numbers are used to describe the degree of uncertainty. In this chapter, fuzzy numbers are presented in triangular form. Note that other forms of the membership function can be easily employed by using the same procedures.

Step 1: Compare the ratios of weights between criteria with respect to each cluster using the fuzzy judgments. To satisfy the condition of the fuzzy reciprocal matrix, we assume that $\tilde{a}_{ji} = 1/\tilde{a}_{ij}$ and $\tilde{a}_{ii} = 1$ (Mikhailov 2003). That is, it is assumed that if

* Originally abstracted from Huang, J.J., and G.H. Tzeng. (2007). A constrained fuzzy arithmetic method for the fuzzy analytic network process. Fuzzy Systems and Knowledge Discovery, Fourth International Conference 3, 401–405.

$$\tilde{a}_{ij} = \left(\underline{a}_{ij}, a_{ij}^c, \overline{a}_{ij}\right), \tag{3.3}$$

then

$$\tilde{a}_{ji} = \left(\frac{1}{\overline{a}_{ij}}, \frac{1}{a_{ij}^c}, \frac{1}{\underline{a}_{ij}}\right), \tag{3.4}$$

where $\underline{a}_{ij}$ denotes the infimum, a_{ij}^c denotes the center value, and $\overline{a}_{ij}$ denotes the supremum.

Step 2: Derive the fuzzy local weight vectors. Several methods have been proposed for deriving fuzzy local eigenvectors, e.g., the fuzzy geometric mean method (Buckley 1985), the fuzzy least-square method (Wagenknecht and Hartmann 1983), the FPP method (Mikhailov 2003), and the fuzzy logarithmic least-square method (van Laarhoven and Pedrycz 1983). However, all of these methods have some drawbacks and cannot be employed in this chapter. First, since the FPP method only derives crisp weights, it cannot satisfy our requirement for obtaining fuzzy local weights. Meanwhile, the other methods may result in irrational fuzzy weights, where the infimum is greater than the center value or the center value is greater than the supremum (Chang and Lee 1995). Therefore, in this chapter, a fuzzy and positive eigenvector is derived by directly fuzzifying Saaty's eigenvector method (Csutora and Buckley 2001) as follows.

Let $\tilde{\mathbf{\Lambda}}$ be a fuzzy positive and reciprocal matrix and choose the specific value $\alpha \in [0,1]$. In addition, let $\Gamma(\alpha)$ be a Cartesian product of intervals, i.e., $\Gamma(\alpha) = \Pi\{\tilde{a}_{ij}[\alpha] \mid 1 \le i < j \le m\}$; and $\boldsymbol{v} \in \Gamma(\alpha)$, where $\boldsymbol{v} = (a_{12}, \ldots, a_{1m}, a_{23}, \ldots, a_{m-1,m})$. Then, we can define a positive and reciprocal matrix $\mathbf{\Lambda} = [e_{ij}]$ as follows: (1) $e_{ij} = a_{ij}$ if $1 \le i < j \le m$; (2) $e_{ii} = 1$, $1 \le i \le m$; and (3) $e_{ji} = a_{ij}^{-1}$ if $1 \le i < j \le m$. Let

$$\boldsymbol{w} = \lim_{l \to \infty}\left(\frac{\mathbf{\Lambda}^l \mathbf{1}}{\mathbf{1}'\mathbf{\Lambda}^l \mathbf{1}}\right), \tag{3.5}$$

where $\mathbf{1}' = [1,1,\ldots,1]$ and $\mathbf{\Lambda}$ is any positive reciprocal matrix. Having described a continuous mapping $\Phi_i(v) = w_i$, $1 \le i \le m$ for each α in the range $[0,1]$, we can obtain the following fuzzy eigenvector:

$$\tilde{w}_i[\alpha] = \left[\underline{w}_i(\alpha), \overline{w}_i(\alpha)\right], \quad \forall\, 1 \le i \le m, \tag{3.6}$$

where

$$\underline{w}_i(\alpha) = \min\left\{ w_i(\alpha) \,\middle|\, \sum\nolimits_{i=1}^{m} w_i = 1, \boldsymbol{v} \in \Gamma(\alpha)\right\}, \tag{3.7}$$

$$\overline{w}_i(\alpha) = \max\left\{ w_i(\alpha) \middle| \sum_{i=1}^{m} w_i = 1, v \in \Gamma(\alpha) \right\}. \tag{3.8}$$

Next, we provide an example to show how the local weights can be derived. Let

$$\tilde{A} = \begin{bmatrix} 1 & \left(\frac{1}{4},\frac{1}{3},\frac{1}{2}\right) & (3,4,5) \\ (2,3,4) & 1 & (5,6,7) \\ \left(\frac{1}{5},\frac{1}{4},\frac{1}{3}\right) & \left(\frac{1}{7},\frac{1}{6},\frac{1}{5}\right) & 1 \end{bmatrix}$$

be a positive and reciprocal fuzzy comparison matrix. By employing Equations 3.6 through 3.8 and setting the α-cuts = 0, 0.2, 0.4, 0.6, 0.8, and 1, we can derive the corresponding fuzzy weights for the matrix, as shown in Table 3.1.

From the results in Table 3.1, we can depict the triangular-shaped fuzzy weights of the given example, as shown in Figure 3.5.

Note that, since the bounds of fuzzy eigenvectors may be hard to calculate, some heuristic methods, such as genetic algorithms or simulated annealing algorithms, should be used to handle this problem.

Step 3: Form the fuzzy weighted supermatrix. By using the equations in Step 2, the fuzzy local eigenvectors can be derived to form a fuzzy weighted supermatrix based on the network structure of the problem, as shown in Figure 3.6.

Now, we can obtain the fuzzy global priorities by raising the fuzzy weighted supermatrix to its limiting power in the following step.

Step 4: Raise the fuzzy weighted supermatrix until the convergent condition is satisfied. In this step, the fuzzy weighted supermatrix is raised to its limiting power to obtain the fuzzy global weights. Note that the supermatrix should always follow the properties of the stochastic matrix in each power.

TABLE 3.1
The Fuzzy Weights of the Given Example with α-cuts

α-cuts	$\tilde{w}_1[\alpha]$	$\tilde{w}_2[\alpha]$	$\tilde{w}_3[\alpha]$
α-cut = 0.0	[0.2088,0.3556]	[0.5500,0.7143]	[0.0651,0.1169]
α-cut = 0.2	[0.2206,0.3376]	[0.5696,0.7002]	[0.0686,0.1096]
α-cut = 0.4	[0.2327,0.3202]	[0.5887,0.6870]	[0.0724,0.1029]
α-cut = 0.6	[0.2450,0.3034]	[0.6074,0.6730]	[0.0764,0.0966]
α-cut = 0.8	[0.2576,0.2868]	[0.6259,0.6588]	[0.0806,0.0907]
α-cut = 1.0	[0.2706,0.2706]	[0.6442,0.6442]	[0.0852,0.0852]

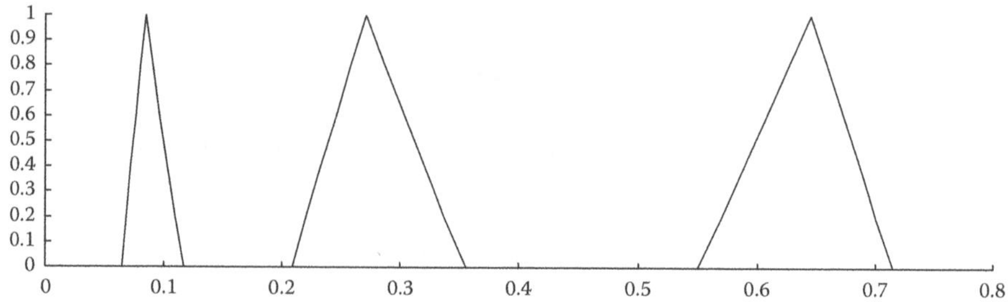

FIGURE 3.5 The triangular-shaped fuzzy weights of the given example.

However, it is clear that we cannot use standard fuzzy arithmetic operations to raise the fuzzy supermatrix to its limiting power, due to the problem of convergence. Hence, constrained fuzzy arithmetic operations (Klir and Pan 1997, 1998) are employed to avoid the convergent problem in raising the fuzzy supermatrix to its limiting power. The concepts of the constrained fuzzy arithmetic operations can be described as follows.

Suppose two fuzzy numbers can be represented as $\tilde{G}$ and $\tilde{Q}$. Then the constrained fuzzy arithmetic operations can be defined by:

$$\left(\tilde{G}*\tilde{Q}\right)[\alpha]=\left\{g*q \mid g,q\in\left(\tilde{G}[\alpha]\times\tilde{Q}[\alpha]\right)\cap R[\alpha]\right\}, \tag{3.9}$$

where * denotes the four basic arithmetic operations on fuzzy numbers, and R is the constraint. Using the concept of constrained fuzzy arithmetic operations, we can derive the global weights of the fuzzy supermatrix as follows.

$$
{}^{S}\tilde{W} = \begin{array}{cc}
 & \begin{array}{cccc} \quad C_1 & \quad C_2 & \cdots & \quad C_m \\ e_{11}\cdots e_{1n_1} & e_{21}\cdots e_{2n_2} & \cdots & e_{m1}\cdots e_{mn_m} \end{array} \\
\begin{array}{cc} C_1 & \begin{array}{c} e_{11} \\ e_{12} \\ \vdots \\ e_{1n_1} \end{array} \\ C_2 & \begin{array}{c} e_{21} \\ e_{22} \\ \vdots \\ e_{2n_2} \end{array} \\ \vdots & \vdots \\ C_m & \begin{array}{c} e_{m1} \\ e_{m2} \\ \vdots \\ e_{mn_m} \end{array} \end{array} &
\begin{bmatrix} \tilde{W}_{11} & \tilde{W}_{12} & \cdots & \tilde{W}_{1m} \\ \tilde{W}_{21} & \tilde{W}_{22} & \cdots & \tilde{W}_{2m} \\ \vdots & \vdots & \vdots & \vdots \\ \tilde{W}_{m1} & \tilde{W}_{m2} & \cdots & \tilde{W}_{mm} \end{bmatrix}
\end{array}
$$

FIGURE 3.6 A fuzzy weighted supermatrix.

Assume a transition probability matrix of an m-state fuzzy Markov chain can be described as:

$$\Pi = \begin{bmatrix} \pi_{11} & \pi_{12} & \cdots & \pi_{1m} \\ \pi_{21} & \pi_{22} & \cdots & \pi_{2m} \\ \vdots & \vdots & & \vdots \\ \pi_{m1} & \pi_{m2} & \cdots & \pi_{mm} \end{bmatrix}, \tag{3.10}$$

where $\tilde{\pi}_{ij}$ denotes the transition probability. Then, the constraint of the transition matrix can be described by the following equation (Buckley and Eslami 2002):

$$S = \left\{ \pi' = (\pi_1, \pi_2, \ldots, \pi_m) \middle| \pi_i \geq 0, \sum_{i=1}^{m} \pi_i = 1 \right\}, \tag{3.11}$$

where π_i denotes the ith entity in the vector $\boldsymbol{\pi}$ and the α-cut domain can be defined by

$$\widetilde{\text{Dom}}_i[\alpha] = \left(\prod_{j=1}^{m} \tilde{\pi}_{ij}[\alpha] \right) \cap S, \quad 0 \leq \alpha \leq 1; 1 \leq i \leq m, \tag{3.12}$$

and

$$\widetilde{\text{Dom}}[\alpha] = \prod_{i=1}^{m} \widetilde{\text{Dom}}_i[\alpha], \quad 0 \leq \alpha \leq 1, \tag{3.13}$$

where $\tilde{\pi}_{ij}[\alpha]$ denotes the α cut fuzzy probability of the ith row and the jth column.

Next, the fuzzy steady-state probabilities can be derived by a specific function of the α-cut domain as shown in Equation 3.14.

$$\tilde{\pi}_{ij}^{(k)}[\alpha] = f_{ij}^{(k)}\left(\widetilde{\text{Dom}}[\alpha]\right), \tag{3.14}$$

where (k) denotes the limiting power that makes the transition probabilities into the steady-state probabilities. Note that Equation 3.14 can be explained by the steady-state probabilities being some function of the transition probability in the transition matrix. Since $f_{ij}^{(k)}(\cdot)$ is continuous and $\widetilde{\text{Dom}}[\alpha]$ is a closed and bounded range, it is clear that $\tilde{\pi}_{ij}^{(k)}[\alpha]$ is also a closed and bounded interval range.

Finally, the fuzzy steady-state probabilities can be expressed using the α-cut as

$$\tilde{\pi}_i^{(k)}[\alpha] = \left[\underline{\pi}_i^{(k)}(\alpha), \overline{\pi}_i^{(k)}(\alpha) \right], \quad \forall\, 1 \leq i \leq m, \tag{3.15}$$

where

$$\underline{\pi}_i^{(k)}[\alpha] = \min\left\{f_{ij}^{(k)}(\pi) \middle| \pi \in \widetilde{\text{Dom}}[\alpha]\right\}, \tag{3.16}$$

and

$$\overline{\pi}_i^{(k)}[\alpha] = \max\left\{f_{ij}^{(k)}(\pi) \middle| \pi \in \widetilde{\text{Dom}}[\alpha]\right\}. \tag{3.17}$$

The computational procedures for finding fuzzy steady-state probabilities are described as follows. First, determine $\alpha \in [0,1]$ and compute the intervals $I_{ij} = \tilde{\pi}_{ij}[\alpha]$. Then, choose a crisp value $z_{ij} \in I_{ij}$ such that $\sum_{i=1}^{m} z_{ij} = 1, \forall i, j = 1,\ldots,m$. Next, we calculate $f_{ij}^{(k)}(\widetilde{\text{Dom}[\alpha]})$ by the process of randomly choosing the z_{ij} in I_{ij} so that all the column sums are equal to one. Finally, we can find the (approximate) intervals as shown in Equations 3.15 through 3.17. Note that in the above procedures, π_{ij} is estimated by z_{ij}. Next, we use a simple example to explain how to derive the fuzzy steady-state priorities in the fuzzy Markov chain according to the equations in Step 4.

Let a 2×2 generalized fuzzy matrix be expressed as:

$$\tilde{\Pi} = \begin{bmatrix} \pi_{11} & \pi_{12} \\ \pi_{21} & \pi_{22} \end{bmatrix},$$

where $\tilde{\pi}_{11} = (0.3, 0.5, 0.7)$, $\tilde{\pi}_{12} = (0.2, 0.3, 0.4)$, $\tilde{\pi}_{21} = (0.3, 0.5, 0.7)$, and $\tilde{\pi}_{22} = (0.6,$ $\tilde{\pi}_{22} = (0.6, 0.7, 0.8)$. Next, select $\alpha = 0$ and $\alpha = 1$ and perform the random process of estimating $f_{ij}^{(k)}(\widetilde{\text{Dom}[\alpha]})$ 1000 times, respectively, to estimate the end points of $\tilde{\pi}_1^{(k)}$ and $\tilde{\pi}_2^{(k)}$. According to the experiment results, the fuzzy steady-state probabilities are $\tilde{\pi}_1^{(k)}[0] = [2/9, 4/7]$ and $\tilde{\pi}_2^{(k)}[0] = [3/7, 7/9]$, and $\tilde{\pi}_1^{(k)}[1] = [3/8, 3/8]$ and $\tilde{\pi}_2^{(k)}[1] = [5/8, 5/8]$.

From a theoretical viewpoint, raising the supermatrix to its limiting power in the ANP is similar to the same properties of Markov chains in that the elements in the supermatrix are similar to the transition probabilities in the transition matrix. Therefore, the concepts of fuzzy Markov chains can be easily applied to the FANP. Using the above concepts, it can be seen that the steady-state priority vector of the fuzzy weighted supermatrix is definitely convergent and rational. In addition, we can obtain fuzzy global weights, rather than crisp global weights, to understand the degree of uncertainty. Note that the proposed method is suitable for fuzzy, interval, crisp, and mixed numbers. Next, we give two numerical examples to illustrate the proposed method.

Example 3.3

The key to developing a successful information system is the alignment of human and technological factors. We assume that the human factor can be measured by

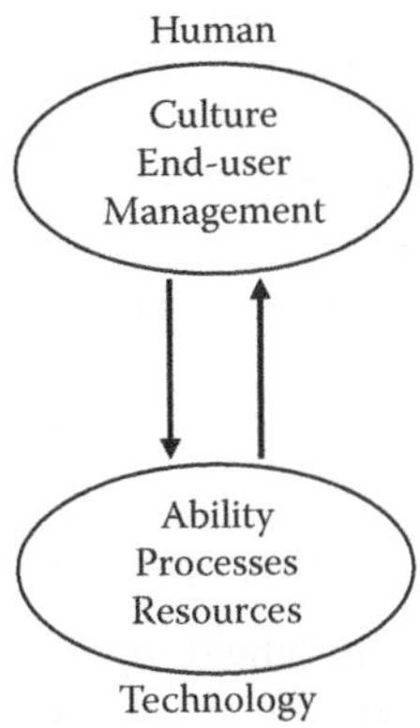

FIGURE 3.7 The network structure of the system development.

the criteria of the business culture, end-user demand, and management. On the other hand, the technological factor can be measured by the criteria of employee ability, processes, and resources. The human and the technology factors are interdependent, as shown in Figure 3.7.

Due to the restrictions of incomplete information and subjective uncertainty, the decision maker adopts fuzzy numbers to judge the ratios of the weights between criteria. For example, in the first matrix below, the question is "For the criterion of culture, how much more importance does one technology criterion have than another?" The other matrices can be easily formed by the same procedures. The next step calculates the fuzzy local weights in each matrix, as shown in the following six matrices:

Culture	**Ability**	**Processes**	**Resources**	**Fuzzy Local Weights**
Ability	[1,1,1]	[3,5,6]	[2,3,4]	[0.5171,0.6370,0.7167]
Processes	[1/6,1/5,1/3]	[1,1,1]	[1/4,1/3,1/2]	[0.0772,0.1047,0.1433]
Resources	[1/4,1/3,1/2]	[2,3,4]	[1,1,1]	[0.1929,0.2583,0.3680]
End-user	**Ability**	**Processes**	**Resources**	**Fuzzy Local Weights**
Ability	[1,1,1]	[3,4,5]	[1,2,3]	[0.3874,0.5580,0.6990]
Processes	[1/5,1/4,1/3]	[1,1,1]	[1/4,1/3,1]	[0.0833,0.1220,0.2451]
Resources	[1/3,1/2,1]	[1,3,4]	[1,1,1]	[0.1800,0.3200,0.4672]
Management	**Ability**	**Processes**	**Resources**	**Fuzzy Local Weights**
Ability	[1,1,1]	[1/7,1/5,1/4]	[1/4,1/3,1/2]	[0.0712,0.1047,0.1518]
Processes	[4,5,7]	[1,1,1]	[2,3,4]	[0.5319,0.6370,0.7300]
Resources	[2,3,4]	[1/4,1/3,1/2]	[1,1,1]	[0.1833,0.2583,0.3510]
Ability	**Culture**	**End-user**	**Management**	**Fuzzy Local Weights**
Culture	[1,1,1]	[2,3,4]	[3,4,7]	[0.4304,0.6337,0.7788]
End-user	[1/4,1/3,1/2]	[1,1,1]	[1/2,1,3]	[0.1042,0.1919,0.3443]
Management	[1/7,1/4,1/3]	[1/3,1,2]	[1,1,1]	[0.0798,0.1744,0.3522]

Processes	Culture	End-user	Management	Fuzzy Local Weights
Culture	[1,1,1]	[1/2,1,2]	[1,2,3]	[0.2051,0.4126,0.6206]
End-user	[1/2,1,2]	[1,1,1]	[1/2,1,3]	[0.1699,0.3275,0.5878]
Management	[1/3,1/2,1]	[1/3,1,2]	[1,1,1]	[0.1168,0.2599,0.4695]
Resources	**Culture**	**End-user**	**Management**	**Fuzzy Local Weights**
Culture	[1,1,1]	[1,2,3]	[1,2,3]	[0.2905,0.4934,0.6392]
End-user	[1/3,1/2,1]	[1,1,1]	[1/3,1/2,1]	[0.1201,0.1958,0.3713]
Management	[1/3,1/2,1]	[1,2,3]	[1,1,1]	[0.1837,0.3108,0.4934]

Now, we can form the fuzzy weighted supermatrix according to the results of the fuzzy eigenvectors above and the network structure shown in Figure 3.7. Since the human and technology factors are interdependent, the fuzzy supermatrix is as follows:

	Culture	End-user	Management	Ability	Processes	Resources
Culture	0	0	0	[0.430,0.634,0.779]	[0.205,0.413,0.621]	[0.291,0.493,0.639]
End-user	0	0	0	[0.104,0.192,0.344]	[0.170,0.327,0.588]	[0.120,0.196,0.371]
Management	0	0	0	[0.080,0.174,0.352]	[0.117,0.260,0.470]	[0.184,0.311,0.493]
Ability	[0.517,0.637,0.717]	[0.387,0.558,0.699]	[0.071,0.105,0.152]	0	0	0
Processes	[0.077,0.105,0.143]	[0.083,0.122,0.245]	[0.532,0.637,0.730]	0	0	0
Resources	[0.193,0.258,0.368]	[0.180,0.320,0.467]	[0.183,0.258,0.351]	0	0	0

Next, we can obtain the steady-state priority vectors of the fuzzy supermatrix. By setting α-cut = 0 and α-cut = 1, we can derive the fuzzy priority vectors shown in Table 3.2.

Example 3.4

Another example of a network structure is shown in Figure 3.8.

There are two ways to deal with the self-feedback effect. One method simply places 1 in diagonal elements, and the other performs pairwise comparison of the criteria with each criterion. Here, we use the first method for simplicity. Using the steps described in Section 3.2, the unweighted supermatrix and the weighted supermatrix can be represented as the following two matrices, respectively:

TABLE 3.2
The Global Fuzzy Weights in Example 3.3

	Culture	End-user	Management	Ability	Processes	Resources
FANP([α] = 0)	[0.1397, 0.3706]	[0.0565, 0.2211]	[0.0491, 0.2375]	[0.1295, 0.3329]	[0.0632, 0.2152]	[0.0888, 0.2063]
FANP([α] = 1)	[0.2722, 0.2722]	[0.1122, 0.1122]	[0.1156, 0.1156]	[0.2481, 0.2481]	[0.1158, 0.1158]	[0.1361, 0.1361]

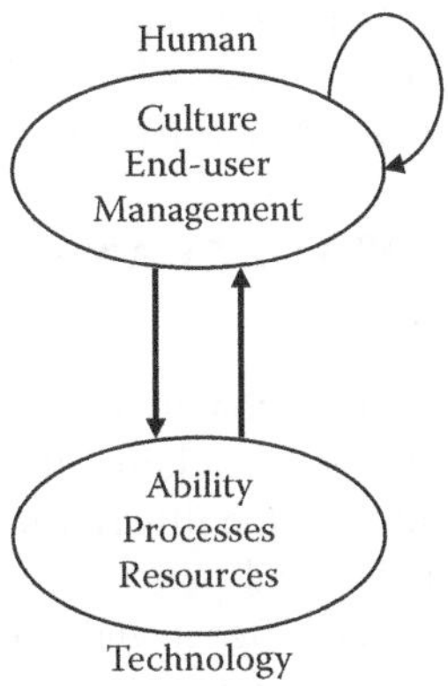

FIGURE 3.8 The network structure of an information system with feedback effects.

	Culture	End-user	Management	Ability	Processes	Resources
Culture	[1, 1, 1]	0	0	[0.430,0.634,0.779]	[0.205,0.413,0.621]	[0.291,0.493,0.639]
End-user	0	[1, 1, 1]	0	[0.104,0.192,0.344]	[0.170,0.327,0.588]	[0.120,0.196,0.371]
Management	0	0	[1, 1, 1]	[0.080,0.174,0.352]	[0.117,0.260,0.470]	[0.184,0.311,0.493]
Ability	[0.517,0.637,0.717]	[0.387,0.558,0.699]	[0.071,0.105,0.152]	0	0	0
Processes	[0.077,0.105,0.143]	[0.083,0.122,0.245]	[0.532,0.637,0.730]	0	0	0
Resources	[0.193,0.258,0.368]	[0.180,0.320,0.467]	[0.183,0.258,0.351]	0	0	0

	Culture	End-user	Management	Ability	Processes	Resources
Culture	[0.5, 0.5, 0.5]	0	0	[0.430,0.634,0.779]	[0.205,0.413,0.621]	[0.291,0.493,0.639]
End-user	0	[0.5, 0.5, 0.5]	0	[0.104,0.192,0.344]	[0.170,0.327,0.588]	[0.120,0.196,0.371]
Management	0	0	[0.5, 0.5, 0.5]	[0.080,0.174,0.352]	[0.117,0.260,0.470]	[0.184,0.311,0.493]
Ability	[0.289,0.319,0.359]	[0.194,0.279,0.350]	[0.036,0.053,0.076]	0	0	0
Processes	[0.039,0.053,0.072]	[0.042,0.061,0.123]	[0.266,0.319,0.365]	0	0	0
Resources	[0.097,0.129,0.184]	[0.090,0.160,0.234]	[0.092,0.129,0.176]	0	0	0

Next, we can derive the fuzzy priorities, as shown in Table 3.3.

TABLE 3.3
The Global Fuzzy Weights in Example 3.4

	Culture	**End-user**	**Management**	**Ability**	**Processes**	**Resources**
FANP([α] = 0)	[0.1810, 0.4984]	[0.0726, 0.3021]	[0.0634, 0.3239]	[0.0878, 0.2208]	[0.0420, 0.1458]	[0.0601, 0.1364]
FANP([α] = 1)	[0.3629, 0.3629]	[0.1496, 0.1496]	[0.1542, 0.1542]	[0.1654, 0.1654]	[0.0772, 0.0772]	[0.0907, 0.0907]

From the results shown in Tables 3.2 and 3.3, we observe that the proposed method can provide fuzzy global weights with a specific α-cut and the ANP can be considered as the special case while α-cut = 0.

3.3 MATRIX METHOD FOR FUZZY ANALYTIC NETWORK PROCESS*

In this section, we relax the assumption of the reciprocal matrix in the FANP. The assumption of the reciprocal matrix claims that if a_{ij} denotes the ratio of the weight that the ith criterion dominates the jth criterion, $a_{ji} = 1/a_{ij}$ should be satisfied. However, the property of the reciprocal matrix is not held in a fuzzy matrix, i.e., $\tilde{a}_{ji} \neq 1/\tilde{a}_{ij}$. Therefore, in this chapter, Cogger and Yu's method (1985) is introduced to show how the eigenvector can be derived when the postulation of reciprocity in a comparison matrix is released as follows.

Let a positive upper triangular comparison matrix

$$\boldsymbol{A} = \left[a_{ij}\right]_{n\times n}, \quad a_{ij} = \begin{cases} a_{ij} & \text{if } i \leq j, \\ 0 & \text{otherwise,} \end{cases} \tag{3.18}$$

where a_{ij} denotes the strength of the ratio of the weight that the criterion i dominates the criterion j.

Let $\boldsymbol{D}$ be the diagonal matrix such that

$$\boldsymbol{D} = \left[d_{ij}\right]_{n\times n}, \quad \text{with } d_{ij} = \begin{cases} n-i+1 & \text{if } i = j, \\ 0 & \text{otherwise.} \end{cases} \tag{3.19}$$

By introducing the weight vector $\boldsymbol{w}' = (w_1, w_2, \ldots, w_n)$, we can obtain

$$\boldsymbol{A}\boldsymbol{w} = \boldsymbol{D}\boldsymbol{w} \tag{3.20}$$

and

$$\left(\boldsymbol{D}^{-1}\boldsymbol{A} - \boldsymbol{I}\right)\boldsymbol{w} = \boldsymbol{0}, \tag{3.21}$$

where

$$d_{ij}^{-1} = \begin{cases} 1/n-i+1 & \text{if } i = j, \\ 0 & \text{otherwise.} \end{cases} \tag{3.22}$$

Next, we incorporate the constraint $\boldsymbol{w}'\boldsymbol{1} = 1$, where $\boldsymbol{1}' = (1,1,\ldots,1)$, into Equation 3.21 and rearrange the matrix such that

$$\boldsymbol{A}^{*}\boldsymbol{w} = \boldsymbol{e}, \tag{3.23}$$

* Originally abstracted from Huang, J.J. (2008). A matrix method for the fuzzy analytic network process. *International Journal of Uncertainty, Fuzziness, and Knowledge-Based Systems* 16 (6): 863–78.

where

$$A^* = \begin{bmatrix} 1-n & a_{12} & \cdots & a_{1n-1} & a_{1n} \\ 0 & 2-n & \cdots & a_{2n-1} & a_{2n} \\ \vdots & \vdots & \ddots & \vdots & \vdots \\ 0 & 0 & \cdots & -1 & a_{n-1n} \\ 1 & 1 & \cdots & 1 & 1 \end{bmatrix}, w = \begin{bmatrix} w_1 \\ w_2 \\ \vdots \\ w_n \end{bmatrix}, \text{ and } e = \begin{bmatrix} 0 \\ 0 \\ \vdots \\ 1 \end{bmatrix}.$$

Finally, since A^* is the non-singular matrix, the local weight vector can be derived as:

$$w = (A^*)^{-1} e, \tag{3.24}$$

where $(A^*)^{-1}$ is the inverse of A^* and $A^*(A^*)^{-1} = I$. Next, we give an example to show how a local weight vector can be derived using Cogger and Yu's method.

Let a positive upper triangular comparison matrix and diagonal matrix D be represented, respectively, as

$$A = \begin{bmatrix} 1 & 3 & 5 \\ 0 & 1 & 2 \\ 0 & 0 & 1 \end{bmatrix} \text{ and } D = \begin{bmatrix} 3 & 0 & 0 \\ 0 & 2 & 0 \\ 0 & 0 & 1 \end{bmatrix}.$$

Then, we can obtain

$$A^* = \begin{bmatrix} -2 & 3 & 5 \\ 0 & -1 & 2 \\ 1 & 1 & 1 \end{bmatrix}.$$

The weight vector can be derived as:

$$w = \begin{bmatrix} -0.1765 & 0.1176 & 0.6471 \\ 0.1176 & -0.4118 & 0.2353 \\ 0.0588 & 0.2941 & 0.1176 \end{bmatrix} \begin{bmatrix} 0 \\ 0 \\ 1 \end{bmatrix} = \begin{bmatrix} 0.6471 \\ 0.2353 \\ 0.1176 \end{bmatrix},$$

where

$$(A^*)^{-1} = \begin{bmatrix} -0.1765 & 0.1176 & 0.6471 \\ 0.1176 & -0.4118 & 0.2353 \\ 0.0588 & 0.2941 & 0.1176 \end{bmatrix}.$$

Therefore, the work of finding weights in the AHP is transformed to calculate the last column vector of the inverse of the matrix A^*.

Once we derive all local weight vectors in the AHP, we can form the supermatrix according to the particular network structure. Next, for simplicity, we rewrite the general form of the supermatrix with the following matrix:

$$\mathbf{\Pi} = \begin{bmatrix} \pi_{11} & \pi_{12} & & \pi_{1m} \\ \pi_{21} & \pi_{22} & & \pi_{2m} \\ & & & \\ \pi_{m1} & \pi_{m2} & & \pi_{mm} \end{bmatrix},$$

where $\sum_{i=1}^{m} \pi_{ij} = 1$, $\pi_{ij} \geq 0$, $\forall j = 1,\ldots,m$.

Since $\mathbf{\Pi}$ can be viewed as a transition matrix of a Markov chain and every entry of $\mathbf{\Pi}^{(\kappa)}$ is positive (i.e., $\mathbf{\Pi}$ is regular), there is a unique column matrix $\boldsymbol{\pi}$ satisfying $\mathbf{\Pi}\boldsymbol{\pi} = \boldsymbol{\pi}$, and the entries of $\boldsymbol{\pi}$ are positive and sum to 1, where $\boldsymbol{\pi}$ can be regarded as the global weight vector in the ANP. Therefore, to derive the steady-state process of a supermatrix, we can solve the following system of linear equations:

$$\begin{cases} \pi_1 = \pi_{11}\pi_1 + \pi_{12}\pi_2 + \cdots + \pi_{1m}\pi_m, \\ \pi_2 = \pi_{21}\pi_1 + \pi_{22}\pi_2 + \cdots + \pi_{2m}\pi_m, \\ \quad \vdots \\ \pi_m = \pi_{m1}\pi_1 + \pi_{m2}\pi_2 + \cdots + \pi_{mm}\pi_m. \end{cases} \tag{3.25}$$

By moving the right side of Equation 3.8 to the left side, we can rewrite Equation 3.25 as:

$$\begin{cases} (1-\pi_{11})\pi_1 - \pi_{12}\pi_2 - \cdots - \pi_{1m}\pi_m = 0, \\ -\pi_{21}\pi_1 + (1-\pi_{22})\pi_2 - \cdots - \pi_{2m}\pi_m = 0, \\ \quad \vdots \\ -\pi_{m1}\pi_1 - \pi_{m2}\pi_2 - \cdots + (1-\pi_{mm})\pi_m = 0. \end{cases} \tag{3.26}$$

Since the last equation of the above linear system is superfluous, we replace it with the constraint $\boldsymbol{\pi}'1 = 1$, where $\mathbf{1}' = (1,1,\ldots,1)$. Then, Equation 3.26 can be represented as the following matrix form:

$$\boldsymbol{B}^{*}\pi = \boldsymbol{e}, \tag{3.27}$$

where

$$\boldsymbol{B}^{*} = \begin{bmatrix} 1-\pi_{11} & -\pi_{12} & & -\pi_{1m} \\ -\pi_{21} & 1-\pi_{22} & & -\pi_{2m} \\ & & & \\ 1 & 1 & & 1 \end{bmatrix}, \quad \boldsymbol{\pi} = \begin{bmatrix} \pi_1 \\ \pi_2 \\ \\ \pi_m \end{bmatrix}, \quad \text{and} \quad \boldsymbol{e} = \begin{bmatrix} 0 \\ 0 \\ \\ 1 \end{bmatrix}.$$

Finally, if $\boldsymbol{B}^*$ is the non-singular matrix, the global weight vector can be derived as

$$\pi = \left(\boldsymbol{B}^*\right)^{-1}\boldsymbol{e}, \tag{3.28}$$

where $(\boldsymbol{B}^*)^{-1}$ denotes the inverse of $\boldsymbol{B}^*$ and $\boldsymbol{B}^*(\boldsymbol{B}^*)^{-1} = \boldsymbol{I}$.

Similar to the result of the AHP, the work of finding weights of criteria in the ANP is transformed to calculate the last column vector of the inverse of the matrix $\boldsymbol{B}^*$. In order to demonstrate the proposed method, we give an example as follows.

Let a supermatrix be formed as:

$$\Pi = \begin{bmatrix} 0.1290 & 0.6223 & 0.5171 & 0.0657 \\ 0.6066 & 0.0000 & 0.1243 & 0.2146 \\ 0.1984 & 0.1307 & 0.0000 & 0.1869 \\ 0.0660 & 0.2470 & 0.3586 & 0.5327 \end{bmatrix}.$$

Then, we can obtain

$$\boldsymbol{B}^* = \begin{bmatrix} 1-0.1290 & -0.6223 & -0.5171 & -0.0657 \\ -0.6066 & 1-0.0000 & -0.1243 & -0.2146 \\ -0.1984 & -0.1307 & 1-0.0000 & -0.1869 \\ 1 & 1 & 1 & 1 \end{bmatrix}.$$

Finally, we can derive the global weight vector as

$$\pi = \begin{bmatrix} 1.3169 & 0.5824 & 0.4565 & 0.2968 \\ 0.4256 & 1.0144 & 0.0847 & 0.2615 \\ -0.0074 & -0.0424 & 0.8429 & 0.1480 \\ -1.7351 & -1.5544 & -1.3841 & 0.2937 \end{bmatrix} \begin{bmatrix} 0 \\ 0 \\ 0 \\ 1 \end{bmatrix} = \begin{bmatrix} 0.2968 \\ 0.2615 \\ 0.1480 \\ 0.2937 \end{bmatrix},$$

where

$$\left(\boldsymbol{B}^*\right)^{-1} = \begin{bmatrix} 1.3169 & 0.5824 & 0.4565 & 0.2968 \\ 0.4256 & 1.0144 & 0.0847 & 0.2615 \\ -0.0074 & -0.0424 & 0.8429 & 0.1480 \\ -1.7351 & -1.5544 & -1.3841 & 0.2937 \end{bmatrix}.$$

We can summarize the characteristics of the proposed method as follows. First, the proposed method does not need to hold the property of the reciprocal matrix in the AHP. The first property makes it possible to naturally extend the AHP to the FAHP. Second, instead of solving the limiting power of a supermatrix, a global weight vector can be derived by solving the particular matrix problem. The second property avoids the convergent problem of the fuzzy supermatrix. Next, we will describe how to derive the procedures of the FANP as follows.

In order to consider the ANP under fuzzy environments, fuzzy numbers are used to compare the ratio of weights between criteria. In this chapter, a fuzzy number is presented as the triangular form. Other forms of fuzzy numbers can be easily employed using the same procedures.

Let a fuzzy positive upper triangular comparison matrix

$$\tilde{\boldsymbol{A}} = \left[\tilde{a}_{ij}\right]_{n\times n}, \quad \tilde{a}_{ij} = \begin{cases} \tilde{a}_{ij} & \text{if } i \le j, \\ 0 & \text{otherwise,} \end{cases} \tag{3.29}$$

where $\tilde{a}_{ij}$ denotes the strength of the ratio of the weight that the criterion i dominates the criterion j.

Then, the matrix of $\tilde{\boldsymbol{A}}^*$ can be represented as

$$\tilde{\boldsymbol{A}}^* = \begin{bmatrix} 1-n & \tilde{a}_{12} & \cdots & \tilde{a}_{1n-1} & \tilde{a}_{1n} \\ 0 & 2-n & \cdots & a_{2n-1} & \tilde{a}_{2n} \\ \vdots & \vdots & \ddots & \vdots & \vdots \\ 0 & 0 & \cdots & -1 & \tilde{a}_{n-1n} \\ 1 & 1 & \cdots & 1 & 1 \end{bmatrix}. \tag{3.30}$$

It is clear that if we can derive the inverse of $\tilde{\boldsymbol{A}}^*$, we can obtain local fuzzy weights in the AHP. Therefore, to find the inverse of a fuzzy matrix, at least two methods can be used: the linear programming approach and Cramer's rule. Next, we briefly introduce the above methods as follows.

Let us first consider the crisp case and $\boldsymbol{C}$ be a non-singular matrix. The inverse of $\boldsymbol{C}$, denoted by $\boldsymbol{C}^{-1}$, holds the following property

$$\boldsymbol{C} \times \boldsymbol{C}^{-1} = \boldsymbol{I}, \tag{3.31}$$

or

$$\begin{pmatrix} c_{11} & \cdots & c_{1n} \\ \vdots & \ddots & \vdots \\ c_{n1} & \cdots & c_{nn} \end{pmatrix} \begin{pmatrix} c'_{11} & \cdots & c'_{1n} \\ \vdots & \ddots & \vdots \\ c'_{n1} & \cdots & c'_{nn} \end{pmatrix} = \begin{pmatrix} 1 & \cdots & 1 \\ \vdots & \ddots & \vdots \\ 1 & \cdots & 1 \end{pmatrix}. \tag{3.32}$$

To derive $\boldsymbol{C}^{-1}$, we can rewrite Equation 3.32 and solve the following system of linear equations synchronously:

$$\begin{cases} c_{11}c'_{11} + c_{12}c'_{21} + \cdots + c_{1n}c'_{n1} = 1, \\ c_{11}c'_{12} + c_{12}c'_{22} + \cdots + c_{1n}c'_{n2} = 1, \\ \vdots \\ c_{n1}c'_{1n} + c_{n2}c'_{2n} + \cdots + c_{nn}c'_{nn} = 1, \end{cases} \tag{3.33}$$

where $\boldsymbol{I}$ denotes the identity matrix.

For example, to derive the last column vector of $\boldsymbol{C}^{-1}$, we can solve the following linear system:

$$\begin{cases} c_{11}c'_{1n} + c_{12}c'_{2n} + \cdots + c_{1n}c'_{nn} = 1, \\ c_{21}c'_{1n} + c_{22}c'_{2n} + \cdots + c_{2n}c'_{nn} = 1, \\ \vdots \\ c_{n1}c'_{1n} + c_{n2}c'_{2n} + \cdots + c_{nn}c'_{nn} = 1. \end{cases} \tag{3.34}$$

In addition, we can also solve Equation 3.34 using Cramer's rule such that

$$c'_{jn} = \frac{|\boldsymbol{C}_j|}{|\boldsymbol{C}|}, \tag{3.35}$$

where $\boldsymbol{C}_j$ is $\boldsymbol{C}$ with its jth column replaced by $\mathbf{1}' = (1,1,\ldots,1)$. Other column vectors of $\boldsymbol{C}^{-1}$ can be derived using the same procedure.

For fuzzy numbers, we can solve the following linear programming problem for deriving local fuzzy weights in the AHP:

$$\max/\min c'_{ij} \tag{3.36}$$

$$\begin{aligned} s.t.\quad & c_{11}c'_{11} + c_{12}c'_{21} + \cdots + c_{1n}c'_{n1} = 1, \\ & c_{11}c'_{12} + c_{12}c'_{22} + \cdots + c_{1n}c'_{n2} = 1, \\ & \vdots \\ & c_{n1}c'_{1n} + c_{n2}c'_{2n} + \cdots + c_{nn}c'_{nn} = 1, \\ & c_{ij} \in \tilde{c}_{ij}[\alpha], c'_{ij} \in [0,1], \quad \forall i,j = 1,\ldots,n, \end{aligned}$$

where $\tilde{c}'_{ij}[\alpha] = [\min c'_{ij}, \max c'_{ij}]$ denotes the fuzzy element and $[\alpha]$ is the α-cut operation.

Or, by using Cramer's rule, we can directly fuzzify Equation 3.36 to derive the inverse of a fuzzy matrix. For example, to derive $\tilde{c}'_{jn}[\alpha] = [c'_{jn1}(\alpha), c'_{jn2}(\alpha)]$, we can calculate

$$c'_{jn1}(\alpha) = \min\left\{ \frac{|\boldsymbol{C}_j|}{|\boldsymbol{C}|} \middle| c \in \tilde{c}[\alpha] \right\}, \tag{3.37}$$

and

$$c'_{jn2}(\alpha) = \max\left\{ \frac{|\boldsymbol{C}_j|}{|\boldsymbol{C}|} \middle| c \in \tilde{c}[\alpha] \right\}. \tag{3.38}$$

Since Equations 3.36 through 3.38 may be hard to evaluate, some heuristic algorithms (e.g., genetic algorithm, ant algorithm, or simulated annealing) can be used to obtain approximate solutions.

Next, an example is given to show how the inverse of a fuzzy matrix, can be derived in the AHP. Assume that the upper triangular fuzzy comparison matrix can be given by the decision maker as:

$$\tilde{A} = \begin{bmatrix} 1 & (2,3,4) & (1/6,1/5,1/4) & (6,7,8) \\ 0 & 1 & (1/8,1/7,1/6) & (1,2,3) \\ 0 & 0 & 1 & (8,9,9) \\ 0 & 0 & 0 & 1 \end{bmatrix}.$$

Then, $\tilde{A}^*$ can be formed as:

$$\tilde{A}^* = \begin{bmatrix} -3 & (2,3,4) & (1/6,1/5,1/4) & (6,7,8) \\ 0 & -2 & (1/8,1/7,1/6) & (1,2,3) \\ 0 & 0 & -1 & (8,9,9) \\ 1 & 1 & 1 & 1 \end{bmatrix}.$$

Next, by solving Equation 3.36 and setting the α-cuts = 0, 0.2, 0.4, 0.6, 0.8, and 1, we can derive the fuzzy local weights, i.e., the last column vector of $(\tilde{A}^*)^{-1}$, as shown in Table 3.4.

To check the fuzzy local weights visually, we can depict the triangular-shaped fuzzy local weights of the given example as shown in Figure 3.9.

On the other hand, to find the global weight vector in the FANP can also be viewed as the problem of calculating the inverse of a fuzzy matrix, i.e., to solve $(\tilde{B}^*)^{-1}$. Therefore, the method of finding global weights in the FANP is similar to the above procedures. Next, an application is used to demonstrate the proposed method in Section 3.4.

TABLE 3.4
The Local Fuzzy Weights

α-cuts	$\tilde{w}_1[\alpha]$	$\tilde{w}_2[\alpha]$	$\tilde{w}_3[\alpha]$	$\tilde{w}_4[\alpha]$
0	[0.2248,0.3578]	[0.0668,0.1439]	[0.4601,0.6307]	[0.0536,0.0763]
0.2	[0.2359,0.3426]	[0.0737,0.1347]	[0.4786,0.6152]	[0.0551,0.0730]
0.4	[0.2472,0.3274]	[0.0806,0.1259]	[0.4974,0.5998]	[0.0567,0.0699]
0.6	[0.2587,0.3122]	[0.0874,0.1174]	[0.5164,0.5847]	[0.0583,0.0670]
0.8	[0.2704,0.2971]	[0.0943,0.1092]	[0.5356,0.5697]	[0.0600,0.0643]
1.0	[0.2821,0.2821]	[0.1013,0.1013]	[0.5549,0.5549]	[0.0617,0.0617]

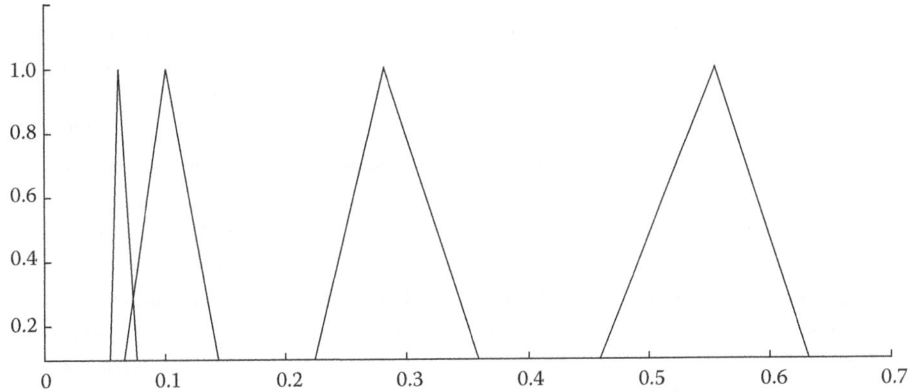

FIGURE 3.9 The triangular-shaped fuzzy local weights.

Example 3.5

Consider the market share of a food company can be evaluated by three clusters: Advertising ability (C_1), Quality ability (C_2), and Attraction ability (C_3). Each cluster can be divided into three criteria, including Creativity, Promotion, Frequency, Nutrition, Taste, Cleanliness, Price, Location, and Reputation, respectively. The decision maker wants to determine the weights of the criteria using the ANP so that he/she can allocate the appropriate budgets for obtaining the maximum market share. Due to the restrictions of incomplete information and human subjective judgments, the decision maker employs fuzzy numbers to judge the ratios of the weights between the criteria. The network structure adopted in this application to deal with the problem of the market share is depicted as shown in Figure 3.10.

In order to calculate all the fuzzy local vectors, we should first compare the relative fuzzy ratios of the weights between the criteria, and then the corresponding fuzzy local weights with α-cut = 0 can be derived using Equation 3.36. The results can be given as shown in the following nine matrices:

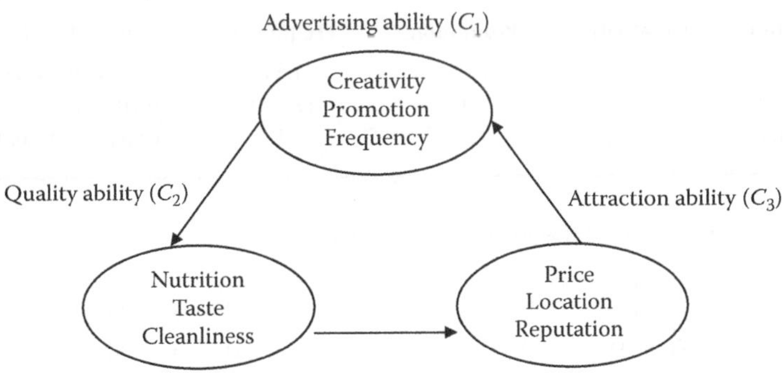

FIGURE 3.10 The network structure in the application.

Creativity	**Nutrition**	**Taste**	**Cleanliness**	**Fuzzy Local Vector**
Nutrition	1	(1/3,1/2,1)	(3,4,5)	(0.2632,0.3514,0.4737)
Taste	0	1	(4,5,6)	(0.4211,0.5405,0.6316)
Cleanliness	0	0	1	(0.0800,0.1081,0.1395)
Promotion	**Nutrition**	**Taste**	**Cleanliness**	**Fuzzy Local Weights**
Nutrition	1	(3,4,5)	(4,5,6)	(0.6250,0.6842,0.7333)
Taste	0	1	(1,2,2)	(0.1333,0.2105,0.2500)
Cleanliness	0	0	1	(0.0909,0.1053,0.1818)
Frequency	**Nutrition**	**Taste**	**Cleanliness**	**Fuzzy Local Vector**
Nutrition	1	(1/3,1/2,1)	(1/3,1/2,1)	(0.1429,0.2000,0.3333)
Taste	0	1	(1/2,1,1)	(0.2222,0.4000,0.4286)
Cleanliness	0	0	1	(0.3333,0.4000,0.5714)
Nutrition	**Price**	**Location**	**Reputation**	**Fuzzy Local Vector**
Price	1	(1,2,3)	(1/5,1/4,1/3)	(0.1358,0.2131,0.3023)
Location	0	1	(1/6,1/5,1/4)	(0.1053,0.1312,0.1695)
Reputation	0	0	1	(0.5581,0.6557,0.7407)
Taste	**Price**	**Location**	**Reputation**	**Fuzzy Local Vector**
Price	1	(2,3,4)	(1/4,1/3,1/2)	(0.2000,0.2800,0.3750)
Location	0	1	(1/6,1/5,1/4)	(0.0952,0.1200,0.1538)
Reputation	0	0	1	(0.5000,0.6000,0.6857)
Cleanliness	**Price**	**Location**	**Reputation**	**Fuzzy Local Vector**
Price	1	(3,4,5)	(1/3,1/2,1)	(0.2800,0.3750,0.5000)
Location	0	1	(1/5,1/4,1/3)	(0.0909,0.1250,0.1667)
Reputation	0	0	1	(0.3750,0.5000,0.6000)
Price	**Creativity**	**Promotion**	**Frequency**	**Fuzzy Local Vector**
Creativity	1	(1/4,1/3,1/2)	(2,3,4)	(0.2000,0.2800,0.3750)
Promotion	0	1	(4,5,6)	(0.5000,0.6000,0.6857)
Frequency	0	0	1	(0.0952,0.1200,0.1538)
Location	**Creativity**	**Promotion**	**Frequency**	**Fuzzy Local Vector**
Creativity	1	(1/5,1/4,1/3)	(1/4,1/3,1/2)	(0.0960,0.1220,0.1724)
Promotion	0	1	(1,2,3)	(0.4138,0.5853,0.6779)
Frequency	0	0	1	(0.2105,0.2927,0.4494)
Reputation	**Creativity**	**Promotion**	**Frequency**	**Fuzzy Local Vector**
Creativity	1	(3,4,5)	(4,5,6)	(0.6154,0.6800,0.7273)
Promotion	0	1	(2,3,4)	(0.1818,0.2400,0.3077)
Frequency	0	0	1	(0.0556,0.0800,0.1250)

Then, we can formulate the fuzzy supermatrix as:

$$\tilde{\Pi} = \begin{bmatrix} 0 & 0 & \boldsymbol{W}_{13} \\ \boldsymbol{W}_{21} & 0 & 0 \\ 0 & \boldsymbol{W}_{32} & 0 \end{bmatrix}, \quad \text{where } 0 = \begin{bmatrix} 0 & 0 & 0 \\ 0 & 0 & 0 \\ 0 & 0 & 0 \end{bmatrix},$$

$$\tilde{W}_{13} = \begin{bmatrix} (0.2000, 0.2800, 0.3750) & (0.0960, 0.1220, 0.1724) & (0.6154, 0.6800, 0.7273) \\ (0.5000, 0.6000, 0.6857) & (0.4138, 0.5853, 0.6779) & (0.1818, 0.2400, 0.3077) \\ (0.0952, 0.1200, 0.1538) & (0.2105, 0.2927, 0.4494) & (0.0556, 0.0800, 0.1250) \end{bmatrix},$$

$$\tilde{W}_{21} = \begin{bmatrix} (0.2632, 0.3514, 0.4737) & (0.6250, 0.6842, 0.7333) & (0.1429, 0.2000, 0.3333) \\ (0.4211, 0.5405, 0.6316) & (0.1333, 0.2105, 0.2500) & (0.2222, 0.4000, 0.4286) \\ (0.0800, 0.1081, 0.1395) & (0.0909, 0.1053, 0.1818) & (0.3333, 0.4000, 0.5714) \end{bmatrix},$$

$$\tilde{W}_{32} = \begin{bmatrix} (0.1358, 0.2131, 0.3023) & (0.2000, 0.2800, 0.3750) & (0.2800, 0.3750, 0.5000) \\ (0.1053, 0.1312, 0.1695) & (0.0952, 0.1200, 0.1538) & (0.0909, 0.1250, 0.1667) \\ (0.5581, 0.6557, 0.7407) & (0.5000, 0.6000, 0.6857) & (0.3750, 0.5000, 0.6000) \end{bmatrix}.$$

Let α-cuts = 0, 0.2, 0.4, 0.6, 0.8, and 1.0, and we can obtain the fuzzy global weights by calculating $(\tilde{B}^*)^{-1}$ as shown in Table 3.5.

Next, to show the justification of the proposed method, we find the crisp global weights using the vertices of the fuzzy numbers in the fuzzy supermatrix, and show they belong to the alpha-zero cut of the fuzzy global weights.

Let a crisp supermatrix

TABLE 3.5
The Fuzzy Global Weights in the Application

Fuzzy Global Weights	**α-cut = 0**	**α-cut = 0.2**	**α-cut = 0.4**
Creativity	[0.1257,0.2047]	[0.1345,0.1976]	[0.1433,0.1904]
Promotion	[0.0860,0.1700]	[0.0936,0.1590]	[0.1014,0.1484]
Frequency	[0.0250,0.0651]	[0.0277,0.0596]	[0.0305,0.0542]
Nutrition	[0.1106,0.2005]	[0.1190,0.1905]	[0.1275,0.1805]
Taste	[0.0840,0.1704]	[0.0934,0.1628]	[0.1030,0.1552]
Cleanliness	[0.0323,0.0829]	[0.0351,0.0748]	[0.0380,0.0672]
Price	[0.0547,0.1310]	[0.0611,0.1218]	[0.0676,0.1129]
Location	[0.0289,0.0613]	[0.0313,0.0571]	[0.0339,0.0531]
Reputation	[0.1549,0.2493]	[0.1645,0.2400]	[0.1742,0.2307]
Fuzzy Global Weights	**α-cut = 0.6**	**α-cut = 0.8**	**α-cut = 1.0**
Creativity	[0.1521,0.1831]	[0.1602,0.1757]	[0.1682,0.1682]
Promotion	[0.1094,0.1381]	[0.1177,0.1321]	[0.1260,0.1260]
Frequency	[0.0334,0.0490]	[0.0363,0.0441]	[0.0391,0.0391]
Nutrition	[0.1360,0.1704]	[0.1446,0.1618]	[0.1532,0.1532]
Taste	[0.1128,0.1478]	[0.1230,0.1405]	[0.1331,0.1331]
Cleanliness	[0.0409,0.0601]	[0.0440,0.0536]	[0.0471,0.0471]
Price	[0.0742,0.1042]	[0.0809,0.0959]	[0.0876,0.0876]
Location	[0.0365,0.0492]	[0.0393,0.0456]	[0.0420,0.0420]
Reputation	[0.1841,0.2216]	[0.1940,0.2127]	[0.2038,0.2038]

$$\boldsymbol{\Pi} = \begin{bmatrix} \mathbf{0} & \mathbf{0} & \boldsymbol{W}_{13} \\ \boldsymbol{W}_{21} & \mathbf{0} & \mathbf{0} \\ \mathbf{0} & \boldsymbol{W}_{32} & \mathbf{0} \end{bmatrix},$$

where

$$\mathbf{0} = \begin{bmatrix} 0 & 0 & 0 \\ 0 & 0 & 0 \\ 0 & 0 & 0 \end{bmatrix}, \boldsymbol{W}_{13} = \begin{bmatrix} 0.3750 & 0.1724 & 0.7273 \\ 0.5000 & 0.4138 & 0.1818 \\ 0.1250 & 0.4138 & 0.0909 \end{bmatrix}, \boldsymbol{W}_{21} = \begin{bmatrix} 0.4737 & 0.7333 & 0.3333 \\ 0.4211 & 0.1333 & 0.2222 \\ 0.1052 & 0.1334 & 0.4445 \end{bmatrix}, \text{and}$$

$$\boldsymbol{W}_{32} = \begin{bmatrix} 0.3023 & 0.3750 & 0.5000 \\ 0.1053 & 0.0952 & 0.0909 \\ 0.5924 & 0.5298 & 0.4091 \end{bmatrix}.$$

By calculating the steady-state process of $\boldsymbol{\Pi}$, the global weight vector can be derived by raising $\boldsymbol{\Pi}$ to its limiting power as:

$$\Pi = \begin{bmatrix} 0.1821 & 0.1061 & 0.0451 & 0.1791 & 0.1009 & 0.0534 & 0.1187 & 0.0333 & 0.1814 \end{bmatrix}'.$$

Clearly, it belongs to the alpha-zero cut of the fuzzy global weight vector. Readers can use other vertices of the fuzzy numbers to show that all crisp global weights belong to the alpha-zero cut of the fuzzy global weights.

Although the ANP has been widely used in various applications, it is hard for a decision maker to quantify precise ratios of weights between criteria with incomplete information and subjective uncertainty. In this chapter, the FANP is proposed to extend the conventional ANP and the fuzzy judgments are used to compare the relative ratios of weights between criteria. Compared with the crisp ANP, the advantages of the proposed methods are as follows. First, because of the restrictions of incomplete information and subjective uncertainty, fuzzy numbers are more suitable for judging the ratios of weights between criteria. Secondly, fuzzy global weights can help decision makers understand the uncertainty degrees of problems. Finally, it should be noted that the crisp ANP is a special case of the proposed method when α-cut = 1.

4 Simple Additive Weighting Method

In this chapter, the simple additive weighting method (SAW) and fuzzy simple additive weighting method (FSAW) are introduced. SAW can be considered the most intuition and easy way to deal with multiple criteria decision-making MCDM problems, because the linear additive function can represent the preferences of decision makers (DM). This is true, however, only when the assumption of preference independence (Keeney and Raiffa 1976) or preference separability (Gorman 1968) is met.

4.1 SIMPLE ADDITIVE WEIGHTING METHOD

Churchman and Ackoff (1954) first utilized the SAW method to cope with a portfolio selection problem. The SAW method is probably the best known and widely used method for multiple attribute decision making MADM. Because of its simplicity, SAW is the most popular method in MADM problems and the best alternative can be derived by the following equation:

$$A^* = \left\{u_i(\boldsymbol{x}) \mid \max_i u_i(\boldsymbol{x}) \mid i = 1,2,\ldots,n\right\}, \tag{4.1}$$

or the gaps of alternatives can be improved to build a new best alternative A^* for achieving aspired/desired levels in each criterion.

Also

$$u_i(\boldsymbol{x}) = \sum_{j=1}^{n} w_j r_{ij}(\boldsymbol{x}), \tag{4.2}$$

where $u_i(\boldsymbol{x})$ denotes the utility of the ith alternative and $i = 1,2,\ldots,n$; w_j denotes the weights of the jth criterion; $r_{ij}(\boldsymbol{x})$ is the normalized preferred ratings of the ith alternative with respect to the jth criterion for all commensurable units; and all criteria are assumed to be independent. In addition, the normalized preferred ratings ($r_{ij}(\boldsymbol{x})$) of the ith alternative with respect to the jth criterion can be defined by:

Form 1

- For benefit criteria (larger is better), $r_{ij}(\boldsymbol{x}) = x_{ij}/x_j^*$, where $x_j^* = \max_i x_{ij}$ or let x_j^* be the aspired/desired level, and it is clear $0 \le r_{ij}(\boldsymbol{x}) \le 1$.

- For cost criteria (smaller is better), $r_{ij}(\boldsymbol{x}) = (1/x_{ij})/(1/x_j^*) = (\max_i x_j^*)/(x_{ij})$ or let x_j^* be the aspired/desired level.

Form 2

- For benefit criteria (larger is better), $r_{ij} = (x_{ij} - x_j^-)/(x_j^* - x_j^-)$, where $x_j^* = \max_i x_{ij}$ and $x_j^- = \min_i x_{ij}$ or let x_j^* be the aspired/desired level and x_j^- be the worst level.
- For cost criteria (smaller is better), $r_{ij} = (x_j^- - x_{ij})/(x_j^- - x_j^*)$.

Therefore, the synthesized performance is

$$p_i = \sum_{j=1}^{m} w_j r_{ij}$$

where p_i is a synthesizing performance value of the ith alternative; w_j denotes the weights of the jth criterion; r_{ij} is the normalized preferred ratings of the ith alternative with respect to the jth criterion for becoming the commensurable units; and the criteria are assumed to be independent of each other. If the units of the performance matrix are the commensurable units, we do not need to transfer the data matrix into the normalized preferred rating scales. Next, a simple example is given to demonstrate the procedures of SAW in determining the preferred order of alternatives.

Example 4.1

Assume the bank evaluation problem can be described as follows. Suppose the criteria of evaluating banks can be represented by investment income (x_1), number of customers (x_2), brand image (x_3), and branch numbers (x_4). Let the five banks and the corresponding evaluation ratings be described as shown in Table 4.1.

First, the normalized preferred ratings should be calculated, as shown in Table 4.2, to transform the scale into [0,1].

TABLE 4.1
The Decision Table in Example 4.1

Bank	x_1	x_2	x_3	x_4
A	2,500 (million)	160,000	6	12
B	2,300 (million)	120,000	8	17
C	1,900 (million)	150,000	5	18
D	3,100 (million)	100,000	7	14
E	2,800 (million)	130,000	7	10
Weights	0.300	0.200	0.250	0.250

TABLE 4.2
The Decision Table of Normalized Preferred Ratings in Example 4.1

Bank	r_1	r_2	r_3	r_4
A	0.806	1.000	0.750	0.667
B	0.742	0.750	1.000	0.944
C	0.613	0.938	0.625	1.000
D	1.000	0.625	0.875	0.778
E	0.903	0.813	0.875	0.556
Weights	0.300	0.200	0.250	0.250

Next, the utility of alternative A can be obtained as:

$$u_A(\boldsymbol{x}) = 0.806 \times 0.30 + 1.000 \times 0.20 + 0.750 \times 0.25 + 0.667 \times 0.25 = 0.796.$$

With the same procedure as above, the utilities of other alternatives can also be obtained as:

$$u_B(\boldsymbol{x}) = 0.859,\ u_C(x) = 0.778,\ u_D(x) = 0.838, \text{ and } u_E(x) = 0.791.$$

On the basis of the utilities above, therefore, it can be shown that the preferred order of alternatives can be expressed as:

$$B \succ D \succ A \succ E \succ C$$

On the basis of the results above, it can be seen that alternative A should be the optimal bank.

4.2 FUZZY SIMPLE ADDITIVE WEIGHTING

In practice, for fuzzy multiattribute decision making (FMADM) problems, if we assume that there is a mutually independent relationship among the criteria, after calculating the relative weights and the performance score of each criterion with respect to each alternative, we can use the FSAW method to aggregate the fuzzy preferred ratings to rank the order of alternatives. The procedure of SAW for FMADM can be summarized as follows:

Step 1: Calculate the relative fuzzy weight $\tilde{w}_j$ of the jth attribute. The fuzzy relative weights can be obtained/assigned using triangle or interval value by the subjective/perceptive judgment of DM or evaluators.

Step 2: Obtain the fuzzy decision matrix whose elements are composed of a set of fuzzy comparable ratings $\tilde{r}_{ij}(\boldsymbol{x})$ for the jth attribute with respect to the ith alternative. If the raw decision matrix is comprised of $\tilde{x}_{ij}$ for the jth attribute with respect to the ith alternative, in order to reduce the influence of the dimension, we can extend the Hwang and Yoon (1981) method to transfer the fuzzy raw data $\tilde{x}_{ij}$ to non-dimension data $\tilde{r}_{ij}(\boldsymbol{x})$ according to the following principle:

Form 1

Case 1. If the criteria are defined by benefit criteria (the larger $\tilde{x}_j$, the greater preference), then the transformed outcome $\tilde{x}_{ij}$ is $\tilde{r}_{ij}(\boldsymbol{x}) = \tilde{x}_{ij}/\tilde{x}_j^*$, where $\tilde{x}_j^* = \max_i \tilde{x}_{ij}$, or let $\tilde{x}_j^*$ be the aspired/desired level and it is clear that $0 \leq \tilde{r}_{ij}(\boldsymbol{x}) \leq 1$.

Case 2. If the criteria are defined by cost criteria (the smaller $\tilde{x}_j$, the greater preference), then the transformed outcome $\tilde{x}_{ij}$ is $\tilde{r}_{ij}(\boldsymbol{x}) = (1/\tilde{x}_{ij})/(1/\tilde{x}_j^*) = \min_i \tilde{x}_{ij}/\tilde{x}_{ij}$ or let $\tilde{x}_j^*$ be the aspired/desired level.

Form 2

- For benefit criteria (larger is better), $\tilde{r}_{ij}(\boldsymbol{x}) = (\tilde{x}_{ij} - \tilde{x}_j^-)/(\tilde{x}_j^* - \tilde{x}_j^-)$, where $\tilde{x}_j^* = \max_i \tilde{x}_{ij}$ and $\tilde{x}_j^- = \min_i \tilde{x}_{ij}$, or let $\tilde{x}_j^*$ be the aspired/desired level and let $\tilde{x}_j^-$ be the worst level.
- For cost criteria (smaller is better), $\tilde{r}_{ij}(\boldsymbol{x}) = (\tilde{x}_j^- - \tilde{x}_{ij})/(\tilde{x}_j^- - \tilde{x}_j^*)$.

Step 3: Synthesize the fuzzy value $\tilde{u}_i(\boldsymbol{x})$ for the ith alternative, which is a summation of multiplying the relative fuzzy weight $\tilde{w}_j$ and non-dimension comparable data $\tilde{r}_{ij}(\boldsymbol{x})$ as follows: $\tilde{u}_i(\boldsymbol{x}) = \sum_j \tilde{w}_j \tilde{r}_{ij}(\boldsymbol{x})$, where $\tilde{u}_i(\boldsymbol{x})$ is a synthesizing fuzzy performance value of the ith alternative, $\tilde{w}_j$ denotes the weights of the jth criterion, and $\tilde{r}_{ij}(\boldsymbol{x})$ is the normalized preferred ratings of the ith alternative with respect to the jth criterion for becoming the commensurable units, and it is assumed that the criteria are independent of each other. If the units of the performance matrix are the commensurable units, we do not need to transfer the data matrix into the normalized preferred rating scales, as for a satisfactory scale for a performance matrix by linguistics (natural language).

Step 4: Select the best alternative defined by $\tilde{A}^* = \{\tilde{u}_i(\boldsymbol{x}) \mid \max_i \tilde{u}_i(\boldsymbol{x})\}$ or improve the gaps of alternatives to build a new best alternative $\tilde{A}^*$ for achieving aspired/desired levels.

It should be highlighted that for operations of fuzzy numbers, refer to Chapter 1.4. Furthermore, it can be seen that since the final rating of each alternative is also a fuzzy number, a defuzzified method, such as the center of area (CoA) method, can be used for DM to determine the best non-fuzzy performance (BNP) value of alternatives. Next, an extended example from Example 4.1 is given to demonstrate the procedures of FSAW.

Example 4.2

Extending the problem of Example 4.1 to consider the criteria of evaluating banks can be represented by investment income (x_1), number of customers (x_2), brand image (x_3), and branch numbers (x_4) in the fuzzy environment. Instead of assuming fuzzy numbers in ratings and weights, only fuzzy weights are considered in this example for simplicity. However, the procedures can be easily extended to consider the general form of FSAW.

Let the normalized preferred ratings of the alternatives and the fuzzy weights be described as shown in Table 4.3.

Then, the utility of Bank A can be calculated as:

$$\tilde{u}_A(\boldsymbol{x}) = 0.806 \times (0.20, 0.30, 0.40) + 1.000 \times (0.15, 0.20, 0.35)$$
$$+ 0.750 \times (0.10, 0.25, 0.30) + 0.667 \times (0.15, 0.25, 0.30) = (0.486, 0.796, 1.098).$$

With the same procedure above, the utilities of other alternatives can be obtained as:

$$\tilde{u}_B(\boldsymbol{x}) = (0.503, 0.859, 1.143);\ \tilde{u}_C(\boldsymbol{x}) = (0.476, 0.778, 1.061);$$

$$\tilde{u}_D(\boldsymbol{x}) = (0.498, 0.838, 1.115);\ \tilde{u}_E(\boldsymbol{x}) = (0.473, 0.791, 1.075).$$

Next, in order to provide concrete information for DM to determine the preferred order of alternatives, the CoA method is used to calculate the defuzzified utilities of alternatives as:

$$u_A^* = 0.793;\ u_B^* = 0.835;\ u_C^* = 0.772;\ u_D^* = 0.817;\ u_E^* = 0.780.$$

TABLE 4.3
The Decision Table of the Normalized Preferred Ratings in Example 4.2

Bank	r_1	r_2	r_3	r_4
A	0.806	1.000	0.750	0.667
B	0.742	0.750	1.000	0.944
C	0.613	0.938	0.625	1.000
D	1.000	0.625	0.875	0.778
E	0.903	0.813	0.875	0.556
Weight	(0.20,0.30,0.40)	(0.15,0.20,0.35)	(0.10,0.25,0.30)	(0.15,0.25,0.30)

According to the defuzzified utilities of alternatives, we can conclude that the preferred order of alternatives can be expressed as:

$$B \succ D \succ A \succ E \succ C.$$

On the basis of the results above, it can be seen that alternative A is the optimal bank.

From the procedures above, it can be seen why SAW or FSAW is so popular in dealing with MADM problems. However, it should be kept in mind that SAW or FSAW only work when the assumption of preference independence (Keeney and Raiffa 1976) or preference reparability (Gorman 1968) is satisfied. Next, we briefly describe the concepts of preference independence as follows (Keeney and Raiffa 1976).

Let two vector attributes be **y** and **z**, where their consequences can be expressed by (y, z). Then, **y** is preference independence of **z**, if preferences for consequences (y, z') with z' fixed do not influence the amount z' and can be mathematically defined by:

$$\left[u\left(y', z^0 \geq y'', z^0\right)\right] \Rightarrow \left[u\left(y', z \geq y'', z\right)\right], \ \forall z.$$

Therefore, if $\{\boldsymbol{x}_1, \boldsymbol{x}_i\}$ is preference independence of $\bar{\boldsymbol{x}}_{1i}, i = 2,3,\ldots,n$, then the value function of $\boldsymbol{x}$ can be expressed as:

$$u(\boldsymbol{x}) = \sum_{i=1}^{n} u_i(\boldsymbol{x}_i).$$

However, if the characteristic of preference independence is not satisfied, we should consider other methods that can account for the interaction effect between attributes, such as the fuzzy integral (refer to Chapter 9), to calculate the utility of an alternative.

4.3 FUZZY SIMPLE ADDITIVE WEIGHTING FOR THE BEST PLAN: EXAMPLE OF AN ENVIRONMENT-WATERSHED PLAN

The purpose of this subsection is to establish a hierarchical structure for tackling the evaluation problem of the best plan alternative of an environment-watershed as an example (Chen, Tzeng, and Ding 2008). Multiple criteria decision making is an analytic method to evaluate the advantages and disadvantages of alternatives based on multiple criteria in a fuzzy environment. This subsection focuses mainly on the evaluation problem. The typical multiple criteria evaluation problem examines a set of feasible alternatives and considers more than one criterion to improve or determine a best alternative for implementation. The contents include three parts: building hierarchical structure of evaluation criteria, determining the evaluation criteria weights, and getting the performance value.

4.3.1 Building a Hierarchical Structure of Environment-Watershed Evaluation Criteria

For example, we take an environment-watershed plan as an explanation (Chen et al. 2011). What is the watershed? Component landforms that commonly occur in a watershed include stream channels, floodplains, stream terraces, alluvial valley bottoms, alluvial fans, mountain slopes, and ridge tops (Petersen 1999). Environment-watershed plan measurements involve a number of complex factors however, including engineering of management, ecological restoration, environmental construction, and environmental conservation issues. In the past, a plan dimension index could be based, simply, on the aggregate environment engineering of catastrophe rate for a period of time or landing cycles, but this may have been incomplete. Yeh (2005) suggested that merging ecological engineering measures into the framework of watershed management would become one of the most crucial research topics for our local authority institutions. At the moment, we need to consider many factors/criteria for the environment-watershed plan index, focused on reducing catastrophe, promoting human safety, increasing comfortable interest, ecological systems, and sustainable environment. Chen and Lin (2005) suggested 4 dimensions and 26 criteria. While many studies provide useful methodologies and models based on problem-solving procedures, they have mainly been applied to the field of environment-watershed plan management in Taiwan and the rest of the world. A watershed plan, restoration, and management have a specific hydrologic function and ecological potential. Inventory, evaluation, and restoration of watershed plans are based on geomorphic, hydrologic, and ecological principles. That is nature approach to watershed plans that works with nature to restore degraded watersheds (Petersen 1999). The operation procedures of several key model components, participation of the local community, utilization of geographical information systems, investigation and analysis of the ecosystem, habitat, and landscape, and allocation of ecological engineering measures, are illustrated in detail for a better understanding of their roles in the model (Yeh and Lin 2005; Özelkan and Duckstein 2002). In the Austrian Danube case study, there are 12 alter natives and 33 criteria. The criteria include three main conflicting types of interest: economy, ecology, and sociology. Apart from calamities, which still account for environment-watershed plans in natural catastrophes, engineering design error and incident data, maintenance, and operational deficiencies are typically cited as causes of failed plans. It has been suggested that "proactive" plan measures be instituted, especially during monitoring of design errors related to human error.

Environment-watershed problems in the world statistics describe from natural disasters and artificially jamming two levels, in the first the typhoon, torrential rain and earthquake cause the flood to overflow, violent perturbation of landslide, potential debris flow torrent and so on. In addition the reason why space and water environmental demand increase in artificial disturbances because of population expansion, so that the changes of land pattern utilizing and terrain features, moreover carryout the transition of developing and also leading to the fact road water and soil conservation is destroyed, the environment falls in the destruction, biological habitat in destroyed, rivers and creeks of the quality had polluted, threatened fish species, loss of forest cover, erosion and urban growth, among others things. How can

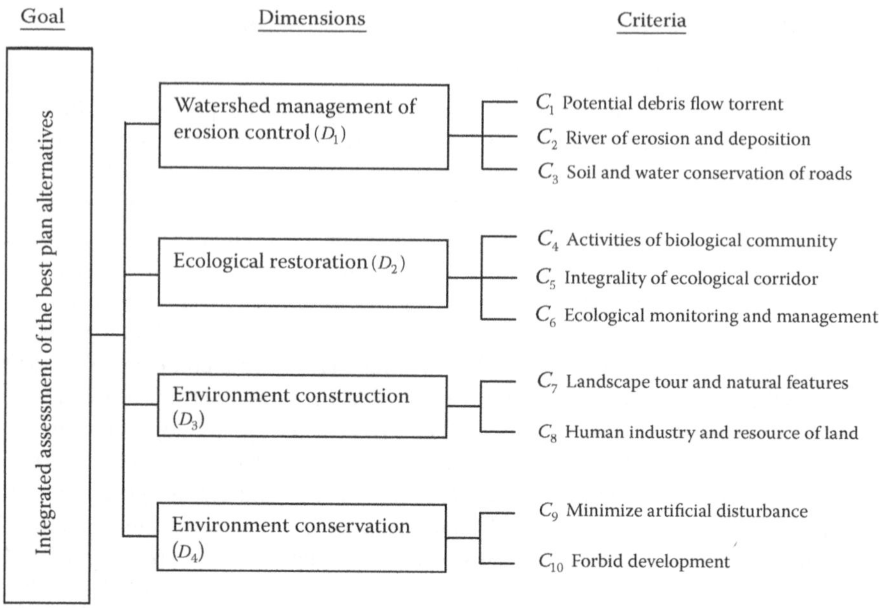

FIGURE 4.1 The hierarchical structure for the best plan alternatives assessment.

we do for solving environment-watershed problems? Firstly from the environment-watershed survey data found characteristic values to improve stabilize the river canal shape, increase the activity of biological community, habitat mold and regeneration, structure integrity of ecological corridor, and to create peripheral landscapes and natural environment features, develope from tour facilities and resources of humane industry, repeat structure nature of beautiful material, and raise property of tourism. However, in areas of steep slopes, erosion and environmental preservation, artificial disturbance should be minimized or not allowed. In summary, we need to consider intact factors/criteria that include four dimensions and ten factors/criteria, i.e., (a) watershed management and erosion control, (b) ecological restoration, (c) environmental construction, and (d) environmental conservation. Based on these, ten evaluation criteria for the hierarchical structure were used in this study.

The hierarchical structure adopted in this study to deal with the problems of environment-watershed plan assessment is shown in Figure 4.1.

4.3.1.1 Determining the Evaluation Criteria Weights

Since the criteria for the best plan evaluation have diverse significance and meanings, we cannot assume that each evaluation criterion is of equal importance. There are many methods that can be employed to determine weights (Hwang and Yoon 1981) such as the eigenvector method, weighted least-square method, entropy method, analytic hierarchy process (AHP), and linear programming technique for multidimensional analysis of preference (LINMAP). The selection of the method depends on the nature of the problems. To evaluate the best plan is a complex and wide-ranging problem, requiring the most inclusive and flexible method. The AHP developed by Saaty (1980, 1996) is a very useful decision analysis tool in dealing with multiple criteria decision problems and has been successfully applied to many construction industry decision areas (Hsieh, Lu, and Tzeng 2004; McIntyre and Parfitt 1998; Cheng et al. 2004; Hastak 1998; Cheung et al. 2001; Fong and Choi 2000). However, in the operation process of applying the AHP method, it is easier and more humanistic for evaluators to assess "criterion A is much more important than criterion B" than to consider "the importance of principle A and principle B is seven to one." Hence, Buckley (1985) extended Saaty's AHP to the case where the evaluators are allowed to employ fuzzy ratios in place of exact ratios to handle the difficulty of assigning exact ratios when comparing two criteria and deriving the fuzzy weights of criteria by the geometric mean method. Therefore, in this study, we employ Buckley's method, FAHP, to fuzzify hierarchical analysis by allowing fuzzy numbers for the pairwise comparisons and find the fuzzy weights. In this section, we briefly review concepts for fuzzy hierarchical evaluation.

4.3.2 Fuzzy Numbers

Fuzzy numbers are a fuzzy subset of real numbers, representing the expansion of the idea of the confidence interval. According to the definition of Laarhoven and Pedrycz (1983), a triangular fuzzy number (TFN) should possess the following basic features.

A fuzzy number $\tilde{A}$ on $\mathbb{R}$ is a TFN if its membership function $x \in \tilde{A}$, $\mu_{\tilde{A}}(x)$: $\mathbb{R} \rightarrow [0,1]$ is equal to

$$\mu_{\tilde{A}}(x) = \begin{cases} (x-l)/(m-l), & l \le x \le m \\ (u-x)/(u-m), & m \le x \le u, \\ 0, & \text{otherwise} \end{cases} \tag{4.3}$$

where l and u stand for the lower and upper bounds of the fuzzy number $\tilde{A}$, respectively, and m for the modal value (see Figure 4.2). The TFN can be denoted by $\tilde{A} = (l, m, u)$ and the following is the operational law of two TFNs $\tilde{A}_1 = (l_1, m_1, u_1)$ and $\tilde{A}_2 = (l_2, m_2, u_2)$.

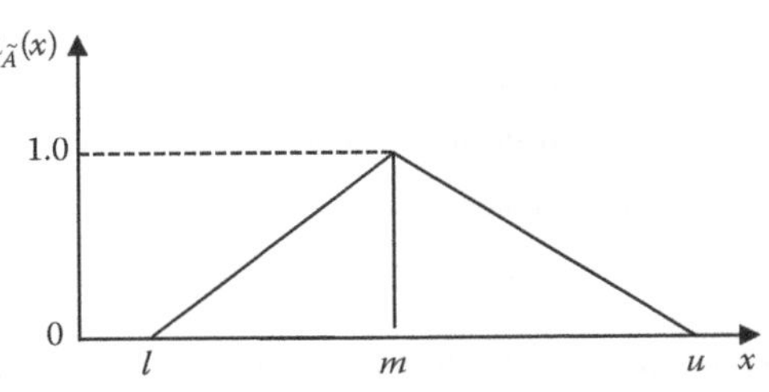

FIGURE 4.2 The membership function of the triangular fuzzy number.

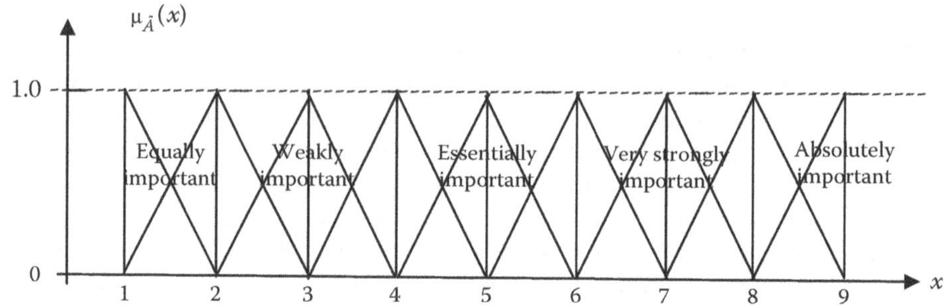

FIGURE 4.3 Membership functions of linguistic variables for comparing two criteria.

4.3.3 Linguistic Variables

According to Zadeh (1975), it is very difficult for conventional quantification to reasonably express those situations that are overtly complex or hard to define; so the notion of a linguistic variable is necessary in such situations. A linguistic variable is a variable whose values are words or sentences in a natural or artificial language. Here, we use this kind of explanation to compare two building the best plan in evaluation criteria by five basic linguistic terms, as "absolutely important," "very strongly important," "essentially important," "weakly important," and "equally important" with respect to a fuzzy five-level scale (see Figure 4.3) (Chiou and Tzeng 2001). In this chapter, the computational technique is based on the following fuzzy numbers, defined by Mon, Cheng, and Lin (1994) in Table 4.4. Here each membership function (scale of fuzzy number) is defined by three parameters of the symmetric TFN, the left point, middle point, and right point of the range over which the function is defined. The use of linguistic variables is currently widespread and the linguistic effect values of the best plan alternatives found in this study are primarily used to assess the linguistic ratings given by the evaluators. Furthermore, linguistic variables are used as a way to measure the performance value of the best plan alternative for each criterion

TABLE 4.4
Membership Function of Linguistic Scales (Example)

Fuzzy Number	Linguistic Scales	Scale of Fuzzy Number
$\tilde{1}$	Equally important (Eq)	(1,1,2)
$\tilde{3}$	Weakly important (Wq)	(2,3,4)
$\tilde{5}$	Essentially important (Es)	(4,5,6)
$\tilde{7}$	Very strongly important (Vs)	(6,7,8)
$\tilde{9}$	Absolutely important (Ab)	(8,9,9)

Note: This table synthesizes the linguistic scales defined by Chiou and Tzeng (2001, 2002) and fuzzy number scale used in Mon et al. (1994).

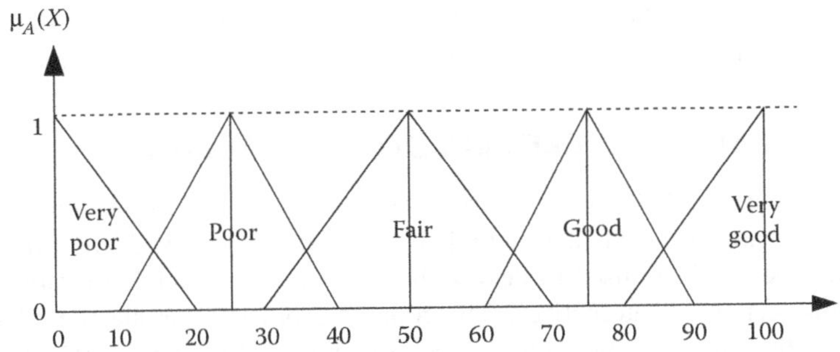

FIGURE 4.4 Membership functions of linguistic variables for measuring the performance value of alternatives (example).

as "very good," "good," "fair," "poor," and "very poor". TFNs, as shown in Figure 4.4 for example, can indicate the membership functions of the expression values.

4.3.4 Fuzzy Analytic Hierarchy Process

The procedure for determining the evaluation criteria weights by FAHP can be summarized as follows:

Step 1: Construct pairwise comparison matrices among all the elements/criteria in the dimensions of the hierarchy system. Assign linguistic terms to the pairwise comparisons by asking which is the more important of each two elements/criteria, such as

$$\tilde{A} = \begin{bmatrix} \tilde{1} & \tilde{a}_{12} & \cdots & \tilde{a}_{1n} \\ \tilde{a}_{21} & \tilde{1} & \cdots & \tilde{a}_{2n} \\ \vdots & \vdots & \ddots & \vdots \\ \tilde{a}_{n1} & \tilde{a}_{n2} & \cdots & \tilde{1} \end{bmatrix} = \begin{bmatrix} \tilde{1} & \tilde{a}_{12} & \cdots & \tilde{a}_{1n} \\ 1/\tilde{a}_{21} & \tilde{1} & \cdots & \tilde{a}_{2n} \\ \vdots & \vdots & \ddots & \vdots \\ 1/\tilde{a}_{n1} & 1/\tilde{a}_{n2} & \cdots & \tilde{1} \end{bmatrix}, \quad (4.4)$$

where $\tilde{a}_{ij}$ measure denotes: let $\tilde{1}$ be (1,1,1), when i equal j (i.e., $i=j$); if $\tilde{1},\tilde{2},\tilde{3},\tilde{4},\tilde{5},\tilde{6},\tilde{7},\tilde{8},\tilde{9}$ measure that criterion i is of relative importance to criterion j and then $\tilde{1}^{-1},\tilde{2}^{-1},\tilde{3}^{-1},\tilde{4}^{-1},\tilde{5}^{-1},\tilde{6}^{-1},\tilde{7}^{-1},\tilde{8}^{-1},\tilde{9}^{-1}$ measure that criterion j is of relative importance to criterion i.

Step 2: Use the geometric mean technique to define the fuzzy geometric mean and fuzzy weights of each criterion by Buckley (1985a, 1985b) as follows:

$$\tilde{r}_i = \left(\tilde{a}_{i1} \otimes \tilde{a}_{i2} \otimes \cdots \otimes \tilde{a}_{in}\right)^{1/n}, \quad \tilde{w}_i = \tilde{r}_i \otimes \left(\tilde{r}_1 \otimes \cdots \otimes \tilde{r}_n\right)^{-1}, \quad (4.5)$$

where $\tilde{a}_{in}$ is the fuzzy comparison value of criterion i to criterion n, thus $\tilde{r}_i$ is the geometric mean of fuzzy comparison value of criterion i to each criterion, and $\tilde{w}_i$ is the fuzzy weight of the ith criterion and can be indicated by a TFN, $\tilde{w}_i = (lw_i, mw_i, uw_i)$. Here lw_i, mw_i, and uw_i stand for the lower, middle, and upper values of the fuzzy weight of the ith criterion.

4.4 FUZZY MULTIPLE CRITERIA DECISION-MAKING

Bellman and Zadeh (1970) were the first to probe into the DM problem under a fuzzy environment-watershed and they heralded the initiation of FMCDM. This analysis method has been widely used to deal with DM problems involving multiple criteria evaluation/selection of alternatives. The following practical applications have been reported in the literature: weapon system evaluation (Mon et al. 1994), technology transfer strategy selection in biotechnology (Chang and Chen 1994), optimization of the design process of truck components (Altrock and Krause 1994), energy supply mix decisions (Tzeng et al. 1994), selection of urban transportation investment alternatives (Teng and Tzeng 1996a), tourist risk evaluation (Tsaur, Tzeng, and Chang 1997), evaluation of electronic marketing strategies in the information service industry (Tang, Tzeng, and Wang 1999), restaurant location selection (Tzeng et al. 2002), and performance evaluation of distribution centers in logistics (Chen, Chang, and Tzeng 2002). These studies show advantages in handling unquantifiable/qualitative criteria and obtained quite reliable results. This study uses this method to evaluate the performance of the best plan alternatives and rank them accordingly. The method and procedures of the FMCDM theory follow.

1. Alternatives Measurement: Using the measurement of linguistic variables to demonstrate the criteria performance/evaluation (effect-values) by expressions such as "very good," "good," "fair," "poor," and "very poor," the evaluators are asked to conduct their subjective judgments and each linguistic variable can be indicated by a TFN within the scale range 0–100, as shown in Figure 4.4. In addition, the evaluators can subjectively assign their personal range of the linguistic variable that can indicate the membership functions of the expression values of each evaluator. Take $\tilde{e}_{ij}^k$ to indicate the fuzzy performance/evaluation value of evaluator k towards alternative i under criterion j, and all of the evaluation criteria will be indicated by $\tilde{e}_{ij}^k = (le_{ij}^k, me_{ij}^k, ue_{ij}^k)$. Since the perception of each evaluator varies according to the evaluator's experience and knowledge, and the definitions of the linguistic variables vary as well, this study uses the notion of average value to integrate the fuzzy judgment values of m evaluators, that is,

$$\tilde{e}_{ij} = (1/m) \otimes \left(\tilde{e}_{ij}^1 \oplus \tilde{e}_{ij}^2 \oplus \cdots \oplus \tilde{e}_{ij}^m\right). \tag{4.6}$$

The sign $\otimes$ denotes fuzzy multiplication, the sign $\oplus$ denotes fuzzy addition, $\tilde{e}_{ij}$ shows the average fuzzy number of the judgment of the DM, which can be displayed by a TFN as $\tilde{e}_{ij} = (le_{ij}, me_{ij}, ue_{ij})$. The end-point values le_{ij}, me_{ij}, and ue_{ij} can be solved by the method put forward by Buckley [15]. that is,

$$le_{ij} = \left(\sum_{k=1}^{m} le_{ij}^k\right) \Bigg/ m; \quad me_{ij} = \left(\sum_{k=1}^{m} me_{ij}^k\right) \Bigg/ m; \quad ue_{ij} = \left(\sum_{k=1}^{m} ue_{ij}^k\right) \Bigg/ m. \tag{4.7}$$

2. Fuzzy Synthetic Decision: The weights of each criterion of building P&D evaluation as well as the fuzzy performance values must be integrated by the calculation of fuzzy numbers, so as to be located at the fuzzy performance value (effect-value) of the integral evaluation. According to each criterion weight $\tilde{w}_j$ derived by FAHP, the criteria weight vector $\tilde{\boldsymbol{w}} = (\tilde{w}_1,\ldots,\tilde{w}_j,\ldots,\tilde{w}_n)^t$ can be obtained, whereas the fuzzy performance matrix $\tilde{\boldsymbol{E}}$ of each of the alternatives can also be obtained from the fuzzy performance value of each alternative under n criteria, that is, $\tilde{\boldsymbol{E}} = (e_{ij})$. From the criteria weight vector $\tilde{w}$ and fuzzy performance matrix $\tilde{\boldsymbol{E}}$, the final fuzzy synthetic decision can be conducted and the derived result will be the fuzzy synthetic decision vector $\tilde{\boldsymbol{r}}$, that is,

$$\tilde{\boldsymbol{r}} = \tilde{\boldsymbol{E}} \circ \tilde{\boldsymbol{w}}. \tag{4.8}$$

The sign "$\circ$" indicates the calculation of the fuzzy numbers, including fuzzy addition and fuzzy multiplication. Since the calculation of fuzzy multiplication is rather complex, it is usually denoted by the approximate multiplied result of the fuzzy multiplication and the approximate fuzzy number $\tilde{r}_i$. The fuzzy synthetic decision of each alternative can be shown as $\tilde{r}_i = (lr_i,\ mr_i,\ ur_i)$, where lr_i, mr_i, and ur_i are the lower, middle, and upper synthetic performance values of the alternative i, that is:

$$lr_i = \sum_{j=1}^{n} le_{ij} \times lw_j,\quad mr_i = \sum_{j=1}^{n} me_{ij} \times mw_j,\quad ur_i = \sum_{j=1}^{n} ue_{ij} \times uw_j. \tag{4.9}$$

3. Ranking the fuzzy number: The result of the fuzzy synthetic decision reached by each alternative is a fuzzy number. Therefore, it is necessary that a non-fuzzy ranking method for fuzzy numbers be employed for comparison of each best plan alternative. In other words, the procedure of defuzzification is to locate the BNP value (Opricovic and Tzeng 2003b,c). Methods of such defuzzified fuzzy ranking generally include mean of maximal (MOM), CoA, and α-cut. To utilize the CoA method to find the BNP is a simple and practical method, and there is no need for the preferences of any evaluators, so it is used in this study.

The *BNP* value of the fuzzy number $\tilde{R}_i$ can be found by the following equation:

$$BNP_i = lr_i + \left[(ur_i - lr_i) + (mr_i - lr_i)\right]/3 \quad \forall i. \tag{4.10}$$

According to the value of the derived *BNP* for each of the alternatives, the ranking of the best plan of each of the alternatives can then proceed.

5 TOPSIS and VIKOR

The Technique for Order Preferences by Similarity to an Ideal Solution (TOPSIS) method was proposed by Hwang and Yoon (1981). The main idea came from the concept of the compromise solution to choose the best alternative nearest to the positive ideal solution (optimal solution) and farthest from the negative ideal solution (inferior solution). Then, choose the best one of sorting, which will be the best alternative.

5.1 TOPSIS

TOPSIS was proposed by Hwang and Yoon (1981) to determine the best alternative based on the concepts of the compromise solution. The compromise solution can be regarded as choosing the solution with the shortest Euclidean distance from the ideal solution and the farthest Euclidean distance from the negative ideal solution. The procedures of TOPSIS can be described as follows.

Given a set of alternatives, $A = \{A_k \mid k = 1,\ldots,n\}$, and a set of criteria, $C = \{C_j \mid j = 1,\ldots,m\}$, where $X = \{x_{kj} \mid k = 1,\ldots,n;\ j = 1,\ldots,m\}$ denotes the set of performance ratings and $\boldsymbol{w} = \{w_j \mid j = 1,\ldots,m\}$ is the set of weights, the information table $I = (A, C, X, W)$ can be represented as shown in Table 5.1.

The first step of TOPSIS is to calculate normalized ratings by

Form 1

$$r_{kj}(\boldsymbol{x}) = \frac{x_{kj}}{\sqrt{\sum_{k=1}^{n} x_{kj}^2}}, \quad k = 1,\ldots,n; j = 1,\ldots,m. \tag{5.1}$$

Form 2

- For benefit criteria (larger is better), $r_{kj}(x) = (x_{kj} - x_j^-)/(x_j^* - x_j^-)$, where $x_j^* = \max_k x_{kj}$ and $x_j^- = \min_k x_{kj}$ or setting x_j^* is the aspired/desired level and x_j^- is the worst level.
- For cost criteria (smaller is better), $r_{kj}(x) = (x_j^- - x_{kj})/(x_j^- - x_j^*)$, and then to calculate weighted normalized ratings by

$$v_{kj}(\boldsymbol{x}) = w_j r_{kj}(\boldsymbol{x}), \quad k = 1,\ldots,n; j = 1,\ldots,m. \tag{5.2}$$

Next the positive ideal point (*PIS*) and the negative ideal point (*NIS*) are derived as:

$$\begin{aligned} PIS = \boldsymbol{A}^+ &= \left\{v_1^+(\boldsymbol{x}), v_2^+(\boldsymbol{x}),\ldots, v_j^+(\boldsymbol{x}),\ldots, v_m^+(\boldsymbol{x})\right\} \\ &= \left\{\left(\max_k v_{kj}(\boldsymbol{x}) \mid j \in J_1\right), \left(\min_k v_{kj} \mid j \in J_2\right) \mid k = 1,\ldots,n\right\}, \end{aligned} \tag{5.3}$$

TABLE 5.1
The Information Table of TOPSIS

Alternatives	C_1	C_2	...	C_m
A_1	x_{11}	x_{12}	...	x_{1m}
A_2	x_{21}	x_{22}	...	x_{2m}
$\vdots$	$\vdots$	$\vdots$	$\vdots$	$\vdots$
A_n	x_{n1}	x_{n2}	...	x_{nm}
W	w_1	w_2	...	w_m

$$NIS = \boldsymbol{A}^- = \left\{v_1^-(\boldsymbol{x}), v_2^-(\boldsymbol{x}), \ldots, v_j^-(\boldsymbol{x}), \ldots, v_m^-(\boldsymbol{x})\right\}$$
$$= \left\{\left(\min_k v_{kj}(\boldsymbol{x}) \mid j \in J_1\right), \left(\max_k v_{kj}(\boldsymbol{x}) \mid j \in J_2\right) \mid k = 1, \ldots, n\right\}, \quad (5.4)$$

where J_1 and J_2 are the benefit and the cost attributes, respectively.

The next step is to calculate the separation from the *PIS* and the *NIS* between alternatives. The separation values can be measured using the Euclidean distance, which is given as:

$$D_k^* = \sqrt{\sum_{j=1}^{m}\left[v_{kj}(\boldsymbol{x}) - v_j^+(\boldsymbol{x})\right]^2}, \quad k = 1, \ldots, n \quad (5.5)$$

and

$$D_k^- = \sqrt{\sum_{j=1}^{m}\left[v_{kj}(\boldsymbol{x}) - v_j^-(\boldsymbol{x})\right]^2}, \quad k = 1, \ldots, n. \quad (5.6)$$

The similarities to the *PIS* can be derived as:

$$C_k^* = D_k^- / \left(D_k^* + D_k^-\right), \quad k = 1, \ldots, n, \quad (5.7)$$

where $C_k^* \in [0,1] \quad \forall k = 1, \ldots, n$.

Finally, the preferred orders can be obtained according to the similarities to the *PIS* (C_k^*) in descending order to choose the best alternatives. Next, a numerical example is introduced to show the procedures of TOPSIS.

Example 5.1

Consider a manager trying to evaluate whether a new facility is needed to replace the current system. Assume three criteria, durability, capability, and reliability, are considered by the manager and the preferred rating of each alternative can be expressed as shown in Table 5.2.

TABLE 5.2
Information Table in Example 5.1

Alternatives	Durability	Capability	Reliability
A_1	5	8	4
A_2	7	6	8
A_3	8	8	6
A_4	7	4	6
Weight	0.3	0.4	0.3

Next, we should first normalize the preferred ratings of each alternative, as shown in Table 5.3, so that the preferred ratings can fall on [0,1] with the same scale, though the scale of the criteria is the same in our example.

By multiplying the weights of each criterion by the corresponding preferred ratings, we can obtain the weighted normalized ratings as shown in Table 5.4.

Then, by using the Euclidean distance, we can calculate the separation from the *PIS* and the *NIS* to each alternative as shown in Table 5.5.

Finally, the similarities of the alternatives to the *PIS* can be derived as

$$C_1^* = 0.5037;\ \ C_2^* = 0.6581;\ \ C_3^* = 0.7482;\ \ C_4^* = 0.3340$$

and the preferred order of the alternatives can be determined as

$$A_3 \succ A_2 \succ A_1 \succ A_4.$$

On the basis of the preferred order of the alternatives, it can be seen that the current systems should be replaced with the new facility and Alternative 1 is the best choice.

5.2 VIKOR

The VIseKriterijumska Optimizacija I Kompromisno Resenje (VIKOR) method was developed for multicriteria optimization of complex systems. It determines the compromise ranking list, the compromise solution, and the weight stability intervals for preference stability of the compromise solution obtained with the initial (given) weights. This method focuses on ranking and selecting from a set of alternatives in the presence of conflicting criteria. It introduces the multicriteria ranking index based on the particular measure of "closeness" to the "ideal" solution (Opricovic 1998).

TABLE 5.3
Normalized Ratings in Example 5.1

Alternatives	r_1	r_2	r_3
A_1	0.37	0.60	0.32
A_2	0.51	0.45	0.65
A_3	0.59	0.60	0.49
A_4	0.51	0.30	0.49

TABLE 5.4
Weighted Normalized Ratings in Example 5.1

Alternatives	v_1	v_2	v_3
A_1	0.11^-	0.24^+	0.10^-
A_2	0.15	0.18	0.19^+
A_3	0.18^+	0.24	0.15
A_4	0.15	0.12^-	0.15

Assuming that each alternative is evaluated according to each criterion function, the compromise ranking could be performed by comparing the measure of closeness to the ideal alternative. The multicriteria measure for compromise ranking is developed from the L_p-*metric* used as an aggregating function in a compromise programming method (Yu 1973; Zeleny 1982). The various k alternatives ($k = 1,\ldots, n$) are denoted as $a_1, a_2,\ldots,a_n$. For alternative a_k, the rating of the jth aspect/criterion is denoted by f_{kj}, i.e., f_{kj} is the value of the jth criterion function for the alternative a_k; m is the number of criteria ($j = 1,2,\ldots,m$).

Development of the VIKOR method started with the following form of L_p-*metric*:

$$L_{p,k} = \left\{ \sum_{j=1}^{n} \left[w_j \left(f_j^* - f_{kj}\right) / \left(f_j^* - f_j^-\right) \right]^p \right\}^{1/p}, \quad 1 \le p \le \infty;\ k = 1,2,\ldots,n. \tag{5.8}$$

Within the VIKOR method, $L_{1,k}$ and $L_{\infty,k}$ are used to formulate ranking measure. The solution obtained by $\min_k S_k$ is with a maximum group utility ("majority" rule, shown as average gap, when $p = 1$) and the solution obtained by $\min_k R_k$ is with a minimum individual regret of the "opponent."

The compromise solution F^c is a feasible solution that is the "closest" to the ideal F^* and compromise means an agreement established by mutual concessions, as illustrated in Figure 5.1 by $\Delta f_1 = f_1^* - f_1^c$ and $\Delta f_2 = f_2^* - f_2^c$.

The compromise ranking algorithm VIKOR has the following steps:

a. Determine the best f_j^* and the worst f_j^- values of all criterion functions, $j = 1,2,\ldots,m$. If the jth function represents a benefit then $f_j^* = \max_k f_{kj}$ or setting f_j^* is the aspired/desired level, $f_j^- = \min_k f_{kj}$ or setting f_j^- is the worst level.

TABLE 5.5
The *PIS* and the *NIS* in Example 5.1

Alternatives	S^+	S^-
A_1	0.1175	0.1193
A_2	0.0635	0.1223
A_3	0.0487	0.1446
A_4	0.1307	0.0655

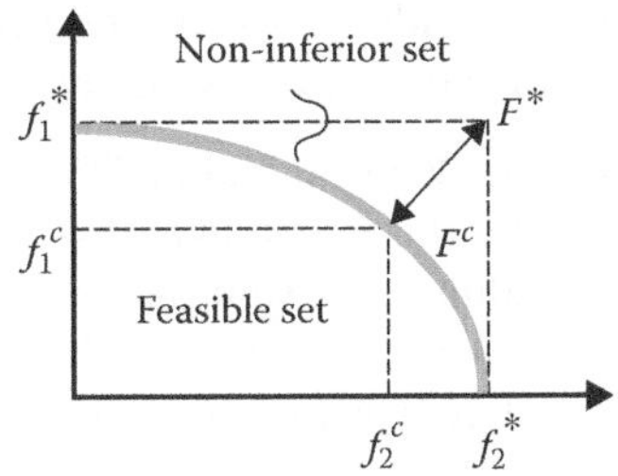

FIGURE 5.1 Ideal and compromise solutions.

b. Compute the values S_k and R_k, $k = 1,2,\ldots,n$, by the relations

$$S_k = \sum_{j=1}^{m} w_j \mid f_j^* - f_{kj} \mid / \mid f_j^* - f_j^- \mid,$$ shown as in the average gap,

$$R_k = \max_j \left\{ \mid f_j^* - f_{kj} \mid / \mid f_j^* - f_j^- \mid j = 1,2,\ldots,m \right\},$$ shown as maximal gap for improvement priority, where w_j are the weights of criteria, expressing their relative importance.

c. Compute the value Q_k, $k = 1,2,\ldots,n$, by the relation

$$Q_k = v(S_k - S^*)/(S^- - S^*) + (1-v)(R_k - R^*)/(R^- - R^*),$$

$k = 1,2,\ldots,m$ (alternatives)

where

$S^* = \min_k S_k$ or let $S^* = 0$ be zero gap, i.e., achieve the aspired level,
$S^- = \max_k S_k$ or let $S^- = 1$ be the worst level,
$R^* = \min R_j$ or let $R^* = 0$, be zero gap, i.e., achieve the aspired level,
$R^- = \max R_j$ or let $R^- = 1$ be the worst level

Therefore, we also can re-write $Q_k = vS_k + (1-v) R_k$, when $S^* = 0$, $S^- = 1$, $R^* = 0$, and $R^- = 1$. v is introduced as the weight of the strategy of "the majority of criteria" (or "the maximum group utility"), here $v = 0.5$.

d. Rank the alternatives, sorting by the values S, R, and Q, in decreasing order. The results are three ranking lists.

e. Propose as a compromise solution the alternative (a'), which is ranked the best by the measure Q (minimum) if the following two conditions are satisfied:

C1. "Acceptable advantage":

$$Q(a'') - Q(a') \geq DQ,$$

where a'' is the alternative with second position in the ranking list by Q; $DQ = 1/(J - 1)$; and J is the number of alternatives.

C2. "Acceptable stability in decision making":
Alternative a' must also be the best ranked by S or/and R. This compromise solution is stable within a decision-making process, which could be: "voting by majority rule" (when $v > 0.5$ is needed), "by consensus" $v \approx 0.5$, or "with vote" ($v < 0.5$). Here, v is the weight of the decision-making strategy "the majority of criteria" (or "the maximum group utility").

If one of the conditions is not satisfied, then a set of compromise solutions is proposed, which consists of:

- Alternative a′ and a″ if only condition C2 is not satisfied, or
- Alternative a', a'',..., $a^{(n)}$ if condition C1 is not satisfied; and $a^{(n)}$ is determined by the relation $Q(a^{(n)} - Q(a')) < DQ$ for maximum n (the positions of these alternatives are "in closeness").

The best alternative, ranked by Q, is the one with the minimum value of Q. The main ranking result is the compromise ranking list of alternatives and the compromise solution with the "advantage rate."

Ranking by VIKOR may be performed with different values of criteria weights, analyzing the impact of criteria weights on the proposed compromise solution. The VIKOR method determines the weight stability intervals, using the methodology presented in Opricovic (1998). The compromise solution obtained with initial weights (w_j, $j = 1,\ldots, m$) will be replaced if the value of a weight is not within the stability interval. The analysis of weight stability intervals for a single criterion is performed for all criterion functions, with the same (given) initial values of weights. In this way, the preference stability of an obtained compromise solution may be analyzed using the VIKOR program.

VIKOR is a helpful tool in multicriteria decision making, particularly in a situation where the decision maker is not able, or does not know, to express his/her preference at the beginning of system design. The obtained compromise solution could be accepted by the decision makers because it provides a maximum "group utility" (represented by min S, Equation 5.1) of the "majority" and a minimum of the individual regret (represented by min R) of the "opponent." The compromise solutions could be the basis for negotiations, involving the decision makers' preference by criteria weights.

Example 5.2

Consider a manager trying to evaluate if a new facility is needed to replace the current system. Assume three criteria, durability, capability, and reliability, are considered by the manager and the preferred ratings of each alternative can be expressed as shown in Table 5.6.

Next, we should first normalize the preferred ratings of each alternative, as shown in Table 5.7, so that the preferred ratings can fall on [0,1] with the same scale, though the scale of the criteria is the same in our example.

TABLE 5.6
Information Table in Example 5.2

Alternatives	Durability	Capability	Reliability
A_1	5	8	4
A_2	7	6	8
A_3	8	8	6
A_4	7	4	6
Weight	0.3	0.4	0.3

From the results of Table 5.7, we can calculate Q_j as:

$$Q_1 = 0.5 \times (0.60 - 0.15)/(0.65 - 0.15) + 0.5 \times (0.30 - 0.15)/(0.40 - 0.15) = 0.75;$$

$$Q_2 = 0.5 \times (0.30 - 0.15)/(0.65 - 0.15) + 0.5 \times (0.20 - 0.15)/(0.40 - 0.15) = 0.25;$$

$$Q_3 = 0.5 \times (0.15 - 0.15)/(0.65 - 0.15) + 0.5 \times (0.15 - 0.15)/(0.40 - 0.15) = 0;$$

$$Q_4 = 0.5 \times (0.65 - 0.15)/(0.65 - 0.15) + 0.5 \times (0.40 - 0.15)/(0.40 - 0.15) = 1.$$

Then, we can rank the alternatives according to the values S_j, R_j, and Q_j with decreasing order as shown in Table 5.8.

On the basis of the preferred order of the alternatives, it can be seen that $A_3 \succ A_2 \succ A_1 \succ A_4$; the current systems should be replaced with the new facility and Alternative 3 is the best choice.

5.3 COMPARING VIKOR AND TOPSIS

From the basic foundation of TOPSIS, we can conclude that the main ideal of TOPSIS comes from the ideal of reference-dependent theory (Kahneman and Tversky 1979). Reference-dependent theory states that consumers evaluate alternatives in terms of gains and losses relative to a subjective reference point (Kahneman and Tversky 1979, 1984; Kahneman, Knetsch, and Thaler 1991; Quattrone and Tversky 1998; Hardie, Johnson, and Fader 1993; Highhouse and Johnson 1996). Therefore, the problem of how to accurately measure the distance from an alternative to the *PIS* and the *NIS* is key to TOPSIS. Although the Euclidean distance is employed in this chapter, the

TABLE 5.7
Information Table in Example 5.2

Alternatives	S_j	R_j
A_1	0.60	0.30
A_2	0.30	0.20
A_3	$0.15 = S^*$	$0.15 = R^*$
A_4	$0.65 = S^-$	$0.40 = R^-$

TABLE 5.8
Sorting S_j, R_j, and Q_j with Decreasing Order

S_j	R_j	Q_j
A_4	A_4	A_4
A_1	A_1	A_1
A_2	A_2	A_2
A_3	A_3	A_3

Minkowski distance of order p (p-norm distance) can also be used. In addition, TOPSIS has recently been widely used for various applications, such as selecting an expatriate host country (Chen and Tzeng 2004), selection of fire station location (Tzeng and Lin 1997), and comparison with other methods (Opricovic and Tzeng 2004).

Unfortunately, Opricovic and Tzeng (2004) found the traditional TOPSIS method cannot be used for ranking purposes. The reasons are explained as follows:

The TOPSIS method introduces an aggregating function for ranking in Equation 5.7. According to the formation of C_r^* (ranking index), alternative a_r is better than a_k ($a_j > a_k$) if $C_r^* > C_k^*$ or $D_r^- /(D_r^* + D_r^-) > D_k^- /(D_k^* + D_k^-)$, which will hold if

1. $D_r^* < D_k^*$ and $D_r^- > D_k^-$; i.e. $C_r^* = D_r^- /(D_r^* + D_r^-) > C_r^* = D_r^- /(D_r^* + D_r^-)$, $a_r > a_k$ in TOPSIS method;
2. $D_r^* > D_k^*$ and $D_r^- > D_k^-$; but $D_r^* < D_k^* D_r^- / D_k^-$; the best choice based on the nearest to the positive ideal solution, $D_k^* < D_r^*$, then $a_k > a_r$.

Condition 1 shows the "regular" situation, when alternative a_r is better than a_k because it is closer to the ideal and farther from the negative ideal. On the contrary, conditional 2 in Equation 5.7 shows that an alternative a_r is farther from the ideal than a_k. Let a_k be that alternative with $D_k^* = D_k^-$ and $C_k^* = 0.5$. In this case, all alternatives a_r with $D_r^* > D_k^*$ and $D_r^- > D_r^*$ are better ranked than a_k, although a_k is closer to the ideal A^*. The distances considered by VIKOR and TOPSIS are illustrated in Figure 5.2. An alternative a_r is better than a_k as a TOPSIS result, but a_k is better than a_r ranked by VIKOR because a_k is closer to the ideal solution. The relative importance of distances D_r^* and D_r^- was not considered within Equation 5.7, although it could be a major concern in decision making.

5.4 FUZZY TOPSIS

Since the preferred ratings usually refer to the subjective uncertainty, it is natural to extend TOPSIS to consider the situation of fuzzy numbers. Fuzzy TOPSIS can be intuitively extended by using the fuzzy arithmetic operations as follows.

Given a set of alternatives, $A = \{A_k \mid k = 1,\ldots,n\}$, and a set of criteria, $C = \{C_j \mid j = 1,\ldots,m\}$, where $X = \{x_{kj} \mid k = 1,\ldots,n; j = 1,\ldots,m\}$ denotes the set of fuzzy ratings and $\tilde{w} = \{\tilde{w}_j \mid j = 1,\ldots,m\}$ is the set of fuzzy weights, the first step of TOPSIS is to calculate normalized ratings by

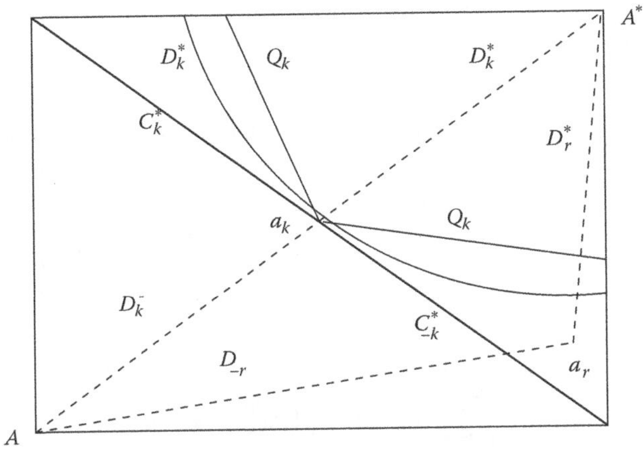

FIGURE 5.2 VIKOR and TOPSIS distances. (From Opricovic, S., and G.H., *European Journal of Operational Research* 156 (2): 445, 2004.)

$$\tilde{r}_{kj}(\boldsymbol{x}) = \frac{\tilde{x}_{kj}}{\sqrt{\sum_{k=1}^{n} \tilde{x}_{kj}^2}}, \quad k = 1,\ldots,n;\ j = 1,\ldots,m \tag{5.9}$$

and then to calculate the weighted normalized ratings by

$$\tilde{v}_{ij}(\boldsymbol{x}) = \tilde{w}_j \tilde{r}_{ij}(\boldsymbol{x}), \quad k = 1,\ldots,n;\ j = 1,\ldots,m. \tag{5.10}$$

Next the *PIS* and the *NIS* are derived as

$$\begin{aligned} PIS = \tilde{A}^+ &= \left\{\tilde{v}_1^+(\boldsymbol{x}), \tilde{v}_2^+(\boldsymbol{x}),\ldots,\tilde{v}_j^+(\boldsymbol{x}),\ldots,\tilde{v}_m^+(\boldsymbol{x})\right\} \\ &= \left\{\left(\max_k \tilde{v}_{kj}(\boldsymbol{x}) \mid j \in J_1\right), \left(\min_k \tilde{v}_{kj}(\boldsymbol{x}) \mid j \in J_2\right) \mid k = 1,\ldots,n\right\}. \end{aligned} \tag{5.11}$$

$$\begin{aligned} NIS = \tilde{A}^- &= \left\{\tilde{v}_1^-(\boldsymbol{x}), \tilde{v}_2^-(\boldsymbol{x}),\ldots,\tilde{v}_j^-(\boldsymbol{x}),\ldots,\tilde{v}_m^-(\boldsymbol{x})\right\} \\ &= \left\{\left(\min_k \tilde{v}_{kj}(\boldsymbol{x}) \mid j \in J_1\right), \left(\max_k \tilde{v}_{kj}(\boldsymbol{x}) \mid j \in J_2\right) \mid k = 1,\ldots,n\right\}, \end{aligned} \tag{5.12}$$

where J_1 and J_2 are the benefit and the cost attributes, respectively.

Similar to the crisp situation, the next step is to calculate the separation from the *PIS* and the *NIS* between the alternatives. The separation values can also be measured using the Euclidean distance given as:

$$\tilde{S}_k^+ = \sqrt{\sum_{k=1}^{m} \left[\tilde{v}_{kj}(\boldsymbol{x}) - \tilde{v}_j^+(\boldsymbol{x})\right]^2}, \quad k = 1,\ldots,n, \tag{5.13}$$

and

TABLE 5.9
Fuzzy Information Table in Example 5.3

Alternatives	Durability	Capability	Reliability
A_1	(4,5,8)	(6,8,9)	(3,4,7)
A_2	(4,7,8)	(3,6,9)	(5,8,9)
A_3	(7,8,8)	(5,8,9)	(5,6,8)
A_c	(5,7,9)	(3,4,7)	(5,6,7)
Weight	(0.2,0.3,0.4)	(0.3,0.4,0.5)	(0.2,0.3,0.4)

$$\tilde{S}_k^- = \sqrt{\sum_{j=1}^{m}\left[\tilde{v}_{kj}(x) - \tilde{v}_j^-(x)\right]^2}, \quad k = 1,\ldots,n, \tag{5.14}$$

where

$$\max\left\{\tilde{v}_{kj}(x)\right\} - \tilde{v}_j^+(x) = \min\left\{\tilde{v}_{kj}(x)\right\} - \tilde{v}_j^-(x) = 0. \tag{5.15}$$

Then, the defuzzified separation values should be derived using one of the defuzzified methods, such as CoA, to calculate the similarities to the *PIS*.

Next, the similarities to the *PIS* are given as

$$C_k^* = D\left(S_k^-\right)/\left[D\left(S_k^+\right) + D\left(S_k^-\right)\right], \quad k = 1,\ldots,n, \tag{5.16}$$

where $C_k^* \in [0,1] \ \forall k = 1,\ldots,n$.

Finally, the preferred orders are ranked according to C_k^* in descending order to choose the best alternatives. Next, a numerical example is considered to demonstrate the procedures of fuzzy TOPSIS.

Example 5.3

On the basis of the problem in Example 5.1, we can use fuzzy numbers to represent the subjective uncertainty of the manager in determining the best alternative. Then the fuzzy information table can be expressed as shown in Table 5.9.

Next, by employing Equation 5.9, we can derive the fuzzy normalized ratings of each alternative as shown in Table 5.10.

TABLE 5.10
Fuzzy Normalized Ratings in Example 5.3

Alternatives	r_1	r_2	r_3
A_1	(0.24,0.37,0.78)	(0.35,0.60,1.01)	(0.19,0.32,0.76)
A_2	(0.24,0.51,0.78)	(0.18,0.45,1.01)	(0.32,0.65,0.98)
A_3	(0.42,0.59,0.78)	(0.29,0.60,1.01)	(0.32,0.49,0.87)
A_c	(0.30,0.51,0.87)	(0.18,0.30,0.79)	(0.32,0.49,0.76)

TABLE 5.11
Fuzzy Weighted Normalized Ratings in Example 5.3

Alternatives	v_1	v_2	v_3
A_1	$(0.048,0.111,0.312)^-$	$(0.105,0.240,0.505)^+$	$(0.038,0.094,0.304)^-$
A_2	(0.048,0.153,0.312)	(0.054,0.180,0.505)	$(0.064,0.195,0.392)^+$
A_3	$(0.084,0.177,0.312)^+$	(0.087,0.240,0.505)	(0.064,0.147,0.348)
A_c	(0.006,0.153,0.348)	$(0.054,0.120,0.395)^-$	(0.064,0.147,0.304)

Then the fuzzy weighted normalized ratings can be obtained, as shown in Table 5.11, using Equation 5.10.

According to Equations 5.11 through 5.15, we can calculate the fuzzy *PIS* and the fuzzy *NIS* and the defuzzified *PIS* and *NIS* can also be derived by using the CoA method as shown in Table 5.12.

Finally, the similarities to the *PIS* can be calculated based on Equation 5.16 as:

$$C_1^* = 0.5040; C_2^* = 0.5629; C_3^* = 0.5657; C_4^* = 0.3906.$$

From the similarities above, we can determine the preferred order of the alternatives as:

$$A_3 \succ A_2 \succ A_1 \succ A_4.$$

On the basis of the preferred order of the alternatives, it can be seen that the current systems should be replaced with the new facility and Alternative 1 is the best choice under the circumstances of the subjective uncertainty.

TABLE 5.12
The *PIS* and the *NIS* in Example 5.3

Alternatives	$\tilde{S}_k^+$	$D(S_k^+)$	$\tilde{S}_k^-$	$D(S_k^-)$
A_1	(0.0000,0.1207,0.4412)	0.1873	(0.0000,0.1200,0.4510)	0.1903
A_2	(0.0000,0.0646,0.5226)	0.1957	(0.0000,0.1248,0.6312)	0.2520
A_3	(0.0000,0.0480,0.5313)	0.1931	(0.0000,0.1469,0.6076)	0.2515
A_4	(0.0000,0.1315,0.6065)	0.2460	(0.0000,0.0676,0.4055)	0.1577

6 ELECTRE Method

Roy (1968) and Benayoun et al. (1966) originally used the concept of outranking relations to introduce the ELimination Et Choice Translating REality (ELECTRE) method. Since then various ELECTRE models have been developed based on the nature of the problem statement (to find a kernel solution or to rank the order of alternatives), the degree of significance of the criteria to be taken into account (true or pseudo), and the preferential information (weights, concordance index, discordance index, veto effect).

6.1 ELECTRE I

The ELECTRE I model was first developed by Roy (1968) to find the kernel solution in a situation where true criteria and restricted outranking relations are given. That is, ELECTRE I cannot derive the ranking of alternatives but the kernel set. In ELECTRE I, two indices called the concordance index and the discordance index are used to measure the relations between objects. For the concordance index, $C(a,b)$ measures how much a is at least as good as b. On the other hand, the discordance index, $D(a,b)$ measures the degree to which b is strictly preferred to a. The concordance index and the discordance index in ELECTRE I can be defined by

$$C(a,b)=\frac{\sum_{i\in Q(a,b)} w_i}{\sum_{i=1}^{m} w_i} \tag{6.1}$$

and

$$D(a,b)=\frac{\max_{i\in R(a,b)}\left[w_i\left(g_i(b)-g_i(a)\right)\right]}{\max_{c,d\in A}\left[w_i\left(g_i(c)-g_i(d)\right)\right]}, \tag{6.2}$$

where $C(a,b)$ and $D(a,b) \in [0,1]$, $g_j(k)$ denote the preferred scores of the jth attribute for the kth alternative, $Q(a,b)$ denotes the set of criteria for which a is equal or preferred to b, $R(a,b)$ is the set of criteria for which b is strictly preferred to a, and A denotes the set of all alternatives.

For comparing alternatives a and b, we can determine the relation between a and b as the following rules:

If $C(a,b) > C^*$ and $D(a,b) < D^*$ then a outranks b, otherwise a does not outrank b.
If $C(b,a) > C^*$ and $D(b,a) > D^*$ then b outranks a, otherwise b does not outrank a.

Then, the outrank relation between a and b can be derived by referring to Table 6.1.

TABLE 6.1
Outrank Relation between *a* and *b*

Outrank Relation	*a* does not Outrank *b*	*a* Outranks *b*
b does not outrank *a*	Incomparable	*a* outranks *b* ($a \succ b$)
b outranks *a*	*b* outranks *a* ($b \succ a$)	Indifference ($a - b$)

Example 6.1

Suppose a multiattribute decision problem for mounting a global positioning system (GPS) is considered. Four alternatives are considered according to four criteria: reliability, functionality, service, and accuracy. The corresponding preferred ratings of each alternative can be defined by a five-point ordinal scale: Very Dissatisfied (VD), Dissatisfied (D), Unsatisfied (U), Satisfied (S), and Very Satisfied (VS), as shown in Table 6.2.

According to Equations 6.1 and 6.2, the indices of concordance and discordance can be derived as shown in Tables 6.3 and 6.4, respectively.

Next, by setting $C = 0.7$ and $D = 0.7$, the outranking relations for Example 6.1 can be built as shown in Figure 6.1.

According to Figure 6.1, it can be seen that Alternative 1 should be the kernel solution. However, it is hard to identify which is better between Alternative 2 and Alternative 4. The shortcomings of ELECTRE I are clear. First, it can only find the kernel solution and cannot rank the order among alternatives. Second, since the final results are varied with the vote threshold, how to determine the appropriate threshold remains unknown. Third, if the kernel set contains any circuits, the kernel set is not unique and may not exit.

6.2 ELECTRE II

ELECTRE II was proposed by Roy and Bertier (1973) to overcome ELECTRE I's inability to produce a ranking of alternatives. Instead of simply finding the kernel set, ELECTRE II can order alternatives by introducing the strong and the weak outranking relations. Furthermore, an additional constraint that a is preferred to b should meet both $C(a,b) > C^*$ and $C(a,b) \geq C(b,a)$. It is clear that this constraint prevents the circuits in the kernel set.

TABLE 6.2
Decision Table for the GPS Problem

Alternatives	Reliability	Functionality	Service	Accuracy
Alternative 1	U	D	S	VS
Alternative 2	D	VS	D	S
Alternative 3	D	VD	S	D
Alternative 4	U	VS	U	S
Weights	0.35	0.15	0.20	0.30

TABLE 6.3
Value of Concordance Index for the GPS Problem

Concordance Index	Alternative 1	Alternative 2	Alternative 3	Alternative 4
Alternative 1	1.00	0.85	0.80	0.50
Alternative 2	0.15	1.00	0.45	0.00
Alternative 3	0.00	0.20	1.00	0.20
Alternative 4	0.20	0.00	0.15	1.00

Before describing the procedures of ELECTRE II, the following quantities should be first introduced:

$$I^+(a,b) = \{C_i \mid g_i(a) > g_j(b)\}; \tag{6.3}$$

$$I^=(a,b) = \{C_i \mid g_i(a) = g_i(b)\}; \tag{6.4}$$

$$I^-(a,b) = \{C_i \mid g_i(a) < g_i(b)\}; \tag{6.5}$$

$$W^+(a,b) = \sum_{j \in I^+(a,b)} w_i; \tag{6.6}$$

$$W^=(a,b) = \sum_{j \in I^=(a,b)} w_i; \tag{6.7}$$

$$W^-(a,b) = \sum_{j \in I^-(a,b)} w_i. \tag{6.8}$$

Next, the concordance and the discordance indices for a pair (a, b) can be defined as:

$$C(a,b) = \frac{W^+(a,b) + W^=(a,b)}{W^+(a,b) + W^=(a,b) + W^-(a,b)} \tag{6.9}$$

TABLE 6.4
Value of Discordance Index for the GPS Problem

Discordance Index	Alternative 1	Alternative 2	Alternative 3	Alternative 4
Alternative 1	–	0.42	0.00	0.42
Alternative 2	1.00	–	0.57	0.57
Alternative 3	1.00	0.85	–	1.00
Alternative 4	0.85	0.00	0.57	–

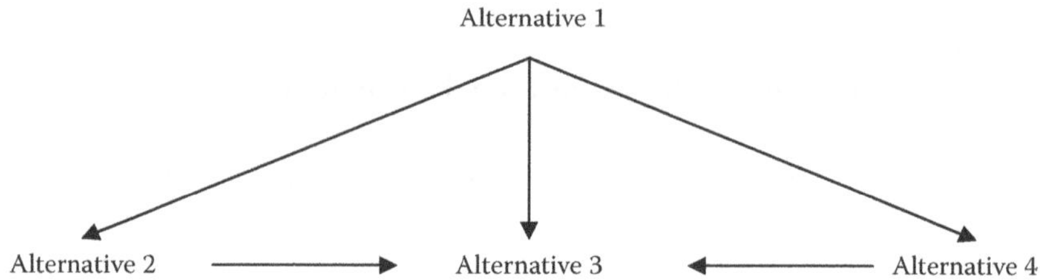

FIGURE 6.1 The outranking for Example 6.1.

and

$$D(a,b) = \frac{\max_{i \in I^-(a,b)} |g_i(a) - g_i(b)|}{\max_{i \in I} (g_i(a), \theta_i)}, \tag{6.10}$$

where θ_i denotes the R-degree parameter used by a decision maker for the ith criterion to represent the degree of attention paid by the decision maker to the ith criterion. From Equations (6.9) and (6.10) it can be seen that $0 \le C(a, b) \le 1$ and $0 \le D(a, b) \le 1$.

The procedure of ELECTRE II can be described as follows. Designate the concordance and the discordance threshold C^+, D^+ for the strong outranking relation, and C^-, D^- for the weak outranking relation where $C^+ > C^-$ and $D^+ < D^-$. The steps of ELECTRE II can be divided into the following three stages (Belton and Stewart 2002).

First stage: determine the descending order.

1. Let Ω be the full set of alternatives.
2. Derive the non-dominated set $\mathbb{N}$ which is not *strongly* outranked by any other alternatives in Ω.
3. Determine the first class of the descending ranking by deriving the set $\mathbb{N}'$ which is not *weakly* outranked by any other alternatives in $\mathbb{N}$.
4. Take off the alternatives in $\mathbb{N}'$ and repeat step 3 until all alternatives have been classified.

Second stage: determine the ascending order.

1. Let Ω be the full set of alternatives.
2. Derive the dominated set $\mathbb{Z}$ which does not *strongly* outrank any other alternatives in Ω.
3. Determine the first class of the ascending ranking by deriving the set $\mathbb{Z}'$ which does not *weakly* outrank any other alternatives in $\mathbb{Z}$.
4. Take off the alternatives in $\mathbb{Z}'$ and repeat step 3 until all alternatives have been classified.

Third stage: determine the final order.

Accounting for the intersection of the descending and ascending orders, we can determine the final order of the alternatives.

TABLE 6.5
Preferred Ratings of Each Alternative in Example 6.2

Preferred Ratings	Quality	Price	Function	Service
Alternative 1	4	2	6	8
Alternative 2	8	9	1	4
Alternative 3	8	2	3	1
Alternative 4	1	2	3	1
Weights	0.20	0.20	0.30	0.30

Example 6.2

Suppose a decision-making problem for purchasing a notebook is considered. In order to choose one of the four alternatives, four criteria, quality, price, function, and service, are considered. The preferred ratings, with respect to each alternative, and the weight, with respect to each criterion, are given as shown in Table 6.5.

In order to calculate the concordance and the discordance indices, the following quantities should be first derived according to Equations 6.6 through 6.8:

$$W^{+}(1,2) = 0.3+0.3 = 0.6; \qquad W^{+}(1,3) = 0.3+0.3 = 0.6;$$
$$W^{+}(1,4) = 0.2+0.3+0.3 = 0.8; \qquad W^{+}(2,1) = 0.2+0.2 = 0.4;$$
$$W^{+}(2,3) = 0.2+0.3 = 0.5; \qquad W^{+}(2,4) = 0.2+0.2+0.3 = 0.7;$$
$$W^{+}(3,1) = 0.2; \qquad W^{+}(3,2) = 0.3;$$
$$W^{+}(3,4) = 0.2; \qquad W^{+}(4,1) = 0;$$
$$W^{+}(4,2) = 0.3; \qquad W^{+}(4,3) = 0.$$

$$W^{=}(1,2) = 0; \qquad W^{=}(1,4) = 0.1;$$
$$W^{=}(2,1) = 0; \qquad W^{=}(2,3) = 0.2;$$
$$W^{=}(2,4) - 0; \qquad W^{=}(3,1) - 0.2;$$
$$W^{=}(3,2) = 0.2; \qquad W^{=}(3,4) = 0.2+0.3+0.3 = 0.8;$$
$$W^{=}(4,1) = 0.2; \qquad W^{=}(4,2) = 0;$$
$$W^{=}(4,3) = 0.2+0.3+0.3 = 0.8.$$

Then, on the basis of Equations 6.9 and 6.10, the concordance and the discordance indices can be calculated as shown in Tables 6.6 and 6.7.

TABLE 6.6
The Concordance Index in Example 6.2

C(*a*, *b*)	Alternative 1	Alternative 2	Alternative 3	Alternative 4
Alternative 1	1	0.6	0.8	0.4
Alternative 2	0.4	1	0.7	0.7
Alternative 3	0.4	0.5	1	1
Alternative 4	0.2	0.3	0.8	1

TABLE 6.7
Discordance Index in Example 6.2

D(*a*, *b*)	Alternative 1	Alternative 2	Alternative 3	Alternative 4
Alternative 1	0	0.78	0.44	0
Alternative 2	0.56	0	0.22	0.22
Alternative 3	0.78	0.78	0	0
Alternative 4	0.78	0.78	0.78	0

Now, let $C^+ = 0.7$ and $C^- = 0.6$ be the strong and weak concordance indices and $D^+ = 0.5$ and $D^- = 0.7$ be the strong and weak discordance indices. Then, the descending and ascending orders can be derived as shown in Table 6.8.

Finally, according to the intersection of the above orders, it can be seen that the following partial order should be indicated as in Table 6.9.

ELECTRE II can be regarded as the extension of ELECTRE I by deriving the partial preorder of alternatives, instead of the subset of kernel solutions. In the next section, ELECTRE III is introduced. Unlike ELECTRE I and ELECTRE II, which use "true" criteria to derive the concordance and the inconcordance indices, ELECTRE III employs pseudocriteria to calculate the partial preorder of alternatives.

6.3 ELECTRE III

Roy (1977, 1978) developed ELECTRE III, extending the crisp outranking relations for modeling decision makers' preferences in fuzzy conditions. Next, we briefly review ELECTRE III (Hokkanen and Salminen 1997a). For a detailed description of these evaluation procedures, refer to Hwang and Yoon (1981), Roy (1991), Tzeng and Wang (1993), Tsaur and Tzeng (1991), and Teng and Tzeng (1994).

Let $A = (a,b,c,\ldots,n)$ be a set of alternatives and $(g_1,g_2,\ldots,g_m)$ a set of criteria for our MCDM problems; $g_j(a_j)$ represents the performance or the evaluation of the alternative $a \in A$ on criterion g_j. Depending on whether the target is to maximize or to minimize the criterion $g_j(a_j)$, the higher or lower it is, the better the alternative meets the criterion in question. Consequently, the multicriteria evaluation of an alternative $a \in A$ will be represented by the vector $g(a) = (g_1(a),g_2(a),\ldots,g_m(a))$.

TABLE 6.8
The Descending and Ascending Orders in Example 6.2

Descending Order	Ascending Order
Alternative 1	Alternative 2
Alternative 2	Alternative 1
Alternative 3	Alternative 3
Alternative 4	Alternative 4

TABLE 6.9
Final Order in Example 6.2

Alternative 1
Alternative 2
Alternative 3
Alternative 4

The evaluation procedures of the ELECTRE III model (refer to Figure 6.2) encompass the establishment of threshold function, disclosure of concordance index and discordance index, confirmation of credibility degree, and the ranking of alternatives. These data are often represented by fuzzy data using a subjective judgment by evaluators or decision-makers, a further description of which follows.

Let $q(g)$ and $p(g)$ represent the indifference threshold and preference threshold, respectively.

If $g(a) \geq g(b)$:

1. $g(a) > g(b) + p(g(b)) \Leftrightarrow aPb,$ (6.11)
2. $g(b) + q(g(b)) < g(a) < g(b) + p(g(b)) \Leftrightarrow aQb,$ (6.12)
3. $g(b) < g(a) < g(b) + q(g(b)) \Leftrightarrow aIb,$ (6.13)

where P denotes strong preference, Q denotes weak preference, I denotes indifference, and $g(a)$ is the criterion value of alternative a.

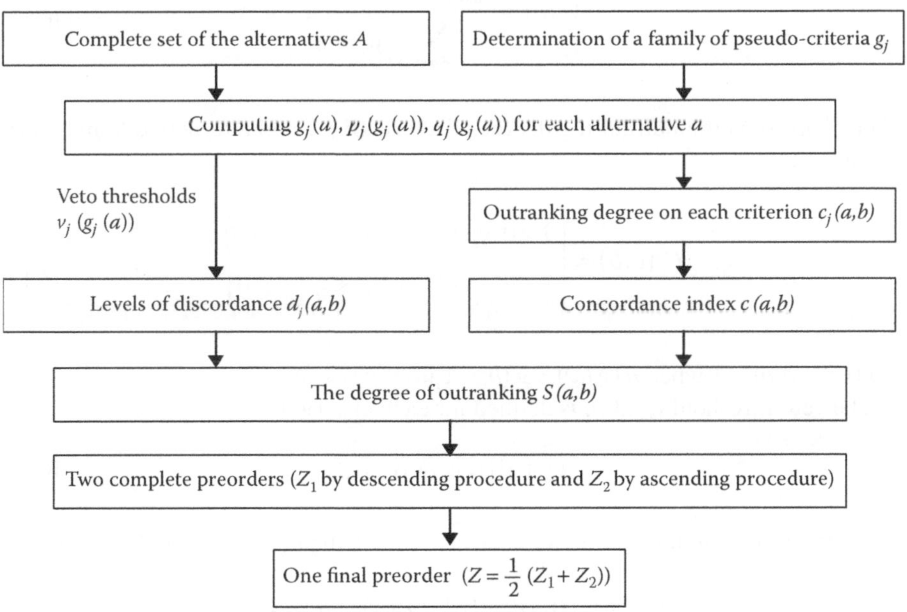

FIGURE 6.2 General structure of ELECTRE III.

The establishment of a threshold function has to satisfy the subsequent constraint equations:

$$g(a) > g(b) \Rightarrow \begin{cases} g(a) + q(g(a)) > g(b) + q(g(b)) \\ g(a) + p(g(a)) > g(b) + p(g(b)) \end{cases}, \tag{6.14}$$

for all criteria, $p(g) > q(g)$.

Furthermore, $p_j(g_j(a))$ and $q_j(g_j(a))$ can be calculated according to Roy's formula:

$$p_j(g_j(a)) = \alpha_p + \beta_p g_j(a); \tag{6.15}$$

$$q_j(g_j(a)) = \alpha_q + \beta_q g_j(a), \tag{6.16}$$

where $p_j(g_j(a))$ and $q_j(g_j(a))$ can be solved in such a way that threshold values are one of the following cases (Roy, Present, and Silhol 1986):

1. Either constant (β equals zero and α has to be determined)
2. Proportional to $g_j(a)$ (β has to be determined and α equals zero); or
3. A form combining these two (both α and β have to be determined)

A concordance index $C(a, b)$ is computed for each pair of alternatives:

$$C(a,b) = \frac{\sum_{i=1}^{m} w_i C_i(a,b)}{\sum_{i=1}^{m} w_i}, \tag{6.17}$$

where $C_i(a, b)$ is the outranking degree of alternative a and alternative b under criterion i, and

$$C_i(a,b) = \begin{cases} 0 & \text{if } g_i(b) - g_i(a) > p_i(g_i(a)) \\ 1 & \text{if } g_i(b) - g_i(a) \le q_i(g_i(a)) \end{cases} \tag{6.18}$$

and $0 < c_i(a,b) < 1$ when $q_i(g_i(a)) < g_i(b) - g_i(a) \le p_i(g_i(a))$.

The veto threshold $v_i(g_i(a))$ is defined for each criterion i:

$$v_i(g_i(a)) = \alpha_v + \beta_v g_i(a). \tag{6.19}$$

A discordance index, $d(a,b)$, for each criterion is then defined as follows:

$$d_i(a,b) = 0 \quad \text{if } g_i(b) - g_i(a) \le p_i(g_i(a)) \tag{6.20}$$

$$d_i(a,b) = 1 \quad \text{if } g_i(b) - g_i(a) > v_i(g_i(a)) \tag{6.21}$$

and $0 < d_i(a,b) < 1$ when $p_i(g_i(a)) < g_i(b) - g_i(a) \le v_i(g_i(a))$.

Finally, the degree of outranking is defined by $S(a,b)$:

$$S(a,b) = \begin{cases} c(a,b) & \text{if } d_j(a,b) \le c(a,b) \quad \forall j \in J \\ c(a,b) \times \prod_{j \in J(a,b)} \dfrac{1 - d_j(a,b)}{1 - c(a,b)} & \text{otherwise} \end{cases}, \tag{6.22}$$

where $J(a,b)$ is the set of criteria for which $d_j(a,b) > c(a,b)$.

The exploiting ranking procedure used in ELECTRE III generally consists of the following steps (Belton and Stewart 2002):

Step 1: *Construct a complete preorder Z_1 by descending distillation procedure.*

1. Determine the maximum value of the credibility index, $\lambda_{max} = \max S(a,b)$, where the maximization is taken over the current set of alternatives under consideration.
2. Set $\lambda^* = \lambda_{max} - (0.3 - 0.15\lambda)$.
3. For each alternative determine its λ-strength, namely, the number of alternatives in the current set to which it is λ-preferred using $\lambda = \lambda^*$.
4. For each alternative determine its λ-weakness, namely, the number of alternatives in the current set which are λ-preferred to it using $\lambda = \lambda^*$.
5. For each alternative determine its qualification, which is its λ-strength minus its λ-weakness.
6. The set of alternatives having the largest qualification is called the first distillate, D_1.
7. If D_1 has more than one member, repeat the process on the set D_1 until all alternatives have been classified; then continue with the original set minus D_1, repeating until all alternatives have been classified.

Step 2: *Construct a complete preorder Z_2 by an ascending distillation procedure.*

This is obtained in the same way as the descending distillation except that at step 6 above, the set of alternatives having the lowest qualification forms the first distillate.

Step 3: *Construct the partial preorder $Z = Z_1 \cap Z_2$ as the final result.*

The final order can be obtained after the downward order and upward order are averaged, that is,

$$Z = \frac{1}{2}(Z_1 + Z_2). \tag{6.23}$$

In addition, in practical problems of fuzzy multiattribute decision making (FMADM), evaluators and decision makers must be anticipated and they necessarily consist of various stakeholders and interest groups. The different backgrounds and positions of the members of these result in greatly varying subjective judgments. For example, the above thresholds (concordance, discordance, and veto) may be presented in fuzzy data; this shows ELECTRE III and IV are more appropriate for the evaluation of real-world problems.

Example 6.3

Suppose a decision-making problem for purchasing a notebook is considered. In order to choose one of the six alternatives, five criteria, quality, price, function, service, and appearance, are considered. The preferred ratings, with respect to each alternative, and the weight, with respect to each criterion, are given as shown in Table 6.10.

For simplicity, we set the indifferent threshold to two, the preference threshold to three, and disable the veto function. Thus, we can calculate the descending and the ascending orders as shown in Table 6.11.

Then, the final orders can be obtained according to the ranking steps of ELECTRE III, as shown in Table 6.12.

From Table 6.12, it can be seen that Alternative 6 should be the best choice and Alternatives 1, 2, and 5 are tied with rank 2. Although ELECTRE III has integrated the concept of fuzzy sets to considered MADM problems under fuzzy environments, the calculation of ELECTRE III is too complex. In addition, we have to quantify the weights of alternatives before we use ELECTRE III. These shortcomings restrict its applications. Next, we will introduce ELECTRE IV to simplify the procedure of ELECTRE and release the requirement of the weights of alternatives.

6.4 ELECTRE IV

Roy and Bouyssou (1983) proposed ELECTRE IV to simplify the procedure of ELECTRE III. The basic difference between ELECTRE III and ELECTRE IV is that ELECTRE IV does not introduce any weight expressing the weights of the criteria, which may be hard to measure in practice. However, this does not mean that

TABLE 6.10
Decision Table in ELECTRE III

Preferred Ratings	Quality	Price	Function	Service	Appearance
Alternative 1	4	2	6	9	9
Alternative 2	10	1	5	9	2
Alternative 3	3	1	9	2	3
Alternative 4	1	4	4	6	4
Alternative 5	4	4	9	5	5
Alternative 6	4	10	8	10	3
Weights	0.3	0.1	0.3	0.1	0.2

TABLE 6.11
Descending and Ascending Orders of ELECTRE III

Descending Order	Ascending Order
Alternative 6	Alternative 1, Alternative 2, Alternative 6
Alternative 5	Alternative 5
Alternative 1, Alternative 2	Alternative 3
Alternative 3, Alternative 4	Alternative 4

the weights of the criteria are assumed to be equal. Therefore, the pseudocriteria are used, as in ELECTRE III.

Five outranking relations are defined in ELECTRE (Roy and Bouyssou 1993):

1. *Quasi-dominance*
   The couple (*b*,*a*) verifies the relation of quasi-dominance if and only if:
   - For every criterion, *b* is either preferred or indifferent to *a*, and
   - If the number of criterion for which the performance of *a* is better than that of *b* (*a* staying indifferent to *b*) is strictly inferior to the number of criteria for which the performance of *b* is better than that of *a*.
2. *Canonic dominance*
   The couple (*b*,*a*) verifies the relation of canonic-dominance if and only if:
   - For no criterion, *a* is strictly preferred to *b*
   - If the number of criteria for which *a* is weakly preferred to *b* is inferior or equal to the number of criteria for which *b* is strictly preferred to *a*, and
   - If the number of criteria for which the performance of *a* is better than that of *b* is strictly inferior to the number of criteria for which the performance of *b* is better than that of *a*.
3. *Pseudo-dominance*
   The couple (*b*,*a*) verifies the relation of pseudo-dominance if and only if:
   - For no criterion, *a* is strictly preferred to *b*, and
   - If the number of criteria for which *a* is weakly preferred to *b* is inferior or equal to the number of criteria for which *b* is strictly or weakly preferred to *a*.
4. *Sub-dominance*
   The couple (*b*,*a*) verifies the relation of sub-dominance if and only if for no criterion, *a* is strictly preferred to *b*.

TABLE 6.12
Final Orders of ELECTRE III

Rank	Final Preorder
1	Alternative 6
2	Alternative 1, Alternative 2, Alternative 5
3	Alternative 3
4	Alternative 4

TABLE 6.13
Decision Table in ELECTRE IV

Preferred Ratings	Quality	Price	Function	Service	Appearance
Alternative 1	4	2	6	9	9
Alternative 2	10	1	5	9	2
Alternative 3	3	1	9	2	3
Alternative 4	1	4	4	6	4
Alternative 5	4	4	9	5	5
Alternative 6	4	10	8	10	3

5. *Veto dominance*
   The couple (*b*,*a*) verifies the relation of veto dominance if and only if:
   - Either for no criterion, *a* is strictly preferred to *b*, or
   - *A* is strictly preferred to *b* for only one criterion but this criterion does not veto the outranking of *a* by *b* and, furthermore, *b* is strictly preferred to *a* for at least half of the criteria.

The partial preorder is performed as in ELECTRE III but is made simpler by the fact that there are only two outranking levels.

Example 6.4

Suppose a decision-making problem for purchasing a notebook is considered. In order to choose one of the six alternatives, five criteria, quality, price, function, service, and appearance, are considered. The preferred ratings, with respect to each alternative, and the weight, with respect to each criterion, are given as shown in Table 6.13. Note that in this example we do not ask the decision maker to quantify weights for the criteria.

For simplicity, we set the indifferent threshold to two, the preference threshold to three, and disable the veto function and pseudo-dominance; we can calculate the descending and ascending orders as shown in Table 6.14.

Then, the final orders can be obtained according to the ranking steps of ELECTRE VI, as shown in Table 6.15.

TABLE 6.14
Descending and Ascending Orders of ELECTRE IV

Descending Order	Ascending Order
Alternative 1, Alternative 6	Alternative 1, Alternative 2, Alternative 6
Alternative 5	Alternative 5
Alternative 2	Alternative 3
Alternative 3, Alternative 4	Alternative 4

TABLE 6.15
Final Orders of ELECTRE IV

Rank	Final Preorder
1	Alternative 1, Alternative 6
2	Alternative 2, Alternative 5
3	Alternative 3
4	Alternative 4

From Table 6.15, we can conclude that Alternative 1 and Alternative 6 are the best choices and Alternative 4 is the worst. Though many ELECTRE models have been developed, researchers usually select ELECTRE III or IV in dealing with FMADM problems in practice. In addition, Roy (1991) summarizes the characteristics of ELECTRE methods that help the researchers to choose the most appropriate one in practical decision-making contexts.

7 PROMETHEE Method

Brans et al. (1984, 1985) consider a new family of outranking methods, called PROMETHEE (Preference Ranking Organization METHods for Enrichment Evaluations) for solving MADM problems. These methods are based on a generalization of the notion of criterion. In this period, a basic concept of fuzzy outranking relation is first considered and built into each criterion by pairwise comparison measures for alternatives to different relation-degrees in each other. These different relation-degrees are then used to set up a partial preorder (PROMETHEE I), a complete preorder PROMETHEE II), or an interval order (PROMETHEE III) on a finite set of feasible solutions. Another method, called PROMETHEE IV, is introduced for the case where the set of feasible solutions is continuous. These results can easily be apprehended by the decision maker, as illustrated in a numerical application.

7.1 THE NOTION OF THE PROMETHEE METHOD

Let a multiattribute decision-making problem be represented as:

$$\text{Max}\left\{g_1(a_i), g_2(a_i), \ldots, g_j(a_i), \ldots, g_n(a_i) | a_i \in \mathbf{A}\right\}, \tag{7.1}$$

where $\mathbf{A} = \{a_i | i = 1,2,\ldots,m\}$ is a set of possible actions (or alternatives) and $g = \{g_j | j = 1,2,\ldots,n\}$ is a set of considered criteria; $g_j(a_i)$ represents performance of action a_i with respect to the jth criterion.

If, for a given pair of alternatives, a and b have $g_j(a) \geq g_j(b)$ for $j = 1,2,\ldots,n$ and at least one inequality is strict, then a dominates b. According to Brans et al. (1984), the PROMETHEE methods belong to the outranking methods consisting in enriching the dominance order. They include three phases:

1. Construction of generalized criteria
2. Determination of an outranking relation on $\mathbf{A}$
3. Evaluation of this relation in order to give an answer (7.1)

In the first phase, a generalized criterion is associated to each criterion g_j by considering a preference function. In the second phase, a multicriteria preference index is defined in order to obtain a valued outranking relation representing the preference of decision makers. The evaluation of outranking relations are obtained by considering for each action a leaving and entering flow.

7.2 PROMETHEE I, II, III, IV

Brans et al. (1984) first suppose that $\boldsymbol{A}$ is a finite set of possible alternatives. A partial preorder (PROMETHEE I) or a complete preorder (PROMETHEE II) on $\boldsymbol{A}$ can first be proposed to the decision maker. PROMETHEE III provides an interval order emphasizing indifference; PROMETHEE IV deals with continuous sets of possible alternatives.

The PROMETHEE methods request additional information but only a few parameters are to be fixed and they all have a real economic significance. Six possible types of generalized criteria can be considered in PROMETHEE methods and, as shown as Table 7.1, each of them can be very easily defined because only one or two parameters are to be fixed:

1. q is a difference threshold. It is the largest value of d below which the decision maker considers there is indifference;
2. p is a strict preference threshold. It is the lowest value of d above which the decision maker considers there is strict preference;
3. σ is a well-known parameter directly connected with the standard deviation of a normal distribution.

Let $\boldsymbol{A}$ be a finite set of alternatives for MCDM problems, and suppose a preference function f_j has been defined for each g_j, for each couple of alternatives $a,b \in \boldsymbol{A}$; i.e., when $a \succ b$ in j criterion, $f_j(a,b) = f_j(d_{ab|j})$ indicates that the degree of alternative a prefers to alternative b (a over b) with different distance of performance value $d_{ab|j} = g_j(a) - g_j(b)$ in j criterion; and $\pi(a,b)$ is a preference index over all the criteria defined by:

$$\pi(a,b) = \frac{1}{n}\sum_{j=1}^{n} f_j(a,b) \tag{7.2a}$$

or

$$\pi(a,b) = \sum_{j=1}^{n} w_j f_j(a,b) \tag{7.2b}$$

where Equation 7.2a shows the criteria are all equal and Equation 7.2b shows the criterion weight is w_j in criterion j and $j = 1,2,\ldots,n$. w_j can be obtained by the analytic hierarchy process (AHP) or the analytic network process (ANP) based on a network relationship map (NRM) from DEMATEL or interpretive structural modeling (ISM) techniques.

The preference index $\pi(a,b)$ gives the intensity of preference of the decision maker for a over b, all criteria being considered. We have $0 \le \pi(a,b) \le 1$.

Moreover, in order to evaluate the alternatives of $\boldsymbol{A}$ by using the outranking relation, they define the following flows:

TABLE 7.1
Generalized Criteria

Types of Criteria	Analytical Definition	Shape	Parameter
Type I: Usual criterion	$H(d)=\begin{cases}0, & d=0;\\ 1, & \lvert d\rvert>0.\end{cases}$	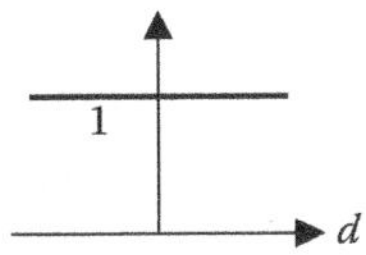	*NA*
Type II: Quasi-criterion	$H(d)=\begin{cases}0, & \lvert d\rvert\le q;\\ 1, & \text{otherwise.}\end{cases}$	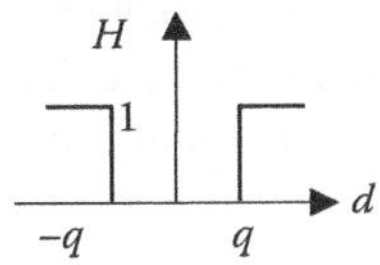	q
Type III: V-sharp criterion	$H(d)=\begin{cases}\dfrac{\lvert d\rvert}{p}, & \lvert d\rvert\le p;\\ 1, & \lvert d\rvert>0.\end{cases}$	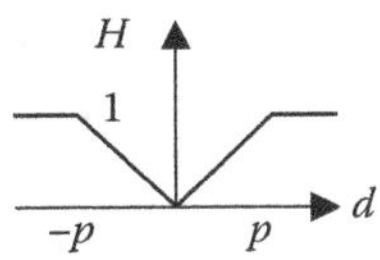	p
Type IV: Level-criterion	$H(d)=\begin{cases}0, & \lvert d\rvert\le q;\\ 1/2, & q<\lvert d\rvert\le p;\\ 1, & \text{otherwise.}\end{cases}$	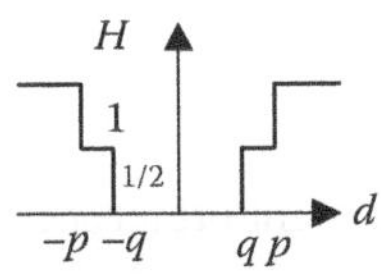	q, p
Type V: Linear criterion	$H(d)=\begin{cases}0, & \lvert d\rvert\le q;\\ \dfrac{\lvert d\rvert-q}{p-q}, & q<\lvert d\rvert\le p;\\ 1, & \text{otherwise.}\end{cases}$	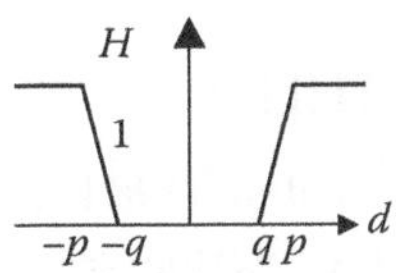	q, p
Type VI: Gaussian criterion	$H(d)=1-\exp\left\{-\dfrac{d^2}{2\sigma^2}\right\}$	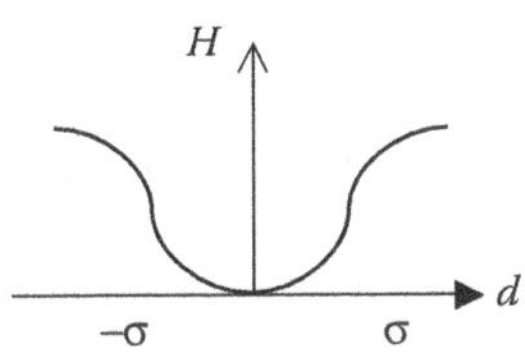	σ

Source: From Brans, J.P., B. Mareschal, and Ph. Vincke, *Operational Research*, Elsevier Science Publishers B.V., North-Holland, 1984b.

1. The leaving flow:

$$\phi^{+}(a)=\sum_{b\in A}\pi(a,b), \tag{7.3}$$

2. The entering flow:

$$\phi^{-}(a)=\sum_{b\in A}\pi(b,a), \tag{7.4}$$

3. The net flow:

$$\phi(a)=\phi^{+}(a)-\phi^{-}(a). \tag{7.5}$$

In PROMETHEE methods, the higher the leaving flow and the lower the entering flow, the better the alternative. The leaving and entering flow induce, respectively, the following preorders on alternatives on $\boldsymbol{A}$:

$$\begin{cases} aP^{+}b & \text{iff } \phi^{+}(a)>\phi^{+}(b); \\ aI^{+}b & \text{iff } \phi^{+}(a)=\phi^{+}(b); \end{cases} \tag{7.6}$$

$$\begin{cases} aP^{-}b & \text{iff } \phi^{-}(a)<\phi^{-}(b); \\ aI^{-}b & \text{iff } \phi^{-}(a)=\phi^{-}(b), \end{cases} \tag{7.7}$$

where P and I represent preference and indifference, respectively.

7.2.1 PROMETHEE I

According to Brans et al. (1984, 1985), PROMETHEE I determines the partial preorder (P^{I},I^{I},R) on the alternatives of $\boldsymbol{A}$ that satisfied the following principle:

$$aP^{I}b\left(a \text{ outranks } b\right), \quad \text{if} \begin{cases} aP^{+}b \text{ and } aP^{-}b \\ aP^{+}b \text{ and } aI^{-}b\,, \\ aI^{+}b \text{ and } aP^{-}b \end{cases} \tag{7.8}$$

$$aI^{I}b\left(a \text{ is indifferent to } b\right), \quad \text{if } aI^{+}b \text{ and } aI^{-}b, \tag{7.9}$$

$$aRb\left(a \text{ and } b \text{ are incomparable}\right), \quad \text{otherwise.} \tag{7.10}$$

From the above equations, we can obtain a partial order for alternatives, while some others are not order (i.e., if aRb cases exist incomparable).

7.2.2 PROMETHEE II

Furthermore, PROMETHEE II gives a complete preorder (P^{II}, I^{II}) induced by the net flow and defined by

$$aP^{II}b\big(a \text{ outranks } b\big), \quad \text{iff } \phi(a) > \phi(b), \tag{7.11}$$

$$aI^{II}b\big(a \text{ is indifferent to } b\big), \quad \text{iff } \phi(a) = \phi(b), \tag{7.12}$$

It seems easier for the decision maker to achieve the decision problem by using the complete preorder in PROMETHEE II instead of the partial one given by PROMETHEE I. However, the partial preorder provides more realistic information by considering only confirmed outranking with respect to the leaving and entering flows. On the other hand, the relation of incomparabilities can also be very useful.

In PROMETHEE I and II, the indifference case between two actions only occurs when the corresponding flows are strictly equal. Nevertheless, due to the continuous character of the generalized criteria (as Table 7.1), it may happen that for two actions a and b the flows are very close to each other, then indifference between a and b is considered.

7.2.3 PROMETHEE III

Based on the above reasons, PROMETHEE III associates, to each action a, an interval $[x_a, y_a]$, and defines a complete interval order (P^{III}, I^{III}) as follows:

$$aP^{III}b\big(a \text{ outranks } b\big) \quad \text{iff } x_a > y_b, \tag{7.13}$$

$$aI^{III}b\big(a \text{ is indifferent to } b\big) \quad \text{iff } x_a \le y_b \text{ and } x_b \le y_a, \tag{7.14}$$

the interval $[x_a, y_a]$ is given by

$$\begin{cases} x_a = \bar{\phi}(a) - \alpha\sigma_a \\ y_a = \bar{\phi}(a) + \alpha\sigma_a \end{cases}, \tag{7.15}$$

where n is the number of actions (or criteria):

$$\bar{\phi}(a) = \frac{1}{n}\sum_{b\in A}\big(\pi(a,b) - \pi(b,a)\big) = \frac{1}{n}\phi(a) \tag{7.16}$$

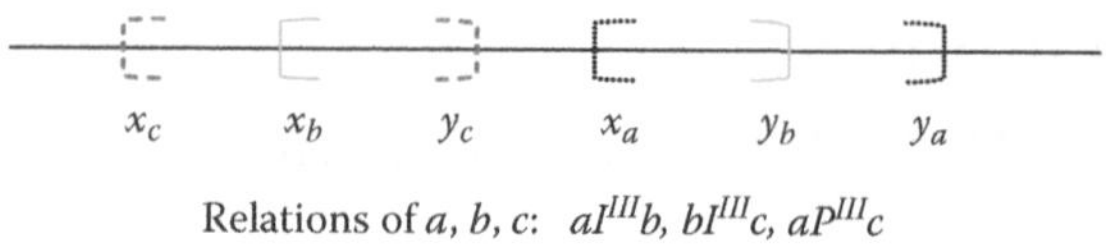

FIGURE 7.1 Diagram of intransitive nature of PROMETHEE III.

$$\sigma_a^2 = \frac{1}{n}\sum_{b \in A}\left(\pi(a,b) - \pi(b,a) - \bar{\phi}(a)\right)^2, \tag{7.17}$$

where $\alpha > 0$ in general.

In other words, $[x_a, y_a]$ is an interval, the center of which is the net mean flow of a and the length of which is proportional to the standard error of the distribution of the numbers $(\pi(a, b) - \pi(b, a))$. In addition, the smaller the value of α, the greater the number of strict outranking; for $\alpha = 0$, (P^{III}, I^{III}) coincides with (P^{II}, I^{II}). It is remarkable that I^{III} is not necessarily transitive while P^{III} is still transitive. For example, for the outranking of three actions a, b, and c, we have $aI^{III}b$ and $bI^{III}c$, but $aP^{III}c$ exists (see as Figure 7.1).

In fact, the choice of α will depend upon the application. For instance, to avoid too many indifferences, it may be requested that the mean length of the intervals be less than the mean distance between two successive mean flows. This leads in general to a value of about 0.15 for α.

In brief, if we utilize PROMETHEE I in our cases, it can help us to determine the partial preorder (P^I, I^I, R) on the set of alternatives $\boldsymbol{A}$; if we use PROMETHEE II, we can obtain a complete preorder (P^{II}, I^{II}) induced by the net flow; furthermore, exploiting PROMETHEE III has the advantage of allowing intransitive indifference and distinguishing incomparability from indifference.

7.2.4 PROMETHEE IV

Furthermore, PROMETHEE IV extends PROMETHEE II to the case of a continuous set of actions (or alternatives) $\boldsymbol{A}$. Such a set arises when the actions are, for instance, percentages, dimensions of a product, compositions of an alloy, investments, and so on.

The generalized criteria of PROMETHEE IV are defined by the above, from preference functions $P_h(a,b)$ such that: $P_h(a,b) = \wp(d)$, where $d_h = f_h(a) - f_h(b)$, $h = 1, 2,...,k$. Besides, the leaving flow, the entering flow, and the net flow for continuous set A are defined as follows:

$$\phi^+(a) = \int_A \pi(a,b)\,db, \tag{7.18}$$

$$\phi^-(a) = \int_A \pi(b,a)\,db, \tag{7.19}$$

TABLE 7.2
Information Table in Example 7.1

Preferred Ratings	Size	Age	Transportation	Facilities	Price
Alternative 1	9	6	4	3	8
Alternative 2	6	9	4	3	3
Alternative 3	2	7	9	6	8
Alternative 4	9	10	4	4	2
Alternative 5	5	9	3	10	1
Weights	0.23	0.18	0.18	0.27	0.14

$$\phi(a) = \phi^{+}(a) - \phi^{-}(a). \tag{7.20}$$

In fact, it is not always easy to integrate the preference index $\pi(a,b)$ into the set A. Brans et al. (1984) suggested simplification of Equations 7.18 through 7.20 as follows:

$$\phi^{+}(a) = \int_{A} P_h(a,b)\,db, \tag{7.21}$$

$$\phi^{-}(a) = \int_{A} P_h(b,a)\,db, \tag{7.22}$$

and to deduce

$$\phi(a) - \frac{1}{k}\sum_{h=1}^{k}\left[\phi_h^{+}(a) - \phi_h^{-}(a)\right]. \tag{7.23}$$

For example, when A is the real interval [0,1], it is possible to obtain the function $\phi(a)$ for the generalized criteria of type I to type V (see Table 7.1) when the functions f_h are piecewise linear or quadratic, showing that a lot of different situations

TABLE 7.3
Leaving, Entering, and Net Flows

Alternatives	$\phi^{+}(a)$	$\phi^{-}(a)$	$\phi(a)$
Alternative 1	0.2727	0.3182	–0.0455
Alternative 2	0.1818	0.3636	–0.1818
Alternative 3	0.4886	0.4318	0.0568
Alternative 4	0.2614	0.2500	0.0114
Alternative 5	0.4205	0.2614	0.1591

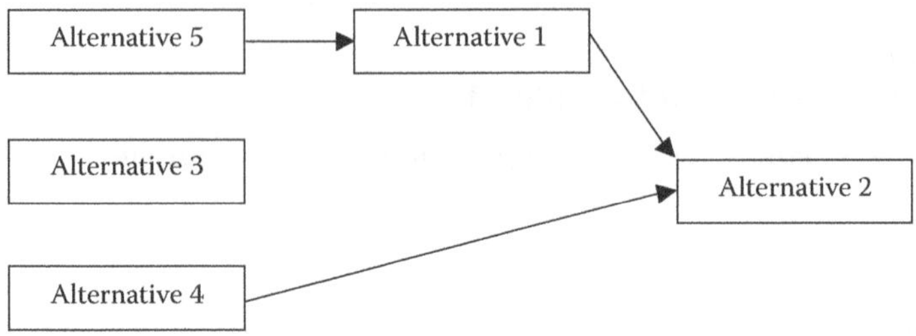

FIGURE 7.2 The partial ranking of PROMETHEE I.

can be considered. However, in more complicated cases a numerical integration may be used.

7.3 EXAMPLE FOR HOUSE SELECTION

Example 7.1

Consider a house-selection problem in Taipei. Fives alternatives with five criteria, size, age, transportation, facilities, and price, are considered and each alternative is evaluated by preferred ratings, as shown in Table 7.2. To choose the best alternative, PROMETHEE I is employed.

The parameters of PROMETHEE I can be set as follows:

Preference function = linear
Indifference threshold = 1
Preference threshold = 2

Then we can calculate the leaving flow, entering flow, and net flow as shown in Table 7.3.

Then, we can depict the partial ranking of PROMETHEE I and the complete ranking of PROMETHEE II as shown in Figures 7.2 and 7.3, respectively.

Comparing the results with PROMETHEE I and PROMETHEE II it can be seen that we can obtain a similar ranking. However, the main difference is that PROMETHEE I can only derive the kernel solutions and PROMETHEE II can obtain the complete ranking.

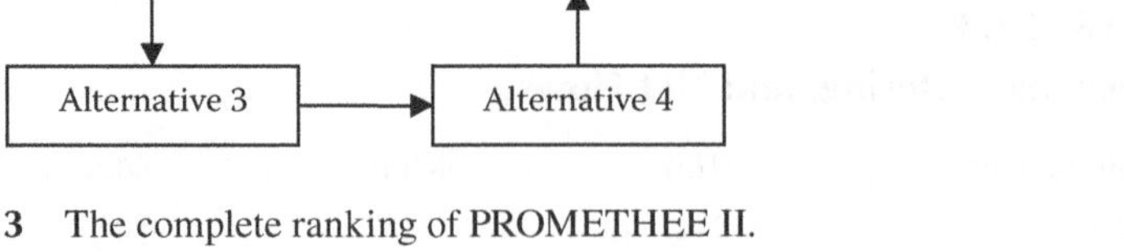

FIGURE 7.3 The complete ranking of PROMETHEE II.

8 Gray Relational Model

The gray system theory proposed by Deng in 1982 is based on the assumption that a system is uncertain and that the information regarding the system is insufficient to build a relational analysis or to construct a model to characterize the system. Gray theory presents a gray relation space and a series of non-functional type models, which are established in this space to overcome the need for a massive number of samples in general statistical methods, or the typical distribution and large amount of calculation work (Tzeng and Tsaur 1994).

8.1 CONCEPTS OF GRAY SYSTEM AND GRAY RELATION

Since there are many abstract systems that cannot be specifically described in this realistic world, we can only reason them out through logic. Then, certain ideas of consciousness or criteria for judgment are exploited to substantiate the structural characteristics of such a system, which will then be displayed through kinds of models. An abstract system of this type is called a "gray system."

The fundamental definition of gray is "information being incomplete or unknown," thus an element from an incomplete message is considered to be a gray element. Furthermore, the relations of incomplete information between systems or elements are taken as being grayish. Should there be an unknown sign or incomplete information in the system, it will be considered as grayish, and such a grayish system is called a gray system. Due to the fact of incomplete information and uncertain relations in this system, it is rather difficult to analyze it with an ordinary method. On the other hand, gray system theory presents such a gray relation space, and serial models of a non-function type are established in the space so as to overcome the obstacles of needing massive numbers of samples in general statistics methods, typical distribution, and much calculation work. The gray relation model is a kind of impact measurement model which takes the measurements of a relation that changes in two systems or between two elements within a system in time, which is called the grade of relation. During the processes of system development, should the trend of change between two elements be consistent, it then enjoys a higher grade of synchronized change and can be considered as having a greater grade of relation; otherwise, the grade of relation would be smaller. Thus, the analysis method, which takes the grade of relation into account, is established using the degree of similarity or of difference of developmental trends among elements to measure the degree of relation among elements.

Fields covered by gray theory include systems analysis, data processing, modeling, prediction, decision making, and control (Deng 1985, 1988, 1989). The relational analysis in gray system theory is a kind of quantitative analysis for the evaluation of alternatives. Similar to fuzzy set theory, gray theory is a feasible mathematical means to deal with systems analysis characterized by poor information. Recently, gray theory has been widely employed in various fields and applications, such as forecasting

(Tseng, Durbin, and Tzeng 2001; Chen et al. 2005b), the problem of selecting an expatriate host country (Chen and Tzeng 2004), artificial neural networks (Hu et al. 2002), and evaluation of pavement condition (Tzeng et al. 2002). In this chapter, we briefly review some relevant definitions and the calculation processes for the gray relational model.

8.2 GRAY RELATION MODEL

Since Deng (1982) put forward the gray system theory, it has already been developed enough to formulate an analysis system based on gray relation space. Gray relation refers to the uncertain relations among things, among elements of systems, or among elements and behaviors. The subsequent text will elaborate on the relevance/similarity definition and calculation processes of the gray relation space.

Definition 8.1

Let X be a factor set of gray relation, $x_0 \in X$ the referential sequence, and $x_i \in X$ the comparative sequence; with $x_0(k)$ and $x_i(k)$ representing, respectively, the numerals at point k for x_0 and x_i. If $\gamma(x_0(k), x_i(k))$ and $\gamma(x_0, x_i)$ are of real numbers, and satisfy the following four gray axioms, then call $\gamma(x_0(k), x_i(k))$ the gray relation coefficient and the grade of gray relation $\gamma(x_0, x_i)$ is the average value of $\gamma(x_0(k), x_i(k))$.

1. Norm interval

$$0 < \gamma(x_0, x_i) \le 1, \quad \forall k; \tag{8.1}$$

$\gamma(x_0, x_i) = 1 \quad \text{iff } x_0 = x_i;$

$\gamma(x_0, x_i) = 0 \quad \text{iff } x_0, x_i \in \phi;$ where ϕ is an empty set.

2. Duality symmetric

$$x, y \in X; \gamma(x, y) = \gamma(y, x) \quad \text{iff } X = \{x, y\}. \tag{8.2}$$

3. Wholeness

$$\gamma(x_i, x_j) \overset{\text{often}}{\neq} \gamma(x_j, x_i) \quad \text{iff } X = \{x_i \mid i = 0, 1, 2, \ldots, n\}, n > 2. \tag{8.3}$$

4. Approachability
   $\gamma(x_0(k), x_i(k))$ decreases along with increasing $|(x_0(k), x_i(k))|$.

Deng also proposed a mathematical equation for the gray relation coefficient as follows:

$$\gamma(x_0(k), x_i(k)) = \frac{\min_i \min_k \Delta_i(k) + \zeta \max_i \max_k \Delta_i(k)}{\Delta_i(k) + \zeta \max_i \max_k \Delta_i(k)}, \tag{8.4}$$

where $\Delta_i(k) = |x_0(k) - x_i(k)|$ and ζ is the distinguished coefficient ($\zeta \in [0,1]$).

Definition 8.2

If $\gamma(x_0, x_i)$ satisfies the four axioms of gray relation, then γ is said to be the gray relational map.

Definition 8.3

If Γ is the entirety of the gray relational map, $\gamma \in \Gamma$ satisfies the four axioms of gray relation, and X is the factor set of gray relation, then (X, Γ) will be called the gray relational space, while γ is the specific map for Γ.

Definition 8.4

Let (X, Γ) be the gray relational space, and if $\gamma(x_0, x_j), \gamma(x_0, x_p), \ldots, \gamma(x_0, x_q)$ satisfy $\gamma(x_0, x_j) > \gamma(x_0, x_p) > \cdots > \gamma(x_0, x_q)$, then we have the gray relational order as $x_j \succ x_p \succ \cdots \succ x_q$.

Once we obtain the degree of the gray relation between the referential sequence and other sequences, the ranking among alternatives can be ordered by

$$\gamma(x_0, x_i) = \sum_{k=1}^{n} w_k \gamma(x_0(k), x_i(k)), \tag{8.5}$$

where w_k denotes the weights of the kth criterion. Weights $\boldsymbol{w} = (w_1, \ldots, w_k, \ldots, w_n)$ can be obtained by the analytic hierarchy process (AHP) (when criteria are independent) or the analytic network process (ANP) (when criteria are dependent on feedback).

8.3 GRAY RELATION FOR MULTIPLE CRITERIA EVALUATION

During the processes of decision making, decision makers always try to use every kind of method, such as investigation, questionnaire, examination, sampling, etc., so as to collect as much practical information as possible, in the hope that the best decision of aspired/desired levels can be reached. Even if such efforts have been made, the hope to have obtained all the necessary information for the decision making remains an impossibility; therefore, decision makers are often compelled to reach their decisions in gray processes.

Viewed from the perspective of multiple criteria decision making, Yu (1990) considered that more extensive decision making should include four basic elements (Table 8.1): (1) the set of substitutive alternatives $\{x_i \mid i = 1,2,\ldots,m\}$ for finding the best alternative; (2) the set of evaluation criteria $\{c_j \mid j = 1, 2,\ldots,n\}$; (3) the anticipated value or outcome matrix $\boldsymbol{X} = [x_i(j)]_{m\times n}$ in regard to the alternatives reckoned by the evaluation criteria; (4) the preference structure of decision making $\{w_j \mid j = 1, 2,\ldots,n\}$. Based on (3), the anticipated values of criteria for each of the alternatives can help to contrive decision matrix or performance matrix $\boldsymbol{X} = [x_i(j)]_{m\times n}$, while the preference structure of the decision maker indicates the preference comparison toward the

TABLE 8.1
Four Basic Elements for Evaluation

		Criteria				
Alternatives		c_1	$\cdots$	c_j	$\cdots$	c_n
		w_1	$\cdots$	w_j	$\cdots$	w_n
x_1		$x_1(1)$	$\cdots$	$x_1(j)$	$\cdots$	$x_1(n)$
$\vdots$		$\vdots$		$\vdots$		$\vdots$
x_i		$x_i(1)$	$\cdots$	$x_i(j)$	$\cdots$	$x_i(n)$
$\vdots$		$\vdots$		$\vdots$		$\vdots$
x_m		$x_m(1)$	$\cdots$	$x_m(j)$	$\cdots$	$x_m(j)$
Aspired value	x^*	$x^*(1)$	$\cdots$	$x^*(j)$	$\cdots$	$x^*(n)$
Worst value	x^-	$x^-(1)$	$\cdots$	$x^-(j)$	$\cdots$	$x^-(n)$

Note: Data matrix: normalization.

outcomes or the weight comparison among criteria by decision makers using AHP or ANP depending on the network relationship map among the criteria. As a result, the four basic elements needed to formulate decision making would be the necessary input to conduct multiple criteria evaluation.

The procedures of calculation are shown as follows:

1. Coefficients of gray relation for aspired values

$$\gamma\left(x^*(j),x_i(j)\right)=\frac{\min_i\min_j|x^*(j)-x_i(j)|+\varsigma\max_i\max_j|x^*(j)-x_i(j)|}{|x^*(j)-x_i(j)|+\varsigma\max_i\max_j|x^*(j)-x_i(j)|}. \tag{8.6}$$

Grade (degree) of gray relation (larger is better)

$$\gamma\left(x^*,x_i\right)=\sum_{j=1}^{n}w_j\gamma\left(x^*(j),x_i(j)\right), \tag{8.7}$$

where the weight w_j can be obtained by AHP or ANP, which depend on the criteria structure. How can we know the criteria structure? Based on the techniques of interpretive structural modeling (ISM), DEMATEL, fuzzy cognitive map (FCM), and so on.

2. Coefficients of gray relation for worst values

$$\gamma\left(x^-(j),x_i(j)\right)=\frac{\min_i\min_j|x^-(j)-x_i(j)|+\varsigma\max_i\max_j|x^-(j)-x_i(j)|}{|x^-(j)-x_i(j)|+\varsigma\max_i\max_j|x^-(j)-x_i(j)|} \tag{8.8}$$

TABLE 8.2
Raw Data in Example 8.1

Raw Data	Quality	Repair Cost	Price	Appearance
Reference point	10	10	10	10
Alternative 1 (A1)	7	4	7	3
Alternative 2 (A2)	5	3	4	9
Alternative 3 (A3)	5	4	8	8
Alternative 4 (A4)	7	7	6	6

Grade (degree) of gray relation (larger is worse, smaller is better)

$$\gamma\left(x^{-},x_i\right)=\sum_{j=1}^{n} w_j\gamma\left(x^{-}(j),x_i(j)\right). \tag{8.9}$$

3. Combining Equations 8.7 and 8.9 for ranking or improving based on the concept of TOPSIS

$$R_i=\frac{\gamma\left(x^{*},x_i\right)}{\gamma\left(x^{-},x_i\right)}. \tag{8.10}$$

8.4 EXAMPLE FOR CAR SELECTION

Example 8.1

Consider the evaluation of the problem of car selection. Four criteria, quality, repair cost, price, and appearance, are considered by a consumer to determine the best alternative. The preferred ratings of alternatives with respect to each criterion are represented as shown in Table 8.2.

Let the normalized transformation be derived by dividing the corresponding reference points (setting aspired/desired levels) of each criterion. Then, we can transfer the raw data above into the following comparative sequence, as shown in Table 8.3.

TABLE 8.3
Comparative Sequence of the Raw Data

Alternatives	Quality	Repair Cost	Price	Appearance
Reference point	1	1	1	1
Alternative 1 (A1)	0.7	0.4	0.7	0.3
Alternative 2 (A2)	0.5	0.3	0.4	0.9
Alternative 3 (A3)	0.5	0.4	0.8	0.8
Alternative 4 (A4)	0.7	0.7	0.6	0.6

TABLE 8.4
Absolute Difference between Reference Points and Sequences

Δ_i	$k = 1$	$k = 2$	$k = 3$	$k = 4$
Reference points	1	1	1	1
$\Delta_{r1}(k)$	0.3	0.6	0.3	0.7
$\Delta_{r2}(k)$	0.5	0.7	0.6	0.1
$\Delta_{r3}(k)$	0.5	0.6	0.2	0.2
$\Delta_{r4}(k)$	0.3	0.3	0.4	0.4

Next, we can calculate the absolute difference between the reference sequence and other sequences, $|x_0(k) - x_i(k)|$, as shown in Table 8.4.

From Table 8.4, it can be seen that $\min_i \min_k \Delta_j(k) = 0.1$ and $\max_i \max_k \Delta_j(k) = 0.7$. Under the assumption of $\zeta = 0.5$, we can derive the grade of gray relation between the reference sequence and other sequences, as shown in Table 8.5. In addition, the weight of each criterion can also be given in Table 8.5.

Finally, we can derive $\gamma(x_0, x_1) = 1/4\sum_{k=1}^{4} w_k \gamma(x_0(k), x_1(k)) = 0.1424$; $\gamma(x_0, x_2) = 0.1598$; $\gamma(x_0, x_3) = 0.1657$; $\gamma(x_0, x_4) = 0.1615$. Therefore, we can derive the preferred order of alternatives as $A_3 \succ A_4 \succ A_2 \succ A_1$.

TABLE 8.5
Grade of Gray Relation between the Reference Sequence and Other Sequences

$\gamma(x_0(k), x_i(k)$	$k = 1$	$k = 2$	$k = 3$	$k = 4$
Reference points	1	1	1	1
$\gamma(x_0(k), x_1(k))$	0.6923	0.4737	0.6923	0.4286
$\gamma(x_0(k), x_2(k))$	0.5294	0.4286	0.4737	1.000
$\gamma(x_0(k), x_3(k))$	0.5294	0.4737	0.8182	0.8182
$\gamma(x_0(k), x_4(k))$	0.6923	0.6923	0.6000	0.6000
Weight	0.3	0.2	0.2	0.3

9 Fuzzy Integral Technique

Multiple attribute decision making (MADM) involves determining the optimal alternative between multiple, conflicting, and interactive criteria (Chen and Hwang 1992). Many methods, which are based on multiple attribute utility theory (MAUT), have been proposed (e.g., the weighted sum and the weighted product methods) to deal with multiple criteria decision-making (MCDM) problems. The concept of MAUT is to aggregate all criteria to a specific unidimension, which is called utility function, to evaluate alternatives. Therefore, the main issue of MAUT is to find a rational and suitable aggregation operator, which can represent the preferences of the decision maker. Although many papers have been proposed to discuss the aggregation operator of MAUT (Fishburn 1970), the main problem of MAUT is the assumption of preferential independence (Hillier 2001; Grabisch 1995).

Preferential independence can be described as the preference outcome of one criterion over another criterion not being influenced by the remaining criteria. However, the criteria are usually interactive in practical MCDM problems. In order to overcome this non-additive problem, the Choquet integral was proposed (Choquet 1953; Sugeno 1974). The Choquet integral can represent a certain kind of interaction between criteria using the concept of redundancy and support/synergy.

9.1 FUZZY INTEGRAL

In 1974, Sugeno introduced the concept of fuzzy measure and fuzzy integral, generalizing the usual definition of a measure by replacing the usual additive property with a weaker requirement, i.e., the monotonicity property with respect to set inclusion. In this chapter, we give an introduction to some notions from the theory of fuzzy measure and fuzzy integral. For a more detailed account, readers can refer to Dubois and Prade (1980), Grabisch (1995), and Hougaard and Keiding (1996).

Definition 9.1

Let X be a measurable set that is endowed with properties of σ-algebra, where $\aleph$ is all subsets of X. A fuzzy measure g defined on the measurable space $(X, \aleph)$ is a set function g: $\aleph \rightarrow [0,1]$ which satisfies the following properties:

1. $g(\psi) = 0$, $g(X) = 1$
2. For all $A, B \in \aleph$, if $A \subseteq B$ then $g(A) \leq g(B)$ (monotonicity)

In view of the definition above, $(X, \aleph, g)$ can be said to be a fuzzy measure space. Furthermore, as a consequence of the monotonicity condition, we can obtain $g(A \cup B) \geq \max\{g(A), g(B)\}$, and $g(A \cap B) \leq \min\{g(A), g(B)\}$. In the case of $g(A \cup B) \geq \max\{g(A), g(B)\}$, the set function g is called a possibility measure (Zadeh 1978), and g is called a necessity measure if $g(A \cap B) \leq \min\{g(A), g(B)\}$.

Definition 9.2

Let $h = \sum_{i=1}^{n} a_i \cdot 1_{A_i}$ be a simple function, where 1_{A_i} is the characteristic function of the set $A_i \in \aleph$, $i = 1,\ldots,n$; the sets A_i are pairwise disjointed and $M(A_i)$ is the measure of A_i. Then the Lebesgue integral of h is

$$\int h \cdot dM = \sum_{i=1}^{n} M(A_i) \cdot a_i. \tag{9.1}$$

Definition 9.3

Let $(X, \aleph, g)$ be a fuzzy measure space. The Choquet integral of a fuzzy measure g: $\aleph \rightarrow [0,1]$ with respect to a simple function h is defined by

$$\int h(x) \cdot dg \cong \sum_{i=1}^{n} \left[h(x_i) - h(x_{i-1})\right] \cdot g(A_i), \tag{9.2}$$

with the same notation above, and $h(x_{(0)}) = 0$.

Let g be a fuzzy measure defined on a power set $P(x)$, which satisfies Definition 9.1 above. The following property is evident:

$$\forall A, B \in P(X), A \cap B = \varnothing, \text{ then,}$$
$$g_\lambda(A \cup B) = g_\lambda(A) + g_\lambda(B) + \lambda g_\lambda(A) g_\lambda(B), \quad \text{for } -1 \leq \lambda < \infty, \tag{9.3}$$

setting $X = \{x_1, x_2, \ldots, x_n\}$, fuzzy density $g_i = g_\lambda(\{x_i\})$ can be formulated as follows:

$$\begin{aligned} g_\lambda(\{x_1, x_2, \ldots, x_n\}) &= \sum_{i=1}^{n} g_i + \lambda \sum_{i_1=1}^{n-1} \sum_{i_2=i_1+1}^{n} g_{i_1} \cdot g_{i_2} + \cdots + \lambda^{n-1} g_1 \cdot g_2 \cdots g_n \\ &= \frac{1}{\lambda} \left| \prod_{i=1}^{n} (1 + \lambda \cdot g_i) - 1 \right| \quad \text{for } -1 \leq \lambda < \infty. \end{aligned} \tag{9.4}$$

Based on the properties above, it can be seen that for an evaluation case with two criteria, A and B, one of the following three cases will be sustained:

Case 1. If $\lambda > 0$, then $g_\lambda(A \cup B) > g_\lambda(A) + g_\lambda(B)$, implying that A and B have a multiplicative effect.

Case 2. If $\lambda = 0$, then $g_\lambda(A \cup B) > g_\lambda(A) + g_\lambda(B)$, implying that A and B have an additive effect.

Case 3. If $\lambda < 0$, then $g_\lambda(A \cup B) > g_\lambda(A) + g_\lambda(B)$, implying that A and B have a substitutive effect.

Next, let h be a measurable set function defined on the fuzzy measurable space $(X,\aleph)$. Assume that $h(x_1) \geq h(x_2) \geq \cdots \geq h(x_n)$, then the fuzzy integral of fuzzy measure $g(\cdot)$ with respect to $h(\cdot)$ can be defined as follows (Ishii and Sugeno 1985):

$$\int h \cdot dg = h(x_n)\cdot g(H_n) + \left[h(x_{n-1}) - h(x_n)\right]\cdot g(H_{n-1}) + \cdots + \left[h(x_1) - h(x_2)\right]\cdot g(H_1)$$
$$= h(x_n)\cdot\left[g(H_n) - g(H_{n-1})\right] + h(x_{n-1})\cdot\left[g(H_{n-1}) - g(H_{n-2})\right]$$
$$+ \cdots + h(x_1)\cdot g(H_1), \tag{9.5}$$

where $H_1 = \{x_1\}$, $H_2 = \{x_1,x_2\},\ldots,H_n = \{x_1,x_2,\ldots,x_n\} = X$. In addition, the corresponding figure can also be depicted as shown in Figure 9.1.

On the basis of Definitions 9.2 and 9.3, it can be seen that the Choquet integral is the Lebesgue integral up to a reordering of the indices. The Choquet integral will reduce to the Lebesgue integral if the fuzzy measure is additive.

Next, the relationships between multiattribute utility function and fuzzy measure are discussed. The general form of multiattribute utility function can be expressed as:

$$u(x_1, x_2, \ldots, x_n) = \sum_{i=1}^{n} w_i u(x_i) + \lambda \sum_{\substack{i=1,\\ j>i}}^{n} w_i w_j u(x_i) u(x_j)$$
$$+ \lambda^2 \sum_{i=1, j>i, l>j}^{n} w_i w_j w_l u(x_i) u(x_j) u(x_l)$$
$$+ \cdots + \lambda^{n-1} w_1 w_2 \ldots w_n u(x_1) u(x_2) \ldots u(x_n), \tag{9.6}$$

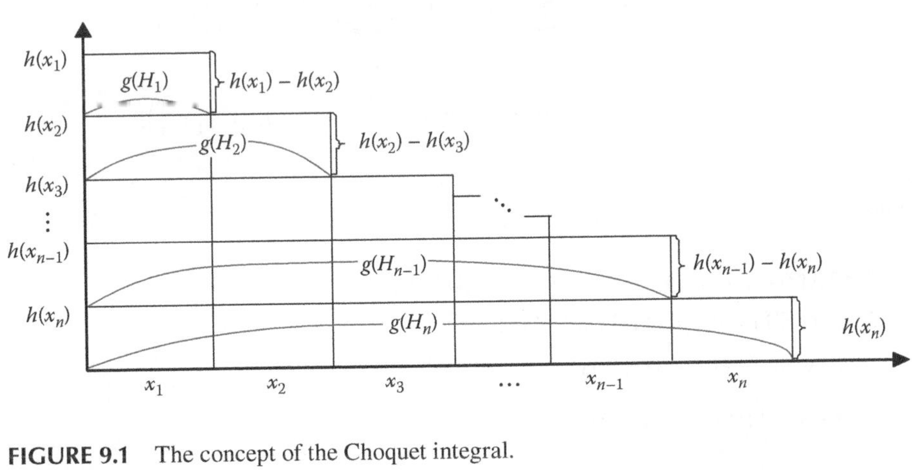

FIGURE 9.1 The concept of the Choquet integral.

where

1. $u(x_1^0, x_2^0, ..., x_n^0) = 0$ and $u(x_1^*, x_2^*, ..., x_n^*) = 1$
2. $u(x_i)$ is a conditional utility function of x_i, i.e., $u(x_i^0) = 0, u(x_i^*) = 1$, $i = 1,2,\ldots,n$
3. $w_i = u(x_i^*, x_{\bar{i}}^0)$
4. λ is a solution of $1+\lambda = \prod_{i=1}^{n}(1+\lambda w_i)$

In addition, if $\sum_{i=1}^{n} w_i = 1$, in other words if $\lambda = 0$, then additive utility function can be written as

$$u(x_1, x_2, ..., x_n) = \sum_{i=1}^{n} w_i u(x_i) \tag{9.7}$$

and if $\sum_{i=1}^{n} w_i \neq 1$, in other words if $\lambda \neq 0$, then multiplicative utility function can be written as:

$$1+\lambda u(x_1, x_2, ..., x_n) = \prod_{i=1}^{n}\left(1+\lambda w_i u(x_i)\right) \tag{9.8}$$

or

$$u(x_1, x_2, ..., x_n) = \frac{1}{\lambda}\left[\prod_{i=1}^{n}\left(1+\lambda w_i u(x_i)\right)-1\right]. \tag{9.9}$$

By contrast, the fuzzy measure can be expressed as:

$$g_\lambda\left(\{x_1, x_2, ..., x_n\}\right) = \sum_{i=1}^{n} g_\lambda\left(\{x_i\}\right) + \lambda \sum_{\substack{i=1,\\ j>i}}^{n} g_\lambda\left(\{x_i\}\right) g_\lambda\left(\{x_j\}\right)$$

$$+\cdots+\lambda^{n-1} g_\lambda\left(\{x_1\}\right) g_\lambda\left(\{x_2\}\right) \ldots g_\lambda\left(\{x_n\}\right), \tag{9.10}$$

where

1. $g_\lambda(\{x_1^*, x_2^*, \ldots, x_n^*\}) = g_\lambda(\{x_1, x_2, \ldots, x_n\}) = 1$
2. $g_\lambda(\{x_i^0\}) = 0,\ g_\lambda(\{x_i^*\}) = 1,\ i = 1,2,\ldots,n$
3. $w_i = u(x_i^*, x_{\bar{i}}^0) = g\lambda(\{x_2\})$
4. $1+\lambda = \prod_{i=1}^{n}(1+\lambda g_\lambda(\{x_i\}))$

TABLE 9.1
Information Table in Example 9.1

Bank	x_1	x_2	x_3	x_4
A	14	11	16	19
B	19	20	10	14
C	19	11	12	17
D	16	18	12	15
E	13	15	16	14
Weight	0.30	0.20	0.25	0.25

In addition, if $\sum_{i=1}^{n} g_\lambda(\{x_i\}) = 1$, then the fuzzy measure can be written as:

$$g_\lambda\left(\{x_1, x_2, \ldots, x_n\}\right) = \sum_{i=1}^{n} g_\lambda\left(\{x_i\}\right) = 1. \tag{9.11}$$

If $\lambda \neq 0$, then this form is called a non-additive fuzzy integral; if $\lambda > 0$, it is called a multiplicative fuzzy measure; and if $-1 < \lambda < 0$, it is called a substitutive fuzzy measure. Next, a numerical example is used to demonstrate the procedures of the Choquet integral.

Example 9.1

Consider the criteria of evaluating banks can be represented as investment income (x_1), loan income (x_2), interest income (x_3), and risk management (x_4). Let the five banks and the corresponding weights and evaluating scores be described as shown in Table 9.1. Since the criteria above are considered as interdependent by decision makers, the Choquet integral is employed to overcome this problem.

By solving Equation 9.3, we can obtain $\lambda = 0.05$, which indicates the multiplicative effect among the criteria. From the result above, next we can derive all the fuzzy measures as shown in Table 9.2.

TABLE 9.2
Fuzzy Measures

Fuzzy Measure	Value	Fuzzy Measure	Value
$g_\lambda\{x_1, x_2\}$	0.5030	$g_\lambda\{x_1, x_3\}$	0.5538
$g_\lambda\{x_1, x_4\}$	0.5538	$g_\lambda\{x_2, x_3\}$	0.4525
$g_\lambda\{x_2, x_4\}$	0.4525	$g_\lambda\{x_3, x_4\}$	0.5031
$g_\lambda\{x_1, x_2, x_3\}$	0.7593	$g_\lambda\{x_1, x_2, x_4\}$	0.7593
$g_\lambda\{x_1, x_3, x_4\}$	0.8107	$g_\lambda\{x_2, x_3, x_4\}$	0.7082

TABLE 9.3
Comparison of Various Aggregated Methods

Bank	Weighted Sum	Rank	Weighted Product	Rank	Fuzzy Integral	Rank
A	15.15	2	7.91	3	15.1883	2
B	15.70	1	7.95	1	15.7522	1
C	15.15	2	7.93	2	15.1797	4
D	15.15	2	7.91	3	15.1809	3
E	14.40	5	7.81	5	14.4107	5

Finally, we can calculate the Choquet integrals to order the alternatives by incorporating the information above. The results, which are compared with the weighted sums and the weighted product methods, can be described as shown in Table 9.3.

On the basis of Table 9.3, it can be seen that the Choquet integral can describe another preference structure by considering the preferential dependence among criteria. It is interesting to clarify which preference structure can represent the preferences of decision makers. In addition to detecting the preference structure using the concepts of preference independence or preference separability, other structural modeling, such as interpretive structural modeling (ISM), decision-making trial and evaluation laboratory (DEMATEL), or fuzzy cognition maps (FCM) can be used. Refer to the Appendix for detailed descriptions of these structural models.

9.2 HIERARCHICAL FUZZY INTEGRAL

In previous discussions of fuzzy integrals, fuzzy measures have been derived according to the concept of λ-measures (Equation 9.4). However, it is not useful in specific MADM problems because decomposable coefficients (e.g., possibility measures or Sugeno's λ-measures) cannot be superadditive for some subsets of criteria and subadditive for other subsets (Grabisch 1995). Hence, the decomposable coefficients can only express either subadditive or superadditive subsets on the whole set of criteria and restrict them from fitting into particular MCDM problems.

To reduce the complexity of identifying coefficients, Sugeno and colleagues proposed the multilevel methods (Sugeno, Fujimoto, and Murofushi 1995; Tanaka and Sugeno 1991) to decompose the Choquet integral into a hierarchical structure according to the condition of inclusion–exclusion covering (IEC) so that the number of estimated coefficients can be significantly reduced. However, their method only provides a way to form the hierarchical Choquet integral, leaving the problem of how to derive coefficients open. In addition, Grabisch (1997) proposed the concept of k-additive measures and Miranda, Grabisch, and Gil (2002) proposed p-symmetric measures to reduce the number of coefficients to $\Sigma_{j=1}^{k}\binom{n}{j}$ and $[(|A_1|+1)] \times \cdots \times [(|A_p|+1)] -2$, where $\{A_1,\ldots,A_p\}$ denotes the partition of criteria, respectively. Although k-additive measures significantly reduce the complexity of identifying coefficients, attempts at determining the appropriate coefficients still overloads the ability of a decision maker in practical problems (e.g., for $n = 10$ and $k = 2$, 55 coefficients should be identified). On the other hand, since p-symmetric measures can only be used in the situation of the particular assumption (i.e., the information about the indifference partition), it restricts the applications of the Choquet integral in dealing with realistic problems.

To identify coefficients in practical problems, two methods have been proposed. The first method, based on minimization of the error criterion, requires numerically independent variables by respondense to derive coefficients, and therefore this method is more suitable for pattern recognition problems. For example, Grabisch (1995) proposed the heuristic least mean squares (HLMS) algorithm, Combarro and Miranda (2006) employed genetic algorithms, and Beliakov (2002) used the least-squares spline method to identify coefficients. On the other hand, the second method, based on constraint satisfaction, identifies coefficients that are based only on the information of revealed preference provided by a decision maker. Hence, this method is more suitable for applications in marketing research and consumer choice. Compared with the first method, the second method has not been very well explored (Marichal and Roubens 2000). The reason is clearly that since the first method identifies coefficients from experimental data, if the data set is large enough, we can always find an optimal method for a satisfactory solution. On the other hand, the result of the second method depends on the information that a decision maker can obtain, i.e., the more complete the information, the more satisfactory the result will be. Therefore, if the information is restricted, we can only reduce the number of estimated coefficients to obtain a satisfactory result.

In this chapter, our focus is the issue of identifying coefficients from the information of revealed preference. Instead of using k-additive or p-symmetric measures, we decompose the Choquet integral into a hierarchical structure according to the concept of preference separability (1968) so that the number of estimated coefficients can be considerably reduced.

Let the preferences over the possible outcomes $X = \prod_{i=1}^{q} X_i$, each $X \subset R^1$, and $Q = \{1,2,\ldots,q\}$ be the index set of the criteria. Given $I \subset Q$ its complement will be denoted by $\bar{I} = Q \backslash I$. In addition, let $\psi = \{I_1, I_2,\ldots,I_m\}$, each $I_k \neq \varnothing$, be a partition of Q.

Definition 9.4

Given the set of revealed preference, $\{\succ\}$, and $I \subset Q$, $I \neq \varnothing$. Let $z \in X_I$ and $w \in X_{\bar{I}}$. It can be defined that z and/or I is preference-separable iff $(z^0, w^0) \succ (z^1, w^0)$ for any z_0, $z_1 \in X_I$ and some $w^0 \in X_{\bar{I}}$ implies that $(z^0, w)\ \psi\ (z^1, w)$ for all $w \in X_{\bar{I}}$.

Definition 9.5

Given $I \subset Q$, $I \neq \varnothing$. It can be said that I is essential if there is some $x_{\bar{I}} \in X_{\bar{I}}$ so that not all elements of X_I are different and I is strictly essential if for each $x_{\bar{I}} \in X_{\bar{I}}$ not all elements of X_I are different.

Definition 9.6

Assume $I_1, I_2 \subset Q$, and I_1 and I_2 overlap iff none of $I_1 \cap I_2$, $I_1 \backslash I_2$, and $I_2 \backslash I_1$ are empty.

Definition 9.7

A topology Γ for a set X is a set of subsets of X such that

i. The empty sets $\varnothing$ and X are in Γ;
ii. The union of arbitrary sets of Γ is inside Γ;
iii. The intersection of any finite number of sets of Γ is in Γ.

If Γ is a topology for X, the pair (X, Γ) is a topological space and the subsets of X in Γ are called open sets.

Definition 9.8

i. A topological space (X, Γ) is connected iff X cannot be partitioned into two non-empty open sets.
ii. The closure of A is the set of all $x \in X$ for which every open set that contains x has a non-empty intersection with A.
iii. (X, Γ) is separable iff X includes a countable subset whose closure is in X.

Assumption 9.1

i. For each (X_i, Γ_i), $i = 1,\ldots,q$, is topologically separable and connected. Thus, (X, Γ), with $X = \prod_{i=1}^{q} X_i$, $\Gamma = \prod_{i=1}^{q} \Gamma_i$ is topologically separable and connected.
ii. $\{\succ\}$ on X is a weak order and for each $x \in X$, $\{x \succ\} \in \Gamma$ and $\{x \prec\} \in \Gamma$.

Definition 9.9

i. Let ψ be a collection of subsets of Q. We say that ψ is complete if (a) $\varnothing$, $Q \in \psi$ and (b) if I_1, $I_2 \in \psi$ overlap then $I_1 \cup I_2$, $I_1 \cap I_2$, $I_1 \backslash I_2$, $I_2 \backslash I_1$, and $(I_1 \backslash I_2) \cup (I_2 \backslash I_1)$ all belong to ψ;
ii. Given I, the completion $\mathbb{C}(\psi)$ of ψ is defined to be the intersection of all the complete collections containing ψ;
iii. Given $T \in \psi$, T is a top element of ψ if $T \neq Q$ and T is not contained by any element of ψ other than Q.

Definition 9.10

Let ψ be a collection of subsets of Q.

i. ψ is connected if for any A, B of ψ, there is a sequence $(I_1,\ldots,I_r)$ of ψ such that $I_1 = A$, $I_r = B$, and I_{k-1} overlaps with I_k, $k = 2,\ldots,r$.
ii. ψ is preference-separable if each element of ψ is preference-separable.

Theorem 9.1 (Gorman 1968)

If (i) ψ is connected and preference-separable, (ii) at least one overlapping pair of elements A, B of ψ exists such that $A\backslash B$ or $B\backslash A$ is strictly essential, (iii) each $\{i\} \subset Q$ is essential, and (iv) Assumption 9.1 is satisfied, then $\mathbb{C}(\psi)$ is preference-separable.

Theorem 9.2 (Gorman 1968)

Assume that Assumption 9.1 holds and $\mathbb{C}(\psi)$ is preference-separable for some preference-separable collection ψ of subsets of Q. Two possible cases in terms of the top elements, $\{T_1,\ldots,T_m\}$ of $\mathbb{C}(\psi)$, should be considered:

Case (i). $\{T_1,\ldots,T_m\}$ do not mutually overlap. Then $\{T_0,T_1,\ldots,T_m\}$, where $T_0 = Q\backslash\left(\bigcup_{i=1}^{m} T_i\right)$, forms a partition of Q and $v\{x\}$ can be written as

$$v(x) = F\left(y_0, v_1(x_1),\ldots,v_m(x_m)\right), \tag{9.12}$$

where $F(x_0,\cdot)$ denotes a strict increase in its components v_i, $i = 1,\ldots,m$, iff each I_i, $i - 1,\ldots,m$, is preference-separable.

Case (ii). Some of $\{T_1,\ldots,T_m\}$ overlap. If each $\{i\}$, $i = Q$, is strictly essential, then $\left\{\bar{T}_1,\ldots,\bar{T}_m\right\}$ forms a partition of Q and $v(x)$ can be written as

$$v(x) = \sum_{i=1}^{m} v_i(x_i), \tag{9.13}$$

if the union of any subsets of ψ is preference-separable.

According to Theorem 9.2, the index set of the criteria Q can be decomposed to a hierarchical structure, which displays independent and non-independent criteria. Therefore, the work to identify fuzzy measures in the original set of the criteria can be reduced dramatically, i.e., we can just focus on the non-independent criteria in the Choquet integral.

Next, a mathematical programme is developed to derive the fuzzy measures based on the viewpoint of statistics as follows. Let $\{\succ\}$ and $\{\sim\}$ be the sets of revealed preference. Assume the preference structure of a decision maker exists and can be represented by a Choquet integral function. Since the information may not be complete, the observed preference cannot reflect the Choquet integral function exactly. Thus, the observed Choquet integral function can be formulated as

$$C(x,\varepsilon) = C(x,g) + \varepsilon, \tag{9.14}$$

where $C(x, g)$ denotes the true Choquet integral function and ε is a random variable with the expected value $E(\varepsilon)$ being equal to zero. Then, because the accurate Choquet integral function is unavailable, the goal is transformed to determine fuzzy measures g such that $C(x, g)$ is most consistent with the sets of revealed preference $\{\succ\}$ and $\{\sim\}$. According to the information of the sets of revealed preference, if $(h^j, h^k) \in \{\succ\}$, it can be seen that

$$\int h_j dg_j - \int h_k dg_k > 0, \quad \text{for all}\left(h^j, h^k\right) \in \{\succ\}. \tag{9.15}$$

On the other hand, if $(h^j, h^k) \in \{\sim\}$, we can expect

$$\int h_j dg_j - \int h_k dg_k = 0 \quad \text{for all}\left(h^j, h^k\right) \in \{\sim\}. \tag{9.16}$$

Therefore, the mathematical program can be developed as

$$\min \sum_{\{\succ\}} \left(y_{jk}\right)^p + \sum_{\{\sim\}} \left(z_{st}\right)^p, \tag{9.17}$$

$$\begin{aligned}
s.t. \quad & \int h_j dg_j - \int h_k dg_k + y_{jk} \geq \delta, \quad \forall\left(h^j, h^k\right) \in \{\succ\}, \\
& \int h_j dg_j - \int h_k dg_k + z_{st} = 0, \quad \forall\left(h^j, h^k\right) \in \{\sim\}, \\
& g\left(\{x_1, x_2, \ldots, x_j\}\right) \geq g\left(\{x_1, x_2, \ldots, x_k\}\right), \\
& g\left(\{\varnothing\}\right) = 0, \; g\left(\{x_1, x_2, \ldots, x_n\}\right) = 1, \quad \forall 1 \leq k < j \leq n,
\end{aligned}$$

where $p \geq 1$ refer to the l_p-norm, $z_{st} \in R$, and δ denotes an arbitrarily small positive constant.

Example 9.2

Let the criteria of purchasing a computer be represented as CUP speed (x_1), memory (x_2), graphics (x_3), display (x_4), storage (x_5), repair cost (x_6), service (x_7), price (x_8), brand (x_9), and appearance (x_{10}). Suppose that the preferred ratings with respect to each criterion are given by the decision maker based on ten-point scales (i.e., excellent = 10, very poor = 1) as shown in Table 9.4.

From the MCDM problem above, it can be seen that $Q = \{1,2,3,4,5,6,7,8,9,10\}$ indicates the index set of the criteria. Assume $\psi = \{(1,2,3), (3,4), (7,8), (8,9)\}$ is a

TABLE 9.4
Decision Table of Purchasing a Computer

Alternative	x_1	x_2	x_3	x_4	x_5	x_6	x_7	x_8	x_9	x_{10}
A	4	2	6	9	9	10	1	5	9	2
B	3	1	8	2	3	7	4	4	6	6
C	4	4	9	5	5	4	10	8	10	3
D	10	1	7	8	10	5	4	8	4	8
E	2	3	2	4	5	6	4	10	5	3
F	1	9	2	3	8	7	8	8	2	2
G	8	7	3	5	6	7	6	3	10	7
H	6	8	8	3	3	9	2	2	8	8
I	10	9	9	6	6	7	5	2	1	8
J	7	8	6	2	9	4	3	3	4	4

partition of the criteria, where (1,2,3) indicates the internal efficiency of the computer, (3,4) denotes the visual capability of the computer (7,8) indicates the service index, and (8,9) is the brand index. Therefore, the completion above can be derived as

$$\mathbb{C}(\psi)=\{\varnothing,(3),(4),(7),(8),(9),(1,2),(7,9),(1,2,4),(7,8,9),(1,2,3,4),Q\}.$$

On the basis of Definition 9.9, the top elements of $C(\psi)$ should be $T_0=\{5,6,10\}$, $T_1=\{1,2,3,4\}$, and $T_2=\{7,8,9\}$.

Next, the top elements, T_1 and T_2, are further decomposed as follows. First, the completion of T_1 can be derived as $\mathbb{C}_1(\psi)=\{\varnothing, (7), (8), (9), (7,8), (8,9), (7,9), (7,8,9)\}$. Then, the top elements with respect to T_1 are $T_{11}=\{1,2\}$, $T_{12}=\{3\}$, and $T_{13}=\{4\}$. On the other hand, the completion of T_2 can also be calculated as $\mathbb{C}_2(\psi)=\{\varnothing, (3), (4), (1,2), (3,4), (1,2,3), (1,2,4), (1,2,3,4)\}$. Therefore, with respect to T_2, the top elements are $T_{21}=\{7\}$, $T_{22}=\{8\}$, and $T_{23}=\{9\}$.

According to the results of decomposition above, the structure of the Choquet integral can be represented by a hierarchical graph as shown in Figure 9.2. From Figure 9.2, it can be seen that the work of identifying fuzzy measures in the original set can be considerably reduced.

Next, Equation 9.17 is employed to derive the fuzzy measures as follows based on the sets of the revealed preference. The revealed preference of the decision problem above can be given by the decision maker as

$$\{\succ\}=\{(A,E),(A,F),(A,J),(B,E),(C,A),(C,B),(C,D),(C,E),(C,F),(C,G),(C,H),$$
$$(C,I),(C,J),(D,A),(D,B),(D,I),(G,E),(H,A),(H,B),(H,E),(H,G),(J,F)\}$$

and

$$\{\sim\}=\{(A,B),(B,J),(D,G)\}.$$

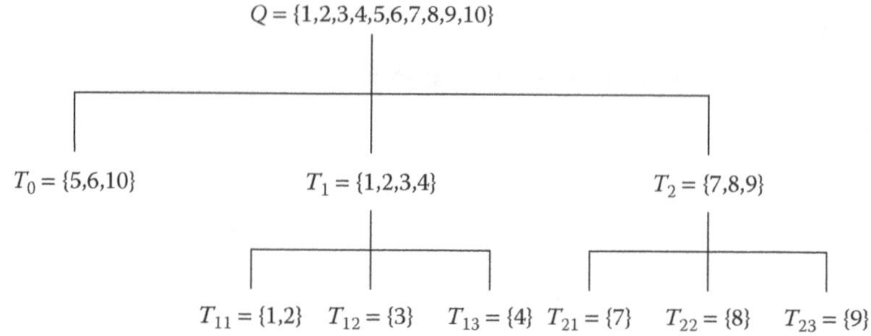

FIGURE 9.2 Hierarchical structure of the Choquet integral.

Note that the property of transitivity in the revealed preference is not necessarily satisfied in order to reflect the realistic MCDM problems. Next, the fuzzy measures can be obtained by

$$\min y_{AE}^{p} + \cdots + y_{JF}^{p} + |z_{AB}|^{p} + \cdots + |z_{EG}|^{p},$$

$$s.t. \quad \int h_A dg_A - \int h_E dg_E + y_{AE} \geq \delta,$$

$$\ldots$$

$$\int h_J dg_J - \int h_F dg_F + y_{JF} \geq \delta,$$

$$\int h_A dg_A - \int h_B dg_B + z_{AB} = 0,$$

$$\ldots$$

$$\int h_D dg_D - \int h_G dg_G + z_{DG} = 0,$$

$$g(\{x_1\}), g(\{x_2\}) \leq g(\{x_1, x_2\}),$$

$$g(\{x_5\}), g(\{x_6\}) \leq g(\{x_5, x_6\}),$$

$$g(\{x_5\}), g(\{x_{10}\}) \leq g(\{x_5, x_{10}\}),$$

$$g(\{x_6\}), g(\{x_{10}\}) \leq g(\{x_6, x_{10}\}),$$

$$g(\{x_5, x_6\}), g(\{x_5, x_{10}\}), g(\{x_6, x_{10}\}) \leq g(\{x_5, x_6, x_{10}\}),$$

$$g(\{\varnothing\}) = 0,\ g(\{x_1, x_2, \ldots, x_{10}\}) = 1.$$

Then, we can derive the fuzzy integral and coefficients of each alternative as shown in Table 9.5.

TABLE 9.5
Results of the Fuzzy Integral and Coefficients

Alternative	A	B	C	D	E
Choquet integral	4.8755	4.8755	5.9172	5.9072	4.8655
Alternative	F	G	H	I	J
Choquet integral	4.8655	5.9072	5.8972	5.8972	4.8755
Coefficients	$g(\{x_1, x_2\})$	$g(\{x_5, x_6\})$	$g(\{x_5, x_{10}\})$	$g(\{x_6, x_{10}\})$	$g(\{x_5, x_6, x_{10}\})$
Value	0.1293	0.2029	0.1526	0.4253	0.4253

From Table 9.5, it can be seen that the results of the fuzzy integral are almost consistent with the sets of the revealed preference, which is proposed by the decision makers. In addition, the optimal fuzzy measures can also be derived using the proposed mathematical programming model so that the results of the fuzzy integral are most consistent with the sets of the revealed preference.

10 Rough Sets

Rough sets, first introduced by Pawlak in 1982 (Pawlak 1982, 1984), are a valuable mathematical tool to deal with vagueness and uncertainty (Pawlak 1997), approaching the fields of artificial intelligence including cognitive sciences, machine learning, knowledge acquisition, decision analysis, knowledge discovery, decision support systems, inductive reasoning, and pattern recognition (Pawlak 1982; Tay and Shen 2002). The starting point of the rough set theory (RST) is the assumption that with every object of interest we associate some information, and objects are similar or indiscernible due to their characters by some information (Pawlak 1997). This kind of indiscernibility relation is the mathematical basis of the RST.

The key concept of RST is the approximative equality of sets in a given approximation space (Pawlak 1982). According to Pawlak (1982, 1984), an approximation space A is an ordered pair (U, R), where U is a certain set called universe, and that equivalence relation $R \subset U \times U$ is a binary relation called the indiscernibility relation. That is, if $x, y \in U$ and $(x, y) \in R$, this means that x and y are indistinguishable in A; equivalence classes of the relation R are called elementary sets (atoms) in A (an empty set is also elementary), and the set of all atoms in A is denoted by U/R.

In the rough set approach, any vague concept is characterized by a pair of precise concepts, that is the lower and upper approximation of the vague concept (Pawlak 1997). Let $X \subseteq U$ be a subset of U, then the lower and upper approximations of X in A are denoted as

$$\underline{A}(X) = \{x \in U \mid [x]_R \subset X\} \tag{10.1}$$

and

$$\overline{A}(X) = \{x \in U \mid [x]_R \cap X \neq \varnothing\}, \tag{10.2}$$

respectively, where $[x]_R$ denotes the equivalence class of the relation R containing element x; in addition, the set

$$BN_A(X) = \overline{A}(X) - \underline{A}(X) \tag{10.3}$$

is called a boundary of X in A (Pawlak 1982). If set X is roughly definable in A, it means that we can describe the set X with some "approximation" by defining its lower and upper approximation in A (Pawlak 1984). The upper approximation $\overline{A}(X)$ means the least definable set in A containing the objects that possibly belong to the concept, whereas the lower approximation $\underline{A}(X)$ means the greatest definable set in A containing the objects that surely belong to the concept (Pawlak 1997).

Using lower and upper approximations of a set, the accuracy and the quality of approximation can be defined, and the knowledge hidden in the data table may be discovered and expressed in the form of decision rules (Mi, Wu, and Zhang 2004). More details of the theory can be found in Pawlak (1982, 1984). The basic concepts of RST and the analytical procedure of data analysis are discussed as follows.

10.1 INFORMATION SYSTEM

Rough set-based data analysis starts from a data table called an information system, which contains data about objects of interest characterized in terms of some attributes (Pawlak 2002a,b). Making decisions in a specific field usually requires structuring and analyzing the information system, which involves data and knowledge. The RST declares that the information system contains information about particular objects in terms of their attributes. The objects can be interpreted as cases, states, processes, and observations, whereas the attributes can be interpreted as features, variables, and characteristic conditions.

In the RST, an information system is used to construct the approximation space. The information system can be viewed as an application of the RST such that each object is described by a set of attributes (Pawlak 1997). According to Pawlak (1984, 1997), an information system is defined by the quadruple $S = (U, Q, V, \rho)$, where universe U is a finite set of objects, Q is a finite set of attributes, $V = \cup_{q \in Q} V_q$ is the set of values of attributes, and V_q is the domain of the attribute q; ρ: $U \times Q \rightarrow V$ is a description function such that $\rho(x, q) \in V_q$ for every $q \in Q$, $x \in U$. In practice, the information system is a finite data table, in which columns are labeled by attributes, rows are labeled by objects, and the entry in column q and row x has the value $\rho(x, q)$; each row in the table represents the information about an object in S (Pawlak 2002a,b).

10.2 INDISCERNIBLE RELATION

It is hard to distinguish objects on the basis that imprecise information is the starting point of RST (Pawlak 1997). In other words, the imprecise information causes the indiscernibility of objects in terms of available data. Let $S = (U, Q, V, \rho)$ be an information system and let $P \subseteq Q$, $x,y \in U$, so that x and y are indiscernible by the set of attributes P in S (denotation $x\tilde{P}y$) iff $r(x,q) = r(x,q)$ for every $q \in P$ (Dimitras et al. 1999). The equivalence classes of relation $\tilde{P}$ (or IND_P) are called P-elementary sets in S, whereas the Q-elementary sets are called atoms in S. These elementary sets represent the smallest discernible groups of objects and the construction of elementary sets is the primary step in performing the classification through rough sets (Walczak and Massart 1999).

Moreover, the indiscernible relation is used to define two main operations on data, namely, the lower and upper approximations of a set. By using the lower and upper approximations of a set, we can define the accuracy and the quality of approximation, which are numbers from interval [0,1] (Pawlak 1984). Further, utilizing the accuracy and the quality of approximation is a favorable way to define exactly how we can describe the classification of objects.

10.3 APPROXIMATION OF SETS AND APPROXIMATION OF ACCURACY

In the RST, the approximations of sets are introduced to deal with the vague concept. Let $P \subseteq Q$ and $Y \subseteq U$. The P-lower approximation of Y is denoted by

$$\underline{P}(Y) = \{x \mid IND_P(x) \subseteq Y\}, \tag{10.4}$$

the P-upper approximation of Y is denoted by

$$\overline{P}(Y) = \{x \mid IND_P(x) \cap Y \neq \varnothing\}, \tag{10.5}$$

and the P-boundary of set Y is the doubtful region denoted by

$$BN_P(Y) = \overline{P}(Y) - \underline{P}(Y). \tag{10.6}$$

If the lower and upper approximations are identical, i.e., $\overline{P}(Y) = \underline{P}(Y)$, the set Y is definable; otherwise, the set Y is undefinable in S. According to Dimitras et al. (1999), the set $\underline{P}Y$ is the set of elements of U, which can be certainly classified as elements of Y by the set of attributes P; the set $\overline{P}Y$ is the set of elements of U, which can be possibly classified as elements of Y by the set of attributes P; and the set $BN_P(Y)$ is the set of elements, which certainly cannot be classified to Y by the set of attributes P.

According to Pawlak (1982), the accuracy of the approximation $\mu_P(Y)$, the quality of classification $\eta_P(\dddot{Y})$, and the accuracy of the classification $\beta_P(\dddot{Y})$ can be measured as follows. To measure the accuracy of the approximation $\mu_P(Y)$ of the set Y by P in S, we can use the way that

$$\mu_P(Y) = \text{card}(\underline{P}Y)/\text{card}(\overline{P}Y), \tag{10.7}$$

where $0 \leq \mu_P(Y) \leq 1$; the Y is definable by P in S if $\mu_P(Y) = 1$, whereas the Y is undefinable by P in S if $\mu_P(Y) < 1$. In addition, let S be an information system, a subset of attributes $P \subseteq Q$; and let $\dddot{Y}$ be the classification of U by P. The subsets $Y_i = \{Y_1, Y_2, \ldots, Y_n\}$, are the classes of the classification $\dddot{Y}$, the P-lower approximation of $\dddot{Y}$ is denoted as $\underline{P}\dddot{Y}$, and the P-upper approximation of $\dddot{Y}$ is denoted as $\overline{P}\dddot{Y}$. Then, the quality of classification $\eta_P(\dddot{Y})$ by P can be measured with the way that

$$\eta_P(\dddot{Y}) = \sum_{i=1}^{n} \text{card}(\underline{P}Y_i)/\text{card}(U). \tag{10.8}$$

As for the accuracy of the classification $\beta_P(\dddot{Y})$ by P, it can be measured with the way that

$$\beta_P(\dddot{Y}) = \sum_{i=1}^{n} \text{card}(\underline{P}Y)_i \Big/ \sum_{i=1}^{n} \text{card}(\overline{P}Y_i). \tag{10.9}$$

10.4 REDUCTION OF ATTRIBUTES

Discovering the dependencies between attributes is important for information table analysis in the rough set approach. In order to check whether the set of attributes is independent or not, every attribute must be checked regarding whether its removal increases the number of elementary sets in an information system (Walczak and Massart 1999). Let $S = (U, Q, V, \rho)$ be an information system and let $P, R \in Q$. Then, the set of attributes P is said to be dependent on the set of attributes R in S (denotation $R \to P$) iff $IND_R \subseteq IND_p$, whereas the set of attributes P, R, are called independent in S iff neither $R \to P$ nor $P \to R$ hold (Pawlak 1982).

Moreover, finding the reduction of attributes is another important thing. Let the minimal subset of attributes $R \subseteq P \subseteq Q$ such that $\eta_P(\ddot{Y}) = \eta_R(\ddot{Y})$ be called the $\ddot{Y}$-reduct of P and be denoted by $RED_{\ddot{Y}}(P)$. Then, the intersection of all $\ddot{Y}$-reducts is called the $\ddot{Y}$-core of P. Especially, the core is a collection of the most relevant attributes in the table (Dimitras et al. 1999) and is the common part of all reducts (Walczak and Massart 1999).

In order to obtain the reducts and their core, there are two popular methods: the indiscernibility relation method and the similarity relation method. The indiscernibility relation method is based on the indiscernibility matrix and the indiscernibility relation concerns mainly qualitative attributes, while the similarity relation method is based on the similarity matrix and the similarity relation concerns mainly quantitative attributes (Greco et al. 2002). Therefore, the indiscernibility relation method is useful for analyzing data containing qualitative attributes with linguistic values, whereas the similarity relation method is useful for analyzing data containing quantitative attributes with continuous values.

10.5 DECISION TABLE AND DECISION RULES

The decision table describes decisions in terms of conditions that must be satisfied in order to carry out the decision specified in the decision table (Pawlak 2002a,b). An information system can be seen as the decision table in the form of $S = (U, C \cup D, V, \rho)$, in which $C \cup D = Q$ and means that condition attributes C and decision attributes D are two disjoint classes of attributes (Greco et al. 2002). Through analyzing the decision table, valuable decision rules can be extracted. To generate decision rules from the data in the decision table, it is required to reduce unnecessary conditions and to minimize superfluous attributes. According to Pawlak (2002a,b), a decision rule in S is an expression $\Phi \to \Psi$, read if Φ then Ψ, where Φ and Ψ are conditions and decisions of the decision rule, respectively; most importantly, $\sigma_s(\Phi, \Psi) = \text{supp}_s(\Phi, \Psi)/\text{card}(U)$ is the strength of the decision rule $\Phi \to \Psi$ in S, where the $\text{supp}_s(\Phi, \Psi)$ is called the support of the rule $\Phi \to \Psi$ in S. Moreover, Dimitras et al. (1999) mentioned that each decision rule is characterized by the strength, which means the number of objects satisfying the condition part of the decision rule and belonging to this decision class; stronger rules are usually more general, i.e., their condition parts are shorter and less specialized.

Especially, the Covering Index (CI) is a rather valuable way of evaluating the quality of the decision rule (Mori et al. 2004). Let the decision attributes D be a singleton $D = \{d\}$, the d-elementary sets in S are denoted by $Y_i = \{Y_1, Y_2, \ldots, Y_n\}$ and called the decision classes of the classification; let the condition attribute $A \subseteq C$ and its domain V_{aj} of the attribute $a_j \in A$. Then, the CI can be expressed as

$$\mathrm{CI}\left(V_{a_j}, Y_i\right) = \mathrm{card}\left(V_{a_j} \wedge Y_i\right) / \mathrm{card}\left(Y_i\right), \tag{10.10}$$

where "$\wedge$" is the operator of conjunction. The CI represents a ratio called the covering ratio, which indicates the degree of decision class, how many objects with the same attribute value match the decision class in contrast with how many objects belong to the same decision class.

10.6 THE ANALYTICAL PROCEDURE OF DATA ANALYSIS

Applications of the RST are varied, such as customer analysis, data analysis and reduction, generation of decision rules, image processing, pattern recognition, knowledge discovery, knowledge representation, and concept naming. There are several kinds of problems that can be solved using the rough set approach, such as: (1) description of a set of objects in terms of the attribute values; (2) dependencies between attributes; (3) reduction of attributes; (4) significance of attributes; and (5) generation of decision rules (Pawlak 1997). Concerning the classification problem, Doumpos and Zopounidis (2002) mention the following: (1) The parametric classification techniques are widely used, such as linear discriminant analysis, quadratic discriminant analysis, logit or probit analysis, and the linear probability model, whereas several alternative non-parametric classification techniques have been developed, such as mathematical programming techniques, multicriteria decision aid methods, neural networks, and machine learning approaches; (2) Most parametric classification techniques have the limitation of statistical assumptions, for example, linear discriminant analysis has these restrictive assumptions, including multivariate normality and equality of dispersion matrices between groups; (3) Among non-parametric classification techniques, the RST, which was developed following the concepts of machine learning, has several distinguishing and attractive features (such as attributes, criteria, variables, etc), including data reduction, handling of uncertainty, ease of interpretation of the developed classification model, etc.

The RST provides a relatively new technique of reasoning from vague and imprecise data. According to Tay and Shen (2002), the rough set approach has several advantages: (1) it can perform the analysis straightforwardly using the original data only and does not need any external information such as probability in statistics or grade of membership in the fuzzy set theory (Krusinska et al. 1992; Dubois and Prade 1992; Skowron and Grzymala-Busse 1993); (2) it is suitable for analyzing not only quantitative attributes but also qualitative ones; (3) it can discover important facts hidden in data and expresses them in the natural language of decision rules; (4) the set of decision rules gives a generalized description of the knowledge contained

TABLE 10.1
Decision Table in Example 10.1

Samples	Length	Width	Color	Shape	Vegetation
Sample 1	1	2	2	1	2
Sample 2	3	2	2	2	1
Sample 3	3	2	3	3	3
Sample 4	2	1	3	1	2
Sample 5	2	3	2	1	3
Sample 6	1	2	2	2	3
Sample 7	2	1	2	1	2
Sample 8	3	3	1	1	1

Length: <3 cm: 1; 3–9 cm: 2; >9 cm: 3.
Width: <3 cm: 1; 3–5 cm: 2; >5 cm: 3.
Color: green: 1; yellow: 2; red: 3.
Shape: rectangle-like: 1; oval-like: 2; diamond-like: 3.

in the information tables; and (5) the results of the rough sets analysis are easy to understand by the natural language.

The central concepts of the RST are the information system, the decision table, indiscernibility, approximation, reducts, and decision rules. In practice, for the analysis of decision tables, the main steps, as mentioned by Walczak and Massart (1999), are: (1) construction of elementary sets; (2) calculation of upper and lower approximations of the elementary sets; (3) finding the core and reducts of attributes; and (4) finding the core and reducts of attribute values. Hence, for the data analysis in the rough set approach, we suggest the three-step analytical procedure: (1) calculating the approximation; (2) finding the reducts of attributes and the core of attributes; and (3) creating the decision rules.

TABLE 10.2
Decision Rules

Rules	Length	Width	Shape	Decision
Rule 1	1	–	1	2
Rule 1′	–	2	1	2
Rule 2	3	2	2	1
Rule 3	3	2	3	3
Rule 4	2	1	1	2
Rule 5	2	3	1	3
Rule 6	1	2	2	3
Rule 7	3	–	1	1
Rule 7′	3	3	–	1

Example 10.1

In this example, we will consider a vegetation classification problem. Eight samples are collected and the vegetations' length, width, color, and shape are recorded, as shown in Table 10.1.

Next, we can calculate the core and reduct of attributes such as Length, Width, and Shape (i.e., Color is a D-superfluous attribute) and derive the decision rules as shown in Table 10.2.

From Table 10.2, we can summarize the decision rules as follows:

Rule 1: If (Length = 3 and Width = 3) or (Length = 3 and Shape = 1) or (Length = 3 and Shape 4 = 2) then Decision = 1.
Rule 2: If (Length = 1 and Shape = 1) or (Width = 1) or (Width = 2 and Shape = 1) then Decision = 2.
Rule 3: If (Length = 2 and Width = 3) or (Length = 1 and Criterion 4 = 2) or (Shape = 3) then Decision = 3.

Rough sets provide a useful and powerful tool to deal with the classification problem of multiple attribute decision making (MADM). However, it should be highlighted that rough sets can only deal with discrete variables and, therefore, continuous variables should be discretized first.

11 Structural Model

Utility independence or utility separability is usually the basic assumption of the multiple attribute decision making (MADM) methods for employing the additive function to represent the preferences of decision makers. However, in the realistic problems, the assumption of utility independence or utility separability seems to be irrational. Therefore, it is interesting to clarify the structure among criteria and then we can determine the appropriate MADM methods based on the results of structural models. Next, three structural models are introduced as follows.

11.1 INTERPRETIVE STRUCTURAL MODELING METHOD

Interpretive structural modeling (ISM), which was proposed by Warfield (1974a, 1974b, 1976), is a computer-assisted methodology to construct and to understand the fundamental relationships of the elements in complex systems or situations. The theory of ISM is based on discrete mathematics, graph theory, social sciences, group decision making, and computer assistance (Warfield 1974a, 1974b, 1976). The procedures of ISM are begun through individual or group mental models to calculate a binary matrix, also called a relation matrix, to present the relations of the elements. The concepts of ISM can be summarized as follows.

A relation matrix can be formed by asking the question "Does the feature e_i inflect the feature e_j?" If the answer is "Yes" then $\pi_{ij} = 1$, otherwise $\pi_{ij} = 0$. The general form of the relation matrix can be presented as follows:

$$\boldsymbol{D} = \begin{array}{c} \\ e_1 \\ e_2 \\ \vdots \\ e_n \end{array} \begin{array}{c} \begin{array}{cccc} e_1 & e_2 & \cdots & e_n \end{array} \\ \left[\begin{array}{cccc} 0 & \pi_{12} & \cdots & \pi_{1n} \\ \pi_{21} & 0 & \cdots & \pi_{2n} \\ \vdots & \vdots & \ddots & \vdots \\ \pi_{n1} & \pi_{n2} & \cdots & 0 \end{array}\right] \end{array},$$

where e_i is the ith element in the system, π_{ij} denotes the relation between the ith and the jth elements ($i, j \in \{1, 2, \ldots, n\}$), and $\boldsymbol{D}$ is the relation matrix.

After constructing the relation matrix, we can calculate the reachability matrix using Equations 11.1 and 11.2 as follows:

$$\boldsymbol{M} = \boldsymbol{D} + \boldsymbol{I}, \tag{11.1}$$

$$\boldsymbol{M}^* = \boldsymbol{M}^k = \boldsymbol{M}^{k+1} \quad k > 1, \tag{11.2}$$

where $\boldsymbol{I}$ is the unit matrix, k denotes the powers, and $\boldsymbol{M}^*$ is the reachability matrix. Note that the reachability matrix is under the operators of the Boolean multiplication and addition (i.e., $1\times 1=1$, $1+1=1$, $1+0=0+1=1$, $1\times 0=0\times 1=0$, $0+0=0$, $0\times 0=0$).

For example

$$M=\begin{pmatrix}1 & 0\\ 1 & 1\end{pmatrix},\quad M^2=\begin{pmatrix}1 & 0\\ 1 & 1\end{pmatrix}.$$

Next we can calculate the reachability set and the priority set based on Equations 11.3 and 11.4, respectively, using the following equations:

$$R(t_i)=\{e_i \mid m_{ji}^*=1\} \tag{11.3}$$

and

$$A(t_i)=\{e_i \mid m_{ij}^*=1\}, \tag{11.4}$$

where m_{ij} denotes the value of the ith row and the jth column of the reachability matrix.

Then, according to Equations (11.3) and (11.4), the levels and the relationships between the elements can be determined and the structure of the elements' relationships can also be expressed using the following equation:

$$R(t_i)\cap A(t_i)=R(t_i). \tag{11.5}$$

Example 11.1

Next, we also use a simple example to demonstrate the steps of ISM in detail (Huang et al. 2005). Assume the ecosystem consists of water (W), fish (F), hydrophytes (H), and fisherman (M), and the relationships of the elements above can be expressed as the relation graph and relation matrix shown in Figure 11.1.

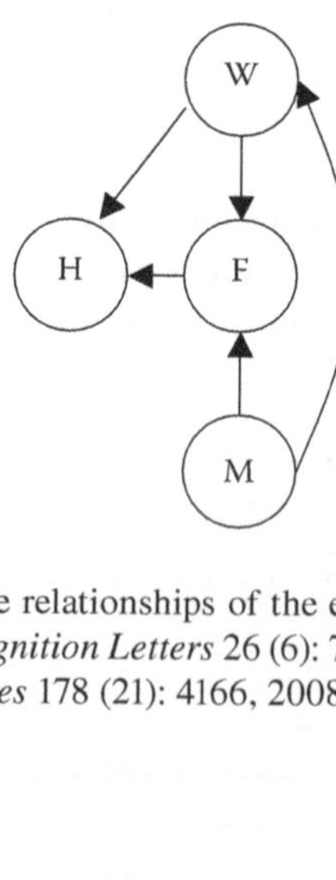

$$D=\begin{array}{c} \\ W\\ F\\ H\\ M\end{array}\begin{array}{c}\begin{array}{cccc}W & F & H & M\end{array}\\ \begin{bmatrix}0 & 1 & 1 & 0\\ 0 & 0 & 1 & 0\\ 0 & 0 & 0 & 0\\ 1 & 1 & 0 & 0\end{bmatrix}\end{array}$$

FIGURE 11.1 The relationships of the elements. (From Huang, J.J., G.H. Tzeng, and C.S. Ong., *Pattern Recognition Letters* 26 (6): 755, 2005d; Yang, J.L., H.N. Chiu, and G.H. Tzeng., *Information Sciences* 178 (21): 4166, 2008.)

TABLE 11.1
The Reachability Set and the Priority Set

e_i	$R(t_i)$	$A(t_i)$	$R(t_i) \cap A(t_i)$
1	1,4	1,2,3	1
2	1,2,4	2,3	2
3	1,2,3,4	3	3
4	4	1,2,3,4	4

Then, the relation matrix adds the identity matrix to form the ***M*** matrix, which can be formed as follows:

$$M = D + I = \begin{bmatrix} 1 & 1 & 1 & 0 \\ 0 & 1 & 1 & 0 \\ 0 & 0 & 1 & 0 \\ 1 & 1 & 0 & 1 \end{bmatrix}.$$

Finally, the reachability matrix can be obtained by powering the matrix, ***M***, to satisfy Equation 11.2.

$$M^* = M^2 = M^{2+k} = \begin{bmatrix} 1 & 1 & 1 & 0 \\ 0 & 1 & 1 & 0 \\ 0 & 0 & 1 & 0 \\ 1 & 1 & 1^* & 1 \end{bmatrix}, \quad k = 1, 2, \ldots, \infty$$

where the symbol (*) indicates the derivative relation, which does not emerge in the original relation matrix. In order to determine the levels of the elements in a hierarchical structure, the reachability set and the priority set are derived based on Equations 11.3 and 11.4 and can be described as shown in Table 11.1.

On the basis of Equation 11.5, it can be seen that the first level is fisherman. The other levels can be determined with the same procedures in turn and can be described as shown in Table 11.2.

The final results of the relationships of the elements, based on the reachability matrix and Table 11.2, can be depicted as shown in Figure 11.2.

TABLE 11.2
Levels in the Ecosystem

Level 1	Fisherman
Level 2	Water
Level 3	Fish
Level 4	Hydrophytes

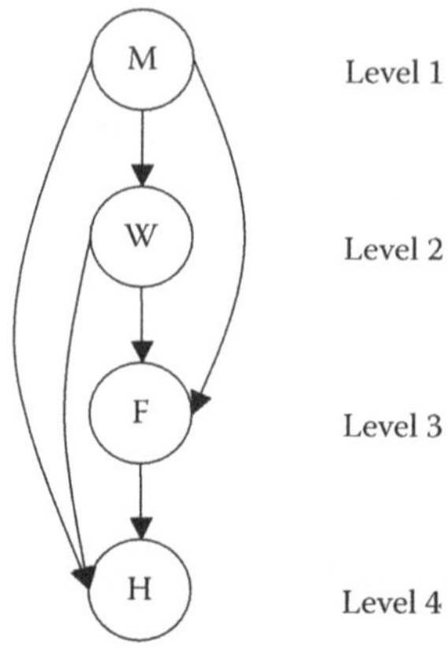

FIGURE 11.2 Hierarchical structure of the elements.

Note that the relation between fisherman and hydrophytes is generated by the reachability matrix. In addition, since hierarchical structural analysis (HSA) and fuzzy ISM (FISM) (Ohuchi and Kaji 1989; Wakabayashi, Itoh, and Ohuchi 1995; Ohuchi, Kase, and Kaji 1988) have been proposed to extend ISM to the feedback structure and the fuzzy environment, we can determine the various network structures in practice.

11.2 DEMATEL METHOD

The Decision Making Trial and Evaluation Laboratory (DEMATEL) method, developed by the Science and Human Affairs Program of the Battelle Memorial Institute of Geneva between 1972 and 1976, was used for researching and solving the complicated and intertwined problem group (Fontela and Gabus 1974, 1976; Warfield 1976). DEMATEL was developed in the belief that pioneering and appropriate use of scientific research methods could improve understanding of the specific problematique, the cluster of intertwined problems, and contribute to identification of workable solutions by a hierarchical structure. The methodology, according to the concrete characteristics of objective affairs, can confirm the interdependence among the variables/attributes and restrict the relation that reflects the characteristic with an essential system and development trend (Hori and Shimizu 1999). Using the DEMATEL method to size and process individual subjective perceptions, brief and impressionistic human insights into problem complexity can be gained. Following the DEMATEL process, the end product of the analysis is a visual representation, an individual map of the mind, according to which the respondent organizes his/her own action in the world, if he/she is to remain internally coherent to respect his/her implicit priorities and to reach his/her secret goals.

The steps of the DEMATEL method can be described as follows:

Step 1: *Calculate the Average Matrix.*

Respondents are asked to indicate the direct influence that they believe each factor exerts on each of the others according to a scale running by integers from 0 to 5. The higher score indicates that the respondent has expressed that the insufficient involvement in the problem of factor i exerts the stronger possible direct influence on the inability of factor j, or, in positive terms, that greater improvement i is required to improve j.

From any group of direct matrices of respondents it is possible to derive an average matrix $\boldsymbol{A}$. Each element of this average matrix will, in this case, be the mean of the same elements in the different direct matrices of the respondents.

Step 2: *Calculate the Initial Direct Influence Matrix.*

The initial direct influence matrix $\boldsymbol{D}$ can be obtained by normalizing the average matrix $\boldsymbol{A}$, in which all principal diagonal elements are equal to zero. Based on matrix $\boldsymbol{D}$, the initial influence that a factor dispatches to and receives from another is shown.

The element of matrix $\boldsymbol{D}$ portrays a contextual relation among the elements of the system and can be converted into a visible structural model, an impact-digraph-map, of the system with respect to that relation. For example, as in Figure 11.2, the respondents are asked to indicate only direct links. In the directed digraph graph represented here, factor i directly affects only factors j and k; indirectly, it also affects first l, m, and n and, secondly, o and q.

Step 3: *Derive the Full Direct/Indirect Influence Matrix.*

There is a continuous decrease of the indirect effects of problems along the powers of matrix $\boldsymbol{D}$, e.g., $\boldsymbol{D}^2$, $\boldsymbol{D}^3$,..., $\boldsymbol{D}^\infty$, thereby guaranteeing convergent solutions to the matrix inversion. In a configuration, such as Figure 11.3, the influence exerted by factor i on factor p will be smaller than that exerted on factor m, and smaller again than that exerted on factor j. This being so, the infinite series of direct and indirect effects can be illustrated. Let the (i, j) element of matrix $\boldsymbol{A}$ be denoted by a_{ij}; the matrix can be found following Equations 11.6 through 11.9.

$$\boldsymbol{D} = s \cdot \boldsymbol{A}, \quad s > 0 \tag{11.6}$$

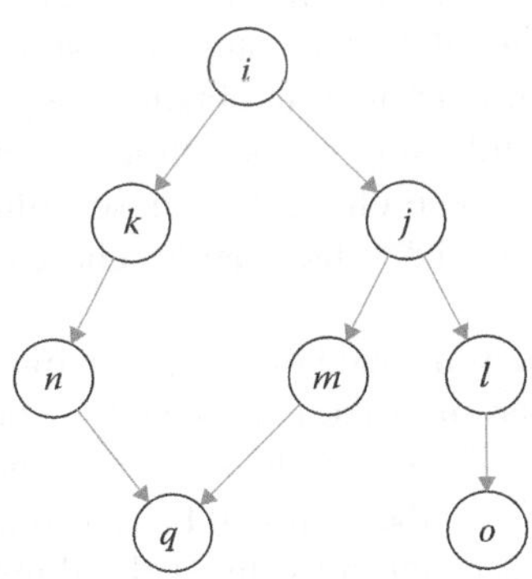

FIGURE 11.3 An example of a direct graph.

or

$$d_{ij} = s \cdot a_{ij}, \quad s > 0 \left(i, j = 1,2,\ldots,n\right), \tag{11.7}$$

$$0 < s < \sup, \quad \sup = \operatorname{Min}\left(\frac{1}{\max\limits_{1\le i\le n}\sum_{j=1}^{n}\left|a_{ij}\right|}, \frac{1}{\max\limits_{1\le j\le n}\sum_{i=1}^{n}\left|a_{ij}\right|}\right), \tag{11.8}$$

and

$$\lim_{m\to\infty} \boldsymbol{D}^m = \left[0\right]. \tag{11.9}$$

The full direct/indirect influence matrix $\boldsymbol{F}$, the infinite series of direct and indirect effects of each factor, can be obtained by the matrix operation of $\boldsymbol{D}$. The matrix $\boldsymbol{F}$ can show the final structure of factors after the continuous process (see Equation 11.10). Let $W_i(f)$ denote the normalized ith row sum of matrix $\boldsymbol{F}$, thus the $W_i(f)$ value means the sum of influence dispatching from factor i to the other factors both directly and indirectly. The $V_i(f)$, the normalized ith column sum of matrix $\boldsymbol{F}$, means the sum of influence that factor i receives from the other factors:

$$\boldsymbol{F} = \sum_{i=1}^{\infty}\boldsymbol{D}^i = \boldsymbol{D}\left(\boldsymbol{I} - \boldsymbol{D}\right)^{-1}. \tag{11.10}$$

Step 4: *Set the Threshold Value and Obtain the Impact-Digraph-Map.*
Setting a threshold value, p, to filter the *obvious* effects denoted by the elements of matrix $\boldsymbol{F}$, is necessary to explain the structure of factors. Based on the matrix $\boldsymbol{F}$, each element, f_{ij}, of matrix $\boldsymbol{F}$ provides the information about a factor i that dispatches influence to factor j, or, in words, factor j receives influence from factor i. If all the information from matrix $\boldsymbol{F}$ converts to the impact-digraph-map, it will be too complex to show the necessary information for decision making. In order to obtain an appropriate impact-digraph-map, setting a threshold value of the influence level is necessary for the decision maker. Only some elements, whose influence level in matrix $\boldsymbol{F}$ is higher than the threshold value, can be chosen and converted into the impact-digraph-map.

The threshold value is decided by a decision maker or, as in this chapter, experts through discussion. Like matrix $\boldsymbol{D}$, the contextual relation among the elements of matrix $\boldsymbol{F}$ can also be converted into a digraph map. If the threshold value is too low, the map will be too complex to show the necessary information for decision making. If the threshold value is too high, many factors will be presented as independent factors without relations to another factor. Each time the threshold value increases, some factors or relationships will be removed from the map. An appropriate threshold value is necessary to obtain a suitable impact-digraph-map and proper information for further analysis and decision making.

After a threshold value and relative impact-digraph-map are decided, the final influence result can be shown. For example, the impact-digraph-map of a factor is the same as Figure 11.2 and eight elements exist in this map. Because of continuous direct/indirect effects between them, finally the effectiveness of these eight elements could be considered to be represented by two independent *final affected elements*: o and q. The other components, not shown in the impact-digraph-map, of a factor can be considered as independent elements because no obvious interrelation with others exists.

Example 11.2

Consider there are five elements and the corresponding direct-relation graph can be represented as shown in Figure 11.4, where the numbers above the arrow denote the pairwise comparison scale.

According to the direct-relation graph, we can formulate the direct-relation matrix as

$$Z = \begin{bmatrix} 0 & 0 & 0 & 0 & 0 \\ 3 & 0 & 0 & 0 & 1 \\ 2 & 0 & 0 & 4 & 1 \\ 4 & 0 & 0 & 0 & 0 \\ 0 & 0 & 1 & 3 & 0 \end{bmatrix}.$$

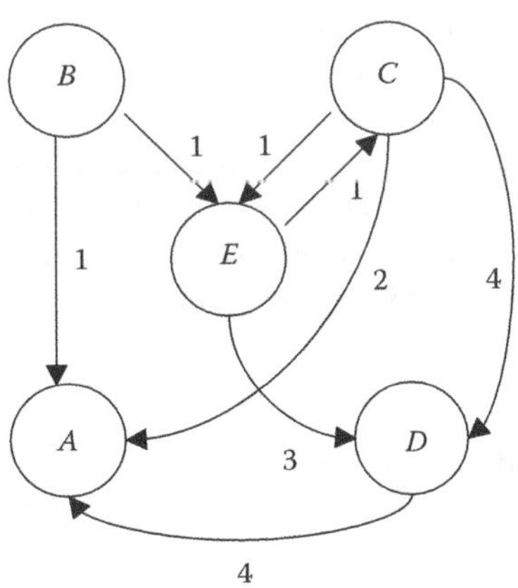

FIGURE 11.4 The direct-relation graph.

Then, we can normalize the direct-relation matrix as

$$X = \begin{bmatrix} 0 & 0 & 0 & 0 & 0 \\ 3/12 & 0 & 0 & 0 & 1/12 \\ 2/12 & 0 & 0 & 4/12 & 1/12 \\ 4/12 & 0 & 0 & 0 & 0 \\ 0 & 0 & 1/12 & 3/12 & 0 \end{bmatrix}.$$

Finally, we can derive the total-relation matrix as

$$T = \begin{bmatrix} 0 & 0 & 0 & 0 & 0 \\ 0.2589 & 0 & 0.0069 & 0.0233 & 0.0839 \\ 0.2867 & 0 & 0.0069 & 0.3566 & 0.0839 \\ 0.3333 & 0 & 0 & 0 & 0 \\ 0.1072 & 0 & 0.0839 & 0.2797 & 0.0069 \end{bmatrix}.$$

From the result of the total-relation matrix, we can derive the indirect relations between elements.

11.3 FUZZY COGNITION MAPS

Fuzzy cognition maps (FCM), which were first proposed by Kosko (1988), extend the original cognitive maps (Axelrod 1976) by incorporating fuzzy measures to provide a flexible and realistic method for extracting the fuzzy relationships among objects in complex systems. Recently, FCM have been widely employed in the applications of political decision making, business management, industrial analysis, and system control (Andreou, Mateou, and Zombanakis 2005; Stylios and Groumpos 2004; Pagageorgiou and Groumpos 2005), except for the area of MCDM. The concepts of FCM can be described as follows.

Consider a 4-tuple $(N, \boldsymbol{E}, \boldsymbol{C}, f)$ where $N = \{N_1, N_2, \ldots, N_n\}$ denotes the set of n objects, $\boldsymbol{E}$ denotes the connection matrix, which is composed of the weights between objects, $\boldsymbol{C}$ is the state matrix, where $\boldsymbol{C}^{(0)}$ is the initial matrix and $\boldsymbol{C}^{(t)}$ is the state matrix at certain iteration t, and f is a threshold function, which indicates the weighting relationship between $\boldsymbol{C}^{(t)}$ and $\boldsymbol{C}^{(t+1)}$. Several formulas have been used as threshold functions such as

$$\text{Hard line function} \quad f(x) = \begin{cases} 1 & \text{if } x \geq 1 \\ 0 & \text{if } x < 1 \end{cases}, \tag{11.11}$$

$$\text{Hyperbolic-tangent function} \quad f(x) = \tanh(x) = (1 - e^{-x})/(1 + e^{-x}), \tag{11.12}$$

and

$$\text{Logistic function} \quad f(x) = 1/(1+e^{-x}). \tag{11.13}$$

The influence of the specific criterion on other criteria can be calculated using the following updating equation:

$$\boldsymbol{C}^{(t+1)} = f\left(\boldsymbol{C}^{(t)}\boldsymbol{E}\right), \boldsymbol{C}^{(0)} = \boldsymbol{I}_{n\times n}, \tag{11.14}$$

where $\boldsymbol{I}_{n\times n}$ denotes the identity matrix.

The vector-matrix multiplication operation to derive successive FCM states is iterated until it converges to a fixed point situation or a limit state cycle. When the state vector remains unchanged for successive iterations, this is called a fixed point situation, and when the sequence of the state vector keeps repeating indefinitely, this is called a limit state cycle.

Example 11.3

In this example, five criteria are used to select the best alternative. In order to derive the local weights, we first compare the importance between the criteria and then employ the eigenvalue method to obtain the eigenvector.

Next, suppose the relationships between the criteria above can be depicted using the FCM as shown in Figure 11.5.

From Figure 11.4, it can be seen that the problem above contains the compound and the interaction effects simultaneously. Next, we present the proposed method to determine the best alternative as follows.

First, on the basis of Figure 11.4, we can formulate the connection matrix as follows:

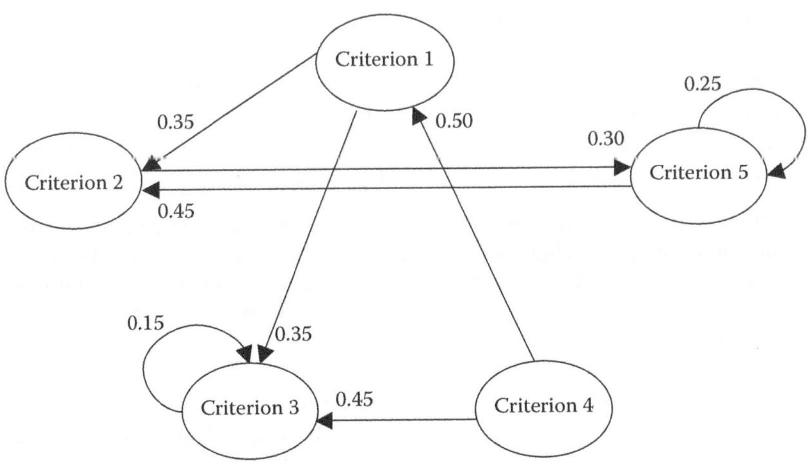

FIGURE 11.5 A fuzzy cognitive map for Example 11.2.

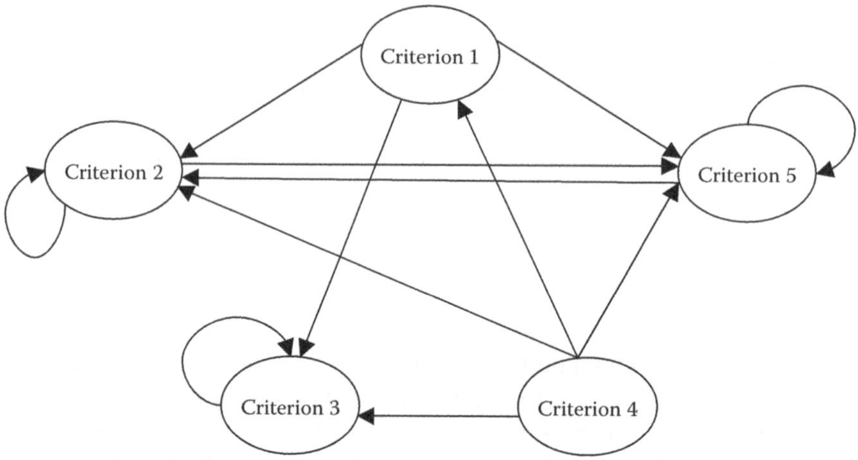

FIGURE 11.6 The relationship between criteria using FCM.

$$
E = \begin{array}{c} \\ C_1 \\ C_2 \\ C_3 \\ C_4 \\ C_5 \end{array}
\begin{array}{c} \begin{array}{ccccc} C_1 & C_2 & C_3 & C_4 & C_5 \end{array} \\
\begin{bmatrix} 0 & 0.35 & 0.35 & 0 & 0 \\ 0 & 0 & 0 & 0 & 0.3 \\ 0 & 0 & 0.15 & 0 & 0 \\ 0.50 & 0 & 0.45 & 0 & 0 \\ 0 & 0.45 & 0 & 0 & 0.25 \end{bmatrix} \end{array}.
$$

Next, by using the pure-linear function, we can calculate the steady-state matrices as follows:

$f(x) = x$	**Criterion 1**	**Criterion 2**	**Criterion 3**	**Criterion 4**	**Criterion 5**
Criterion 1	0	0.4268	0.4118	0	0.1707
Criterion 2	0	0.2195	0	0	0.4878
Criterion 3	0	0	0.1765	0	0
Criterion 4	0.5000	0.2134	0.7353	0	0.0853
Criterion 5	0	0.7317	0	0	0.6260

Note that it can be seen that the results may be different with respect to the different threshold functions. Therefore, another interesting question is which threshold function should be most appropriate for the particular problem. Next, on the basis of the above matrix, we can depict the relationship between criteria as shown in Figure 11.6.

In this chapter, we introduce three structural models, the ISM, DEMATEL, and FCM methods, for recognizing the pattern of elements. This plays a key role in understanding the relationships between criteria and providing the information for the decision maker to use the appropriate MADM methods.

Part II

Applications of MADM

12 AHP: An Application

12.1 INTRODUCTION

When initiating a construction project, most owners must outsource engineering services in order to develop the preliminary plans and the associated design details. In a project life cycle, this planning and design (P&D) phase is most critical to project success. Yet, when selecting an appropriate P&D alternative, most public works owners lack the ability to effectively evaluate the candidates. Substandard P&D work is often a direct result of inadequate tender selection.

The prime issues for tender selection of P&D services for public works projects are threefold. First, the evaluation criteria are generally multiple and often structured in multilevel hierarchies. Secondly, the evaluation process usually involves subjective assessments, resulting in the use of qualitative and imprecise data. Thirdly, other related interest groups' input for the P&D alternative selection process should be considered.

An effective evaluation procedure is essential in promoting decision quality, and for this a governmental agency must be able to respond to these issues and incorporate them into the overall process. This study examines this group decision-making (DM) process and proposes a multicriteria framework for P&D alternative selection in public office buildings.

The fuzzy analytic hierarchy process (FAHP) or fuzzy multiple criteria decision making (FMCDM) analysis has been widely used to deal with DM problems involving multiple criteria evaluation/selection of alternatives. The practical applications reported in the literature (Altrock and Krause 1994; Baas and Kwakernaak 1997; Chang and Chen 1994; McIntyre and Parfitt 1998; Tang et al. 1999; Teng and Tzeng 1996b; Tsaur, Tzeng, and Wang 1997; Tzeng et al. 1994, 2002) have shown advantages in handling unquantifiable/qualitative criteria and obtained quite reliable results. Thus, this study applied fuzzy set theory (Zadeh 1965) to a managerial DM problem of alternative selection, with the intention of establishing a framework of incorporating FAHP and FMCDM, in order to help a government entity select the most appropriate P&D candidate for public building investment.

This study uses the FAHP to determine the criteria weights from subjective judgments of each DM group. Since the evaluation criteria of building P&D have diverse connotations and meanings, there is no logical reason to treat them as if they are each of equal importance. Furthermore, the FMCDM was used to evaluate the synthetic performance of building P&D alternatives, in order to handle qualitative criteria that are difficult to describe in crisp values, thereby strengthening the comprehensiveness and reasonableness of the DM process.

12.2 PLANNING AND DESIGN ALTERNATIVES EVALUATION MODEL

The purpose of this section is to establish a hierarchical structure for tackling the evaluation problem of building a P&D alternative. The contents include three subsections: building hierarchical structure of evaluation criteria, determining the evaluation criteria weights, and obtaining the performance value.

12.2.1 Building Hierarchical Structure of Evaluation Criteria

Multiple criteria decision making (MCDM) is an analytic method to evaluate the advantages and disadvantages of alternatives based on multiple criteria. MCDM problems can be broadly classified into two categories: multiple objective programming and multiple criteria evaluation (Hwang and Yoon 1981). Since this study focuses mainly on the evaluation problem, the second category is emphasized. The typical multiple criteria evaluation problem examines a set of feasible alternatives and considers more than one criterion to determine a priority ranking for alternative implementation. Keeney and Raiffa (1976) suggest that five principles be considered when criteria are being formulated: completeness (the criteria must embrace all of the important characteristics of the DM problems), operational ability (the criteria will have to be meaningful for decision makers and available for open study), decomposability (the criteria can be decomposed from higher hierarchy to lower hierarchy to simplify evaluation processes), non-redundancy (the criteria must avoid duplicate measurement of the same performance), and minimum size (the number of criteria should be as small as possible so as to reduce the needed manpower, time, and cost).

The hierarchical structure adopted in this study to deal with the problems of P&D assessment for public building is shown in Figure 12.1. The key dimensions of the criteria for evaluation and selection of building P&D alternatives were derived through comprehensive investigation and consultation with several experts, including one professor in architectural engineering, one professor in civil engineering, one experienced architect, and five experienced staff in professional services procurement of the Taipei City Public Works Bureau. These individuals were asked to rate the accuracy, adequacy, and relevance of the criteria and dimensions and to verify their "content validity" in terms of building P&D assessment. Synthesizing the literature review (Chen 1978; Wu, Chen, and Zhao 1990), the expert and government staff opinions provided the basis for developing the hierarchical structure used in this study. Furthermore, the five criteria selection principles suggested by Keeney and Raiffa (1976) have been used to formulate the P&D of public building evaluation criteria in this study. There are six dimensions: building lot layout, two-dimensional planning, appearance modeling, electrical and mechanical systems, structural systems, and degree of requirement accomplishment. From these, twenty evaluation criteria for the hierarchical structure were used in this study.

12.2.2 Determining the Evaluation Criteria Weights

Since the criteria of building P&D evaluation have diverse significance and meanings, we cannot assume that each evaluation criteria is of equal importance. There are

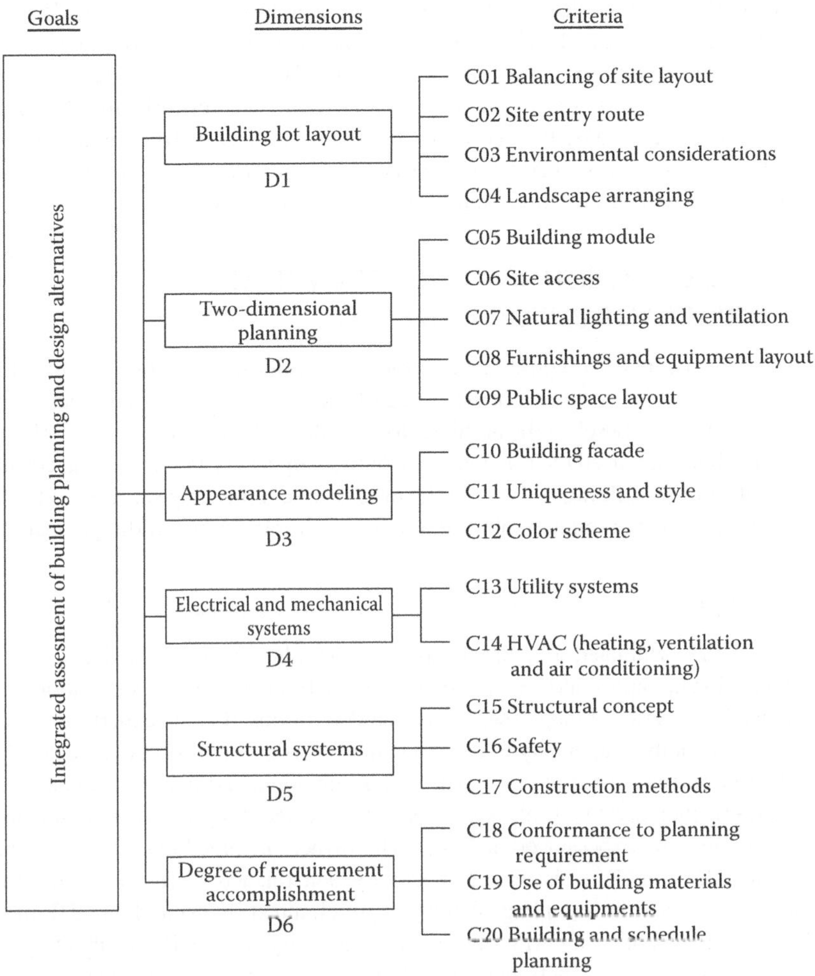

FIGURE 12.1 The hierarchical structure for building planning and design alternatives assessment.

many methods that can be employed to determine weights (Hwang and Yoon 1981), such as the eigenvector method, weighted least-square method, entropy method, analytic hierarchy process (AHP), and linear programming techniques for multidimensional analysis of preference (LINMAP). The selection of method depends on the nature of the problem. To evaluate building P&D is a complex and wide-ranging problem, requiring the most inclusive and flexible method. The AHP developed by Saaty (1977, 1980) is a very useful decision analysis tool in dealing with multiple criteria decision problems and has been successfully applied to many construction industry decision areas (Al-Harbi 2001; Alkhalil 2002; Cheung et al. 2001, 2002; Fong and Choi 2000; Hastak 1998; Mahdi et al. 2002; McIntyre and Parfitt 1998). However, in the operation process of applying the AHP method, it is easier and more humanistic for evaluators to assess "criterion A is much more important than criterion B" than to consider "the importance of principle A and principle B is seven to one." Hence, Buckley (1985b) extended Saaty's AHP to the case where evaluators are allowed to employ fuzzy ratios in place of exact ratios to handle the difficulty of people assigning exact ratios when comparing two criteria and deriving the fuzzy weights of criteria by the geometric mean method. Therefore, in this study, we employ Buckley's method, FAHP, to fuzzify hierarchical analysis by allowing fuzzy numbers for the pairwise comparisons, and find the fuzzy weights.

12.3 CASE OF SELECTING THE ENGINEERING SERVICE FOR PUBLIC BUILDINGS

When a government entity would like to build a new building in Taiwan, it must follow sub-paragraph 9 of first paragraph, article 22 of the Government Procurement Law, to publicly and objectively select the architect to provide professional services for building P&D. Thus, this study used the previous case of the Taipei City Police Bureau constructing a branch station building to exercise the process of engineering service tender selection. In this case, five architects submitted proposals for the new building construction.

12.3.1 Weights Calculation of the Evaluation Criteria

According to the formulated structure of building P&D alternatives evaluation, the weights of the dimension hierarchy and criterion hierarchy can be analyzed. The simulation process was followed by a series of interviews with three DM groups: domain experts (evaluators), superintendents of the Taipei City Police Bureau (owners), and the users of the new building in the future (policemen, users). Each DM group contained five representatives. The domain experts included two professors in architecture and design, two professors in civil engineering, and one experienced architect. The owners included one Director General, three Deputy Director Generals, and one Secretary General; and the five policemen (users) were selected by random sampling. Weights were obtained by using the FAHP method, and then the weights of each DM group and average weights were derived by the geometric mean method suggested by Buckley (1985b). The following example demonstrates the computational procedure of the weights of dimensions for the owners group.

a. According to the interviews with five owners' representatives about the importance of evaluation dimensions, the pairwise comparison matrices of dimensions will be obtained as follows:

	D1	D2	D3	D4	D5	D6
D1	1	*Wk*	*Es*	*Wk*	*Eq*	*Eq*
D2		1	*Wk*	*Eq*	*Eq*	*Wk*
D3			1	*Eq*	*LEs*	*LEs*
D4				1	*LWk*	*LWk*
D5					1	*Eq*
D6						1

Owner no. 1

	D1	D2	D3	D4	D5	D6
D1	1	*LWk*	*Wk*	*Eq*	*Eq*	*LWk*
D2		1	*Wk*	*Wk*	*Wk*	*LWk*
D3			1	*LVs*	*LVs*	*LVs*
D4				1	*LWk*	*LWk*
D5					1	*LWk*
D6						1

Owner no. 2

	D1	D2	D3	D4	D5	D6
D1	1	*LEs*	*Eq*	*Wk*	*Eq*	*LVs*
D2		1	*Wk*	*Ab*	*Vs*	*LEs*
D3			1	*Es*	*Wk*	*LVs*
D4				1	*Eq*	*LAb*
D5					1	*LVs*
D6						1

Owner no. 3

	D1	D2	D3	D4	D5	D6
D1	1	*LWk*	*LEs*	*Wk*	*LVs*	*Es*
D2		1	*Es*	*Vs*	*LAb*	*Vs*
D3			1	*Wk*	*LAb*	*Wk*
D4				1	*LAb*	*Wk*
D5					1	*Ab*
D6						1

Owner no. 4

	D1	D2	D3	D4	D5	D6
D1	1	*Wk*	*Es*	*LWk*	*LVs*	*LEq*
D2		1	*Wk*	*LEs*	*LVs*	*LWk*
D3			1	*LVs*	*LAb*	*LEs*
D4				1	*LWk*	*Wk*
D5					1	*Es*
D6						1

Owner no. 5

where L = less.

b. Applying the fuzzy numbers defined in Table 12.1, the linguistic scales can be transferred to the corresponding fuzzy numbers as follows:

	D1	D2	D3	D4	D5	D6
D1	1	$\tilde{3}$	$\tilde{5}$	$\tilde{3}$	$\tilde{1}$	$\tilde{1}$
D2	$\tilde{3}^{-1}$	1	$\tilde{3}$	$\tilde{1}$	$\tilde{1}$	$\tilde{3}$
D3	$\tilde{5}^{-1}$	$\tilde{3}^{-1}$	1	$\tilde{1}$	$\tilde{5}^{-1}$	$\tilde{5}^{-1}$
D4	$\tilde{3}^{-1}$	$\tilde{1}^{-1}$	$\tilde{1}^{-1}$	1	$\tilde{3}^{-1}$	$\tilde{3}^{-1}$
D5	$\tilde{1}^{-1}$	$\tilde{1}^{-1}$	$\tilde{5}$	$\tilde{3}$	1	$\tilde{1}$
D6	$\tilde{1}^{-1}$	$\tilde{3}^{-1}$	$\tilde{5}$	$\tilde{3}$	$\tilde{1}^{-1}$	1

Owner no. 1

	D1	D2	D3	D4	D5	D6
D1	1	$\tilde{3}^{-1}$	$\tilde{3}$	$\tilde{1}$	$\tilde{1}$	$\tilde{3}^{-1}$
D2	$\tilde{3}$	1	$\tilde{3}$	$\tilde{3}$	$\tilde{3}$	$\tilde{3}^{-1}$
D3	$\tilde{3}^{-1}$	$\tilde{3}^{-1}$	1	$\tilde{7}^{-1}$	$\tilde{7}^{-1}$	$\tilde{7}^{-1}$
D4	$\tilde{1}^{-1}$	$\tilde{3}^{-1}$	$\tilde{7}$	1	$\tilde{3}^{-1}$	$\tilde{3}^{-1}$
D5	$\tilde{1}^{-1}$	$\tilde{3}^{-1}$	$\tilde{7}$	$\tilde{3}$	1	$\tilde{3}^{-1}$
D6	$\tilde{3}$	$\tilde{3}$	$\tilde{7}$	$\tilde{3}$	$\tilde{3}$	1

Owner no. 2

	D1	D2	D3	D4	D5	D6
D1	1	$\tilde{5}^{-1}$	$\tilde{1}$	$\tilde{3}$	$\tilde{1}$	$\tilde{7}^{-1}$
D2	$\tilde{5}$	1	$\tilde{3}$	$\tilde{9}$	$\tilde{7}$	$\tilde{5}^{-1}$
D3	$\tilde{1}^{-1}$	$\tilde{3}^{-1}$	1	$\tilde{5}$	$\tilde{3}$	$\tilde{7}^{-1}$
D4	$\tilde{3}^{-1}$	$\tilde{9}^{-1}$	$\tilde{5}^{-1}$	1	$\tilde{1}$	$\tilde{9}^{-1}$
D5	$\tilde{1}^{-1}$	$\tilde{7}^{-1}$	$\tilde{3}^{-1}$	$\tilde{1}^{-1}$	1	$\tilde{7}^{-1}$
D6	$\tilde{7}$	$\tilde{5}$	$\tilde{7}$	$\tilde{9}$	$\tilde{7}$	1

Owner no. 3

	D1	D2	D3	D4	D5	D6
D1	1	$\tilde{3}^{-1}$	$\tilde{5}^{-1}$	$\tilde{3}$	$\tilde{7}^{-1}$	$\tilde{5}$
D2	$\tilde{3}$	1	$\tilde{5}$	$\tilde{7}$	$\tilde{9}^{-1}$	$\tilde{7}$
D3	$\tilde{5}$	$\tilde{5}^{-1}$	1	$\tilde{3}$	$\tilde{9}^{-1}$	$\tilde{3}$
D4	$\tilde{3}^{-1}$	$\tilde{7}^{-1}$	$\tilde{3}^{-1}$	1	$\tilde{9}^{-1}$	$\tilde{3}$
D5	$\tilde{7}$	$\tilde{9}$	$\tilde{9}$	$\tilde{9}$	1	$\tilde{9}$
D6	$\tilde{5}^{-1}$	$\tilde{7}^{-1}$	$\tilde{3}^{-1}$	$\tilde{3}^{-1}$	$\tilde{9}^{-1}$	1

Owner no. 4

TABLE 12.1
Definition of Fuzzy Numbers

	LAb	*LVs*	*LEs*	*LWk*	*Eq*	*Wk*	*Es*	*Vs*	*Ab*
Fuzzy numbers	$\tilde{9}^{-1}$	$\tilde{7}^{-1}$	$\tilde{5}^{-1}$	$\tilde{3}^{-1}$	$\tilde{1}$	$\tilde{3}$	$\tilde{5}$	$\tilde{7}$	$\tilde{9}$
	$\left(\frac{1}{9},\frac{1}{9},\frac{1}{7}\right)$	$\left(\frac{1}{9},\frac{1}{7},\frac{1}{5}\right)$	$\left(\frac{1}{7},\frac{1}{5},\frac{1}{7}\right)$	$\left(\frac{1}{5},\frac{1}{3},1\right)$	(1,1,3)	(1,3,5)	(3,5,7)	(5,7,9)	(7,9,9)

$$
\begin{array}{c} \\ D1 \\ D2 \\ D3 \\ D4 \\ D5 \\ D6 \end{array}
\begin{array}{c} \begin{array}{cccccc} D1 & D2 & D3 & D4 & D5 & D6 \end{array} \\
\left[\begin{array}{cccccc}
1 & \tilde{3} & \tilde{5} & \tilde{3}^{-1} & \tilde{7}^{-1} & \tilde{1}^{-1} \\
\tilde{3}^{-1} & 1 & \tilde{3} & \tilde{5}^{-1} & \tilde{7}^{-1} & \tilde{3}^{-1} \\
\tilde{5}^{-1} & \tilde{3}^{-1} & 1 & \tilde{7}^{-1} & \tilde{9}^{-1} & \tilde{5}^{-1} \\
\tilde{3} & \tilde{5} & \tilde{7} & 1 & \tilde{3}^{-1} & \tilde{3} \\
\tilde{7} & \tilde{7} & \tilde{9} & \tilde{3} & 1 & \tilde{5} \\
\tilde{1} & \tilde{3} & \tilde{5} & \tilde{3}^{-1} & \tilde{5}^{-1} & 1
\end{array}\right] \end{array}
$$

Owner no. 5

c. Computing the elements of the synthetic pairwise comparison matrix using the geometric mean method suggested by Buckley (1985b), that is: $\tilde{a}_{ij} = (\tilde{a}_{ij}^1 \otimes \tilde{a}_{ij}^2 \otimes \tilde{a}_{ij}^3 \otimes \tilde{a}_{ij}^4 \otimes \tilde{a}_{ij}^5)^{1/5}$, for $\tilde{a}_{12}$ as an example:

$$
\tilde{a}_{12} = \left((1,3,5) \otimes \left(\frac{1}{5},\frac{1}{3},1\right) \otimes \left(\frac{1}{7},\frac{1}{5},\frac{1}{3}\right) \otimes \left(\frac{1}{5},\frac{1}{3},1\right) \otimes (1,3,5) \right)^{1/5}
$$

$$
= \left(\left(1\times\frac{1}{5}\times\frac{1}{7}\times\frac{1}{5}\times 1\right)^{1/5}, \left(3\times\frac{1}{3}\times\frac{1}{5}\times\frac{1}{3}\times 3\right)^{1/5}, \left(5\times 1\times\frac{1}{3}\times 1\times 5\right)^{1/5} \right)
$$

$$
= (0.356, 0.725, 1.528).
$$

The other matrix elements can be obtained by the same computational procedure, therefore the synthetic pairwise comparison matrices of the five representatives will be constructed as follows:

$$
\begin{array}{c} \\ D1 \\ D2 \\ D3 \\ D4 \\ D5 \\ D6 \end{array}
\begin{array}{c} \begin{array}{cccccc} D1 & D2 & D3 & D4 & D5 & D6 \end{array} \\
\left[\begin{array}{cccccc}
1 & (0.356,0.725,1.528) & (1.052,1.179,3.005) & (0.725,1.552,3.272) & (0.415,0.459,1.016) & (0.467,0.750,1.332) \\
(0.654,1.380,2.809) & 1 & (1.246,3.323,5.348) & (1.380,2.068,3.323) & (0.573,0.803,1.310) & (0.491,0.859,1.719) \\
(0.333,0.582,0.951) & (0.187,0.301,0.803) & 1 & (0.517,0.789,1.332) & (0.181,0.524,0.369) & (0.191,0.300,0.467) \\
(0.306,0.644,1.380) & (0.301,0.484,0.725) & (0.751,1.267,1.933) & 1 & (0.245,0.333,0.844) & (0.339,0.644,1.290) \\
(0.985,2.178,2.408) & (0.763,1.246,1.745) & (2.713,3.936,5.512) & (1.185,3.000,4.076) & 1 & (0.859,1.165,2.068) \\
(0.751,1.332,2.141) & (0.582,1.165,2.036) & (2.141,3.328.5.245) & (0.775,1.552,2.954) & (0.484,0.859,1.165) & 1
\end{array}\right] \end{array}.
$$

d. Using Equation 2.7 or 4.10 to obtain the fuzzy weights of dimensions for the owners group, that is:

$$\tilde{r}_1 = \left(\tilde{a}_{11} \otimes \tilde{a}_{12} \otimes \tilde{a}_{13} \otimes \tilde{a}_{14} \otimes \tilde{a}_{15} \otimes \tilde{a}_{16}\right)^{1/6}$$

$$= \left(\left(1\times 0.356\times\cdots\times 0.467\right)^{1/6}, \left(1\times 0.725\times\cdots\times 0.750\right)^{1/6}, \left(1\times 1.528\times\cdots\times 1.332\right)^{1/6}\right)$$

$$= \left(0.612, 0.935, 1.652\right).$$

Likewise, we can obtain the remaining $\tilde{r}_i$, that is,

$$\tilde{r}_2 = \left(0.826, 1.367, 2.197\right),\ \tilde{r}_3 = \left(0.322, 0.468, 0.748\right),\ \tilde{r}_4 = \left(0.423, 0.663, 1.132\right),$$

$$\tilde{r}_5 = \left(1.129, 1.828, 2.409\right),\ \tilde{r}_6 = \left(0.840, 1.379, 2.070\right).$$

The weight of each dimension can be obtained as follows:

$$\tilde{w}_1 = \tilde{r}_1 \otimes \left(\tilde{r}_1 \oplus \tilde{r}_2 \oplus \tilde{r}_3 \oplus \tilde{r}_4 \oplus \tilde{r}_5 \oplus \tilde{r}_6\right)^{-1}$$

$$= \left(0.612, 0.935, 1.652\right) \otimes \left(1/\left(1.652+\cdots+2.070\right), 1/\left(0.935+\cdots+1.379\right),\right.$$

$$\left.\left(1/0.612+\cdots+0.840\right)\right)$$

$$= \left(0.06, 0.141, 0.398\right).$$

Likewise, $\tilde{w}_2 = (0.081, 0.206, 0.529)$, $\tilde{w}_3 = (0.032, 0.071, 0.180)$, $\tilde{w}_4 = (0.041, 0.010, 0.273)$, $\tilde{w}_5 = (0.111, 0.275, 0.580)$, and $\tilde{w}_6 = (0.082, 0.208, 0.499)$.

e. To employ the center of area (CoA) method to compute the best non-fuzzy performance (*BNP*) value of the fuzzy weights of each dimension: Taking the *BNP* value of the weight of building lot layout (D1) for the owners group as an example, the calculation process is as follows.

$$BNP_{w_1} = \left[\left(Uw_1 - Lw_1\right) + \left(Mw_1 - Lw_1\right)\right]/3 + Lw_1$$

$$= \left[\left(0.398 - 0.06\right) + \left(0.141 - 0.06\right)\right]/3 + 0.06 = 0.200.$$

Similarly, the weights for the remaining dimensions and criteria for the owners group can be found as shown in Table 12.2. However, due to limited space, we omit the fuzzy weights of the other two groups and the average of the three, but we list the final *BNP* value of them in Table 12.3.

From the FAHP results, for the owners group, we find the two most important aspects are structural system (0.322) and two-dimensional planning (0.272),

TABLE 12.2
Weights of Dimensions and Criteria for the Owners Group

Dimension and Criteria	Local Weights	Overall Weights	*BNP*
Building Lot Layout	**(0.06,0.141,0.398)**		**0.200**
Balancing of site layout	(0.157,0.384,0.976)	(0.009,0054,0.388)	0.151
Site entry route	(0.153,0.400,0.900)	(0.009,0.056,0.358)	0.141
Matching of environment	(0.063,0.152,0.412)	(0.004,0.021,0.164)	0.063
Landscape arranging	(0.033,0.064,0.182)	(0.002,0.009,0.072)	0.028
Two-Dimensional Planning	**(0.081,0.206,0.529)**		**0.272**
Building module	(0.143,0.384,0.899)	(0.012,0.079,0.476)	0.189
Site access	(0.096,0.242,0.593)	(0.008,0.050,0.314)	0.124
Natural lighting and ventilation	(0.090,0.213,0.571)	(0.007,0.044,0.302)	0.118
Furnishings and equipment layout	(0.032,0.073,0.192)	(0.003,0.015,0.102)	0.040
Public space layout	(0.040,0.088,0.235)	(0.003,0.018,0.124)	0.048
Appearance Modeling	**(0.032,0.071,0.180)**		**0.094**
Building facade	(0.169,0.439,0.946)	(0.005,0.031,0.171)	0.069
Innovation and style	(0.180,0.410,1.001)	(0.006,0.029,0.180)	0.072
Color scheme	(0.077,0.151,0.402)	(0.002,0.011,0.072)	0.028
E&M Systems	**(0.414,0.100,0.273)**		**0.138**
Utility systems	(0.308,0.580,1.077)	(0.013,0.058,0.294)	0.121
HVAC	(0.227,0.420,0.794)	(0.009,0.042,0.217)	0.089
Structural System	**(0.111,0.275,0.580)**		**0.322**
Structure configuration concept	(0.069,0.125,0.297)	(0.008,0.034,0.172)	0.071
Safety	(0.338,0.708,1.302)	(0.037,0.195,0.755)	0.329
Construction methods	(0.096,0.167,0.392)	(0.011,0.046,0.227)	0.095
Degree of Requirement Accomplishment	**(0.082,0.208,0.499)**		**0.263**
Conformance to planning requirements	(0.282,0.627,1.254)	(0.023,0.130,0.625)	0.260
Using of building materials and equipment	(0.119,0.237,0.560)	(0.010,0.049,0.279)	0.113
Budgeting and schedule planning	(0.070,0.136,0.313)	(0.006,0.028,0.156)	0.063

whereas the least important is appearance modeling (0.094). For the users group, the two most important dimensions are structural system (0.446) and E&M systems (0.386), and the least is degree of requirement accomplishment (0.080). However, for the experts group, the two most important dimensions are two-dimensional planning (0.367) and building lot layout (0.327), and the structural system is the least (0.066). These results indicate that both the owners group and users group are worried about the safety of managing the building and living in the building; in addition, the owners group also cares about the two-dimensional planning of building like experts group which will be considering the convenience of their operating. On the contrary, the experts group focuses on the related professional issues for building esthetic aspects and planning of space mechanism, but they deem that the structural design is certain to be safe under professional calculations, so they ranked it as least

TABLE 12.3
Weights of Dimensions and Criteria for Each Group and Average

Dimension and Criteria	Owners	Users	Experts	Average
Building Lot Layout	**0.200**	**0.117**	**0.327**	**0.197**
Balancing of site layout	0.151	0.087	0.210	0.139
Site entry route	0.141	0.059	0.225	0.123
Matching of environment	0.063	0.065	0.080	0.069
Landscape arranging	0.028	0.031	0.041	0.033
Two-Dimensional Planning	**0.272**	**0.100**	**0.367**	**0.215**
Building module	0.189	0.067	0.242	0.145
Site access	0.124	0.044	0.150	0.093
Natural lighting and ventilation	0.118	0.046	0.155	0.094
Furnishings and equipment layout	0.040	0.053	0.041	0.044
Public space layout	0.048	0.027	0.056	0.042
Appearance Modeling	**0.094**	**0.134**	**0.253**	**0.147**
Building facade	0.069	0.114	0.181	0.112
Innovation and style	0.072	0.088	0.156	0.099
Color scheme	0.028	0.036	0.043	0.035
E&M Systems	**0.138**	**0.386**	**0.068**	**0.153**
Utility systems	0.121	0.285	0.062	0.129
HVAC	0.089	0.261	0.040	0.098
Structural System	**0.322**	**0.446**	**0.066**	**0.211**
Structure configuration concept	0.071	0.173	0.025	0.068
Safety	0.329	0.432	0.056	0.199
Construction methods	0.095	0.205	0.027	0.081
Degree of Requirement Accomplishment	**0.263**	**0.080**	**0.170**	**0.153**
Conformance to planning requirements	0.260	0.058	0.174	0.137
Using of building materials and equipment	0.113	0.075	0.033	0.065
Budgeting and schedule planning	0.063	0.033	0.042	0.044

important. As for the criteria hierarchy, both the owners group and users group deem safety to be the most important (0.329, 0.432). This may reflect the urgent need for building safety after the Chi-Chi earthquake of Taiwan in 1999. Building safety was followed in importance by conformance to planning requirements (0.260), building module (0.189), balancing of site layout (0.151), and site entry route (0.141) for the owners group; and utility systems (0.285), HVAC (0.261), construction methods (0.205), and structural concept (0.173) for the users group. For the experts group, the five most important criteria were building module (0.242), site entry route (0.225), balancing of site layout (0.210), building facade (0.181), and conformance to planning requirements (0.174).

12.3.2 Estimating the Performance Matrix

The evaluators can define their own individual range for the linguistic variables employed in this study according to their subjective judgments with a scale of 0–100 (Table 12.4), revealing a degree of variation in their definitions of the linguistic variables. It can be seen in the divergent understandings of the third and fourth evaluator with respect to the same linguistic variable. For each evaluator with the same importance, this study employed the method of average value to integrate the fuzzy/vague judgment values of different evaluators regarding the same evaluation criteria. In other words, fuzzy addition and fuzzy multiplication are used to solve for the average fuzzy numbers of the performance values under each evaluation criterion shared by the evaluators for the five building P&D alternatives.

For alternative A-1, as an example, the average fuzzy performance values of criterion C01 (balancing of site layout) from the experts' judgment can be obtained as follows:

1. The experts assigned their subjective judgments for A-1 under C01 by the expressions "very good (VG)," "good (G)," "fair (F)," "poor (P)," and "very poor (VP)," and corresponding to the linguistic variable of Table 12.4, the fuzzy performance matrix $\tilde{E}_{ij}^{k}$ can be obtained:

$$\begin{matrix} E^1 & E^2 & E^3 & E^4 & E^5 \\ [F & F & F & G & F] \end{matrix} = \begin{matrix} E^1 & E^2 & E^3 & E^4 & E^5 \\ [(35,45,70) & (30,50,70) & (38,48,65) & (80,85,90) & (45,60,75)] \end{matrix}.$$

2. To obtain the fuzzy performance value of A-1 under C01, that is:

$$\tilde{E}_{11} = \left(\left(\sum_{k=1}^{5} LE_{11}^{k} \right) \Big/ 5, \left(\sum_{k=1}^{5} ME_{11}^{k} \right) \Big/ 5, \left(\sum_{k=1}^{5} UE_{11}^{k} \right) \Big/ 5 \right) = (45.6, 57.6, 74.0).$$

The remaining elements of fuzzy performance values of each criterion of experts for each alternative can be obtained by the same procedure, and are shown in Table 12.5.

TABLE 12.4
Subjective Cognition Results of Evaluators for the Five Levels of Linguistic Variables

	Linguistic Variables				
Evaluator	**Very Poor**	**Poor**	**Fair**	**Good**	**Very good**
1	(0,0,20)	(20,30,40)	(35,45,70)	(70,80,90)	(85,100,100)
2	(0,0,25)	(10,30,50)	(30,50,70)	(65,75,85)	(80,100,100)
3	(0,0,19)	(15,27,43)	(38,48,65)	(60,78,90)	(88,100,100)
4	(0,0,40)	(40,50,60)	(60,70,80)	(80,85,90)	(90,100,100)
5	(0,0,15)	(15,30,45)	(45,60,75)	(75,80,90)	(90,100,100)

TABLE 12.5
Average Fuzzy Performance Values of Each Criterion of Experts for Each Alternative

Criterion	A-1	A-2	A-3	A-4	A-5
Balancing of site layout	(45.6,57 6,74)	(35,47.4,62.6)	(36,50.6,67)	(54.6,65.6,79)	(60,71.4,79.6)
Site entry route	(51.6,52 6,77)	(35,47.4,62.6)	(36,50.6,67)	(58.6,67.6,80)	(57,67.4,77.6)
Matching of environment	(52.6,52 6,77)	(42,56.6,73)	(31,42.4,54.6)	(51.6,62.6,77)	(64.6,75.6,84)
Landscape arranging	(41,50.6,62)	(28.6,41.6,56)	(21,31.4,45.6)	(66,76.6,87)	(58.6,67.6,80)
Building facade	(60.6,70 6,82)	(40,54.6,71)	(15,21.4,36.6)	(75,86.6,93)	(57,71.4,79.6)
Innovation and style	(60,70.6,79)	(40,52.6,64)	(17,27.4,41.6)	(70,80.6,89)	(47,60.4,73.6)
Color scheme	(51.6,51 6,72)	(32.6,45.6,60)	(24.6,34.6,52)	(78,91.6,96)	(65.6,78.6,86)
Building module	(53,65.6,80)	(45.6,57.6,74)	(23,36.4,53.6)	(66,78.6,88)	(55,66.6,79)
Site access	(46,60.6,77)	(35.6,47.6,64)	(32,44.4,57.6)	(70,80.6,89)	(62,76.6,86)
Natural lighting and ventilation	(45,58.6,70)	(46.6,58.6,72)	(33,47.6,61)	(38.6,51.6,66)	(63,73.6,85)
Furnishings and equipment layout	(50,62.6,74)	(45,58.6,70)	(31.6,41.6,56)	(44.6,56.6,69)	(70,80.6,89)
Public space layout	(44,57.6,73)	(63,73.6,85)	(28.4,38.6,56)	(50.6,62.6,77)	(60,72.6,84)
Utility systems	(54,67.6,77)	(32,44.4,56.6)	(61.6,70.6,82)	(50,61.4,74.6)	(60,71.4,80.6)
HVAC	(63,73.6,85)	(35,47.4,62.6)	(64.6,74.6,84)	(40,52.4,64.6)	(56,68.4,76.6)
Structure configuration concept	(41.6,54 6,72)	(37.6,50.6,68)	(45.6,57.6,74)	(45.6,57.6,74)	(54.6,66.6,79)
Safety	(41.6,51 6,72)	(37.6,50.6,68)	(53,65.6,80)	(51.6,62.6,77)	(54.6,64.6,78)
Construction methods	(47.6,59 6,75)	(37.6,50.6,68)	(37.6,50.6,68)	(59,71.6,82)	(50,62.4,74.6)
Conformance to planning requirements	(50,63.6,79)	(47.6,59.6,75)	(40,54.6,71)	(63,75.6,86)	(68.6,79.6,87)
Using of building materials and equipment	(54,63.4 75.6)	(47.5,59.6,75)	(63,73.6,85)	(51.6,62.6,77)	(76,89.6,94)
Budgeting and schedule planning	(48.6,59 6,75)	(41.6,54.6,72)	(31.6,44.6,62)	(34.6,47.6,62)	(55,65.4,76.6)

12.3.3 Ranking the Alternatives

From the criteria weights of three DM groups and average of the three obtained by FAHP (Table 12.3) and the average fuzzy performance values of each criterion of experts for each alternative (Table 12.5), the final fuzzy synthetic decision ($\tilde{R}_i$) can then be processed. After the fuzzy synthetic decision is processed, the non-fuzzy values method is then employed, and finally the fuzzy numbers are changed into non-fuzzy values. Though there are methods to rank these fuzzy numbers, this study has employed CoA to determine the *BNP* value, which is used to rank the evaluation results of each P&D alternative.

To take the fuzzy synthetic decision value of alternative A-1 under the weights of the owners group as an example, we can obtain this value as follows.

$$\tilde{R}_1 = (LR_1, MR_1, UR_1) = \left(\sum_{j=1}^{20} LE_{1j} \times Lw_j, \sum_{j=1}^{20} ME_{1j} \times Mw_j, \sum_{j=1}^{20} UE_{1j} \times Uw_j \right)$$

$$= \big((45.6 \times 0.009 + \cdots + 48.6 \times 0.006), (57.6 \times 0.054 + \cdots + 59.6 \times 0.028),$$

$$(74.0 \times 0.388 + \cdots + 75.0 \times 0.156)\big)$$

$$= (9.19, 61.43, 413.86).$$

Next, we use Equation 4.15 to find out its *BNP* value as follows:

$$BNP_1 = \left[(413.86 - 9.19) + (61.43 - 9.19)\right] / 3 + 9.19 = 161.49.$$

Likewise, we can obtain the *BNP* values of other alternatives for comparison purposes. Finally, details of the results are presented in Table 12.6.

As can be seen from the alternative evaluation results in Table 12.6, alternative A-5 is the best alternative considering the weights of the owners group, users group and the average of the three. However, alternative A-4 is the best alternative by the weights of the experts group, which is clearly different from the other two groups. One interesting point that can be observed from Table 12.6 is that the ranking order of the owners group is the same as the average of the three. The results in Table 12.6 reflect the common perception that changes in criteria weights may affect the evaluation outcome to a certain degree. It is clear that most alternatives maintain similar relative rankings under different criteria weights. In addition, obviously, alternative A-1 has the poorest performance rating relative to other alternatives, which is the most common consensus among the three DM groups.

12.4 DISCUSSIONS

According to the results of case simulation, the ranking order of weights of criteria for each DM group in a complete evaluation of the criterion hierarchy, we can see the

TABLE 12.6
Performance Value and Ranking by Various Criteria Weightings

Alternative	Owners Group		Users Group		Experts Group		Compromised	
	BNP_i	Ranking	BNP_i	Ranking	BNP_i	Ranking	BNP_i	Ranking
A-1	161.49	3	164.33	3	149.97	3	135.96	3
A-2	144.55	4	141.59	5	132.15	4	119.76	4
A-3	141.45	5	150.84	4	114.76	5	114.57	5
A-4	170.48	2.2	167.90	2	161.78	1	143.29	2
A-5	175.10	1	173.43	1	160.46	2	145.63	1

Note: Compromised refers to the weights of average of three groups computed by geometric mean.

difference of each DM group in the DM process. In this case of building P&D alternative evaluation, the owners group and users group are both very concerned about the safety of the building structural system, and its importance ratio is much higher than that of the experts group (the weight of the owners group is 0.329, users group is 0.432, experts group is 0.056). This shows that the users and owners of the building are very concerned with the basic security of people and property, but the experts believe that the structural design would never violate the basic design safety coefficients and the structural calculation results should conform to professional standards. On the other hand, the experts group is more concerned about the planning of the space mechanism, because they think that these criteria may identify the design ability of an architect (the first three important criteria are building module 0.242, site entry route 0.225, and balancing of site layout 0.210). However, because the users group and owners group are not aware of deeper professional design concepts, they allocated relatively less importance to these criteria. In the process of obtaining the weights of criteria by FAHP, we can see the different views of participant parties in the evaluation and that these differences are harmonized, making the evaluation results sufficiently represent group DM.

12.5 CONCLUSION

The purpose of this study was to develop a scientific framework for the evaluation of a P&D alternative tender for public building construction. In architectural engineering, preliminary P&D is a highly professional engineering service, which involves an enormous amount of specialized effort. Although judging the quality of the building P&D may be subjective, tender evaluation of the P&D alternative is even more so. In current methods of building P&D tender selection, government agencies rely only on a panel of experts to perform the evaluation, neglecting the fuzziness of subjective judgment and other related interest groups' perception in this process. Thus, an effective evaluation procedure is essential to improve the quality for the decision-making. This work examines this group DM process and proposes a multicriteria framework for building P&D tender selection. To deal with the qualitative attributes in subjective judgment, this work employs FAHP to determine the weights of decision criteria for each relative interest group, including the owners', users', and experts' representatives. Then the FMCDM approach is used to synthesize the group decision. This process enables decision makers to formalize and effectively solve the complicated, multicriteria, and fuzzy/vague perception problem of selection of the most appropriate building P&D alternative. An empirical case study of nine proposed P&D alternatives for a new building project for the Taipei City Police Bureau is used to exemplify the approach. The underlying concepts applied were intelligible to the DM groups, and the computation required is straightforward and simple. It will also assist government agencies in making critical decisions during the selection of building P&D alternatives.

13 VIKOR Technique with Applications Based on DEMATEL and ANP

13.1 INTRODUCTION

Multiple criteria decision making (MCDM) is frequently used to deal with conflict problems in management. Practical problems are often characterized by several non-commensurable and conflicting (competing) criteria, and there may be no solution satisfying all criteria simultaneously. Therefore, using MCDM, a compromise solution for a problem with conflicting criteria can be determined, which can help the decision makers to improve the problems for achieving the final decision. Yu (1973) and Zeleny (1982) proposed the foundation for compromise solutions. The compromise solution is a feasible solution closest to the ideal/aspired level, compromise meaning an agreement established by mutual concessions. The VIKOR technique introduced the multicriteria ranking index based on the particular measure of closeness to the ideal/aspired level solution and was introduced as one applicable technique to implement within MCDM (Tzeng et al. 2002). This method focuses on ranking, improving, and selecting from a set of alternatives in the presence of conflicting criteria to help the decision makers to relax the trade-offs for reaching the aspired levels (Opricovic and Tzeng 2007). Its characteristics are to provide "the maximum group utility" and "the minimum individual regret," so decision makers can accept the VIKOR-proposed compromise solution. The VIKOR method was developed as an MCDM method to solve discrete decision problems with non-commensurable and conflicting criteria (Tzeng et al. 2002, 2005; Opricovic and Tzeng 2002, 2004, 2007). However, few papers discuss conflicting (competing) criteria with dependence and feedback within this compromise solution method. Therefore, this research adopts the VIKOR technique based on the analytic network process (ANP) and DEMATEL (Decision Making Trial and Evaluation Laboratory) methods to achieve the goal of solving this problem involving conflicting criteria with dependence and feedback.

Saaty (1996) proposed a new MCDM method, the ANP, to overcome the problems of interdependence and feedback among criteria and alternatives in the real world. The ANP is an extension of the analytic hierarchy process (AHP), based on concepts of Markov chains, and is a non-linear structure, while the AHP is hierarchical and linear, with the goal at the top and the alternatives at lower levels (Saaty 1999). The ANP method has been applied successfully in many practical decision-making problems (Karsak et al. 2002; Lee and Kim 2000; Meade and Presley 2002; Momoh and Zhu 2003). However, the treatment of inner dependence in those studies involving the ANP was not complete or perfect. Indeed, the DEMATEL method

(Fontela and Gabus 1974, 1976; Warfield 1976) can be applied to build and illustrate the interrelations of a network relation map (NRM) among criteria, to find the central criteria to represent the effectiveness of factors/aspects, and it can also be used as a wise way to handle the inner dependence within a set of criteria (Wu 2007). Furthermore, a novel hybrid model combining the ANP and DEMATEL methods to solve the dependence and feedback problems has been successfully used in various fields (Tzeng et al. 2007; Liou, Tzeng, and Chang 2007; Huang et al. 2007). Thus, this study uses VIKOR, ANP, and DEMATEL to overcome the conflict when there are multiple criteria with dependence and feedback problems.

When dealing with the ANP procedure, this study finds it is not reasonable to use the traditional method to normalize the unweighted supermatrix. In the normalization procedure, the weighted supermatrix is derived by transforming each column to sum exactly to unity (1.00) based on the total influence ratios of NRM. In the traditional method, each criterion in a column is divided by the number of clusters so each column will sum to unity exactly. Using this normalization method implies each cluster has the same weight. However, there are different degrees of influence between the clusters of factors/criteria in the real world, so using the assumption of equal weight for each cluster to obtain the weighted supermatrix is an improvement. Thus, another purpose of this chapter is to adopt the DEMATEL method to normalize the unweighted supermatrix in the ANP to suit the real world. In conclusion, the contribution of this study is to propose a novel model that combines the DEMATEL and ANP with VIKOR technique procedures not only to overcome the conflicting problems of factors/criteria with interdependence and feedback, but also to normalize the unweighted supermatrix in the ANP procedure to suit the real world. In addition, we also demonstrate a numerical example to show the steps of the proposed method with applications thereof. The results show this proposed method not only can deal with the conflicting problems of criteria with interdependence and feedback, but also can improve the normalized supermatrix to suit the real world.

The remainder of this chapter is organized as follows. Section 13.2 describes the hybrid model. A numerical example with applications is illustrated in Section 13.3. Discussions and conclusions are presented in Section 13.4 and 13.5, respectively.

13.2 NOVEL HYBRID MULTIPLE CRITERIA DECISION MAKING MODEL

According to the above descriptions, a VIKOR technique based on DEMATEL and ANP for evaluating and improving problems is proposed. The procedures of this novel hybrid MCDM model, a combination of the DEMATEL and ANP with VIKOR technique procedures, are shown and explained briefly as follows (see Figure 13.1).

We illustrate these methods according to the above model procedures, detailed as follows.

13.2.1 DEMATEL

The Battelle Memorial Institute proposed a DEMATEL method project through its Geneva Research Centre (Gabus and Fontela 1973). The DEMATEL method is used

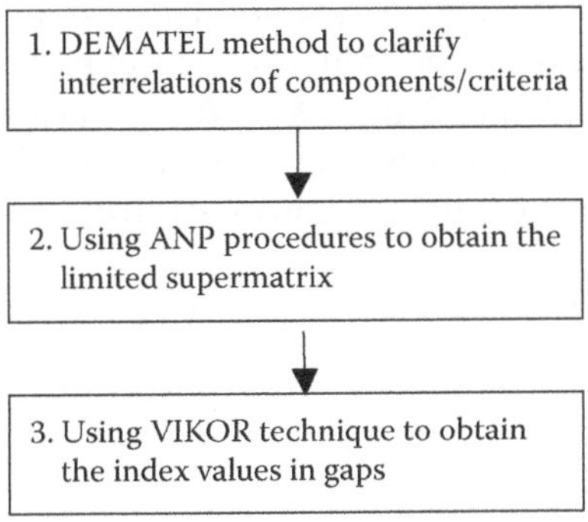

FIGURE 13.1 Hybrid MCDM model procedures.

to construct the interrelations between factors/criteria to build the impact of an NRM (Tzeng et al. 2007; Liou et al. 2007; Huang et al. 2007). The method can be summarized as follows:

Step 1: *Calculate the initial average matrix by scores.*

Assuming the scales 0, 1, 2, 3, and 4 represent the range from "no influence" to "very high influence," respondents are asked to indicate the degree of direct influence each factor/criterion i exerts on each factor/criterion j, which is denoted by a_{ij}, using the assumed scales. Each respondent would produce a direct matrix and an average matrix $\boldsymbol{A}$ is then derived through the mean of the same factors/criteria in the various direct matrices of the respondents. The average matrix $\boldsymbol{A}$ is represented by the following equation:

$$\boldsymbol{A}=\begin{bmatrix} a_{11} & \cdots & a_{1j} & \cdots & a_{1n} \\ \vdots & & \vdots & & \vdots \\ a_{i1} & \cdots & a_{ij} & \cdots & a_{in} \\ \vdots & & \vdots & & \vdots \\ a_{n1} & \cdots & a_{nj} & \cdots & a_{nn} \end{bmatrix} \tag{13.1}$$

Step 2: *Calculate the initial influence matrix.*

The initial direct influence matrix $\boldsymbol{X}(\boldsymbol{X}=[x_{ij}]_{n\times n})$ can be obtained by normalizing the average matrix $\boldsymbol{A}$. Specifically, the matrix $\boldsymbol{X}$ can be obtained through Equations 13.2 and 13.3, in which all principal diagonal criteria are equal to zero.

$$\boldsymbol{X}=s\cdot\boldsymbol{A}. \tag{13.2}$$

$$s=\min\left[1\Big/\max_i\sum_{j=1}^{n}|a_{ij}|,1\Big/\max_j\sum_{i=1}^{n}|a_{ij}|\right]. \tag{13.3}$$

Step 3: *Derive the full direct/indirect influence matrix.*

A continuous decrease of the indirect effects of problems along the powers of $\boldsymbol{X}$, e.g., $\boldsymbol{X}^2, \boldsymbol{X}^3, \ldots, \boldsymbol{X}^k$ and $\lim_{k\to\infty} \boldsymbol{X}^k = [0]_{n\times n}$, when $\boldsymbol{X} = [x_{ij}]_{n\times n}$, $0 \le x_{ij} < 1$, $0 \le (\sum_i x_{ij}, \sum_j x_{ij}) < 1$ and only one column sum $\sum_j x_{ij}$ or one row sum $\sum_i x_{ij}$ equals 1. The total-influence matrix is listed as follows.

$$\boldsymbol{T} = \boldsymbol{X} + \boldsymbol{X}^2 + \cdots + \boldsymbol{X}^k = \boldsymbol{X}(\boldsymbol{I} - \boldsymbol{X})^{-1}, \tag{13.4}$$

[Proof]

$$\begin{aligned}
\boldsymbol{T} &= \boldsymbol{X} + \boldsymbol{X}^2 + \cdots + \boldsymbol{X}^k \\
&= \boldsymbol{X}\left(\boldsymbol{I} + \boldsymbol{X} + \boldsymbol{X}^2 + \cdots + \boldsymbol{X}^{k-1}\right) \\
&= \boldsymbol{X}\left[(\boldsymbol{I} + \boldsymbol{X} + \boldsymbol{X}^2 + \cdots + \boldsymbol{X}^{k-1})(\boldsymbol{I} - \boldsymbol{X})\right](\boldsymbol{I} - \boldsymbol{X})^{-1} \\
&= \boldsymbol{X}\left[\boldsymbol{I} - \boldsymbol{X}^k\right](\boldsymbol{I} - \boldsymbol{X})^{-1} \\
&= \boldsymbol{X}(\boldsymbol{I} - \boldsymbol{X})^{-1}, \quad \text{when } \lim_{k\to\infty} \boldsymbol{X}^k = [0]_{n\times n},
\end{aligned}$$

where $\boldsymbol{T} = [t_{ij}]_{n\times n}$ and $(\boldsymbol{I} - \boldsymbol{X})(\boldsymbol{I} - \boldsymbol{X})^{-1} = \boldsymbol{I}$. In addition, the method presents each row sum and column sum of total matrix $\boldsymbol{T}$.

$$\boldsymbol{r} = (r_i)_{n\times 1} = \left[\sum_{j=1}^{n} t_{ij}\right]_{n\times 1}, \tag{13.5}$$

$$\boldsymbol{c} = (c_j)_{n\times 1} = (c_j)'_{1\times n} = \left[\sum_{i=1}^{n} t_{ij}\right]'_{1\times n} \tag{13.6}$$

where r_i denotes the row sum of the ith row of matrix $\boldsymbol{T}$ and shows the sum of direct and indirect effects of factor/criterion i on the other factors/criteria. Similarly, c_j denotes the column sum of the jth column of matrix $\boldsymbol{T}$ and shows the sum of direct and indirect effects that factor/criterion j has received from the other factors/criteria. In addition, when $i = j$ (i.e., the sum of the row and column aggregates), $(r_i + c_i)$ provides an index of the strength of influences given and received, that is, $(r_i + c_i)$ shows the degree of the central role that factor/criterion i plays in the problem. If $(r_i - c_i)$ is positive, then factor i is affecting other factors, and if $(r_i - c_i)$ is negative, then factor i is influenced by other factors (Tzeng and Chiang 2007; Tamura, Nagata, and Akazawa 2002).

Step 4: *Set a threshold value and obtain the NRM.*

Based on the matrix $\boldsymbol{T}$, each factor t_{ij} of matrix $\boldsymbol{T}$ provides network information about how factor i affects factor j. Setting a threshold value α to filter the minor effects denoted by the factors of matrix $\boldsymbol{T}$ is necessary to isolate the relation structure of the factors. In practice, if all the information from matrix $\boldsymbol{T}$ converts to the NRM, the map would be too complex to show the necessary network information for decision making. In order to reduce the complexity of the NRM, the decision maker sets a threshold value for the influence level: only factors whose influence value in matrix $\boldsymbol{T}$ is higher than the threshold value can be chosen and converted into the NRM. The threshold value can be decided through the brainstorming of experts. When the threshold value and relative NRM have been decided, the NRM can be shown.

13.2.2 Analytic Network Process

The ANP is the general form of the AHP (Saaty 1980) which has been used in MCDM to release the restriction of hierarchical structure. The method can be described in the following steps.

Step 5: *Compare the criteria in the whole system to form the supermatrix.* The original supermatrix of column eigenvectors can be obtained from pairwise comparison matrices of criteria. The relative importance value can be determined using a scale of 1 to 9 to represent equal importance to extreme importance (Saaty 1980, 1996). The general form of the supermatrix can be described as follows:

$$
\boldsymbol{W} = \begin{array}{c c} & \begin{array}{cccc} C_1 & C_2 & & C_n \\ e_{11}\cdots e_{1m_1} & e_{21}\cdots e_{2m_2} & \cdots & e_{n1}\cdots e_{nm_n} \end{array} \\ \begin{array}{c c} C_1 & \begin{array}{c} e_{11} \\ e_{12} \\ \vdots \\ e_{1m_1} \end{array} \\ C_2 & \begin{array}{c} e_{21} \\ e_{21} \\ \vdots \\ e_{2m_2} \end{array} \\ \vdots & \vdots \\ C_n & \begin{array}{c} e_{n1} \\ e_{n2} \\ \vdots \\ e_{nm_n} \end{array} \end{array} & \begin{bmatrix} W_{11} & W_{12} & \cdots & W_{1n} \\ W_{12} & W_{22} & \cdots & W_{2n} \\ \vdots & \vdots & \vdots\vdots\vdots & \vdots \\ W_{n1} & W_{n2} & \cdots & W_{nn} \end{bmatrix} \end{array}, \tag{13.7}
$$

where C_n denotes the nth cluster, e_{nm} denotes the mth criterion in the nth cluster, and W_{ij} is the principal eigenvector of the influence of the criteria in the jth cluster compared to the ith cluster. In addition, if the jth cluster has no influence on the ith cluster, then $W_{ij} = [0]$.

Step 6: *Obtain the weighted supermatrix by multiplying the normalized matrix, which is derived according to the NRM based on the DEMATEL method.* The normalization is used to derive the weighted supermatrix by transforming each column to sum exactly to unity. In the traditional normalized method, each criterion in a column is divided by the number of clusters so each column will sum to unity exactly. Using this normalization method implies each cluster has the same weight. However, we know the effect of each cluster on the other clusters may be different, as described in Section 13.2.1. Therefore, using the assumption of equal weight for each cluster to obtain the weighted supermatrix is irrational. This study adopts the NRM based on the DEMATEL method to solve this problem. First, we use the DEMATEL method (Section 13.2.1) to derive the NRM. Next, this study uses the total-influence matrix $\boldsymbol{T}$ and a threshold value α to generate a new matrix. The values of the clusters in matrix $\boldsymbol{T}$ are reset to zero if their values are less than α, i.e., they have a lower influence on other clusters if their values are less than α, which is decided by decision makers or experts. The new matrix with α-cut is called the α-cut total-influence matrix $\boldsymbol{T}_\alpha$, as in Equation 13.8.

$$\boldsymbol{T}_\alpha = \begin{bmatrix} t_{11}^\alpha & \cdots & t_{1j}^\alpha & \cdots & t_{1n}^\alpha \\ \vdots & & \vdots & & \vdots \\ t_{i1}^\alpha & \cdots & t_{ij}^\alpha & \cdots & t_{in}^\alpha \\ \vdots & & \vdots & & \vdots \\ t_{n1}^\alpha & \cdots & t_{nj}^\alpha & \cdots & t_{nn}^\alpha \end{bmatrix} \rightarrow d_i = \sum_{j=1}^{n} t_{1j}^\alpha, \tag{13.8}$$

where if $t_{ij} < \alpha$, then $t_{ij}^\alpha = 0$, or else $t_{ij}^\alpha = t_{ij}$, and t_{ij} is in the total-influence matrix $\boldsymbol{T}$. The α-cut total-influence matrix $\boldsymbol{T}_\alpha$ needs to be normalized by dividing by the following formula.

$$d_i = \sum_{j=1}^{n} t_{ij}^\alpha. \tag{13.9}$$

Therefore, we could normalize the α-cut total-influence matrix and represent it as $\boldsymbol{T}_s$.

$$\boldsymbol{T}_s = \begin{bmatrix} t_{11}^\alpha / d_1 & \cdots & t_{1j}^\alpha / d_1 & \cdots & t_{1n}^\alpha / d_1 \\ \vdots & & \vdots & & \vdots \\ t_{i1}^\alpha / d_2 & \cdots & t_{ij}^\alpha / d_2 & \cdots & t_{in}^\alpha / d_2 \\ \vdots & & \vdots & & \vdots \\ t_{n1}^\alpha / d_3 & \cdots & t_{nj}^\alpha / d_3 & \cdots & t_{nn}^\alpha / d_3 \end{bmatrix} = \begin{bmatrix} t_{11}^s & \cdots & t_{1j}^s & \cdots & t_{1n}^s \\ \vdots & & \vdots & & \vdots \\ t_{i1}^s & \cdots & t_{ij}^s & \cdots & t_{in}^s \\ \vdots & & \vdots & & \vdots \\ t_{n1}^s & \cdots & t_{nj}^s & \cdots & t_{nn}^s \end{bmatrix}. \tag{13.10}$$

This study adopts the normalized α-cut total-influence matrix $\boldsymbol{T}_s$ (hereafter abbreviated to "the normalized matrix") and the unweighted supermatrix $\boldsymbol{W}$ using Equation 13.11 to calculate the weighted supermatrix $\boldsymbol{W}_w$.

$$\boldsymbol{W}_w = \begin{bmatrix} t_{11}^s \times W_{11} & t_{21}^s \times W_{12} & \cdots & \cdots & t_{n1}^s \times W_{1n} \\ t_{12}^s \times W_{21} & t_{22}^s \times W_{22} & \vdots & & \vdots \\ \vdots & \cdots & t_{ji}^s \times W_{ij} & \cdots & t_{ni}^s \times W_{in} \\ \vdots & & \vdots & & \vdots \\ t_{1n}^s \times W_{n1} & t_{2n}^s W_{n2} & \cdots & \cdots & t_{nn}^s \times W_{nn} \end{bmatrix}. \tag{13.11}$$

Step 7: *Limit the weighted supermatrix by raising it to a sufficiently large power k, as Equation 13.12, until the supermatrix has converged and become a long-term stable supermatrix to get the global priority vectors or called weights.*

$$\lim_{k\to\infty} \boldsymbol{W}_w^{k}. \tag{13.12}$$

If the limiting supermatrix is not the only one, such as if there are N supermatrices, the average of the values is obtained by adding the N supermatrices and dividing by N.

In order to illustrate clearly the procedures of the ANP and DEMATEL methods, this study proposes a case (Case 1). We assume Case 1 has three factors: Cluster 1, Cluster 2, and Cluster 3 (here, "factor" could be "element," "cluster," or "criterion"; however, in order to illustrate the following steps in the ANP procedures, we replace "factors" with "clusters"). First, we operate from Step 1 to Step 4 above to derive the total-influence matrix $\boldsymbol{T}$; then we set a threshold value, α, to filter the minor effects in the criteria of matrix $\boldsymbol{T}$, as in Equation 13.13. If the circled parts are higher than the value of α in the following equation, then their NRM can be shown, as in Figure 13.2.

$$\boldsymbol{T} = \begin{array}{c} \\ C_1 \\ C_2 \\ C_3 \end{array} \begin{array}{c} \begin{array}{ccc} C_1 & C_2 & C_3 \end{array} \\ \begin{bmatrix} \textcircled{t_{11}} & \textcircled{t_{12}} & t_{13} \\ t_{21} & \textcircled{t_{22}} & \textcircled{t_{23}} \\ \textcircled{t_{31}} & \textcircled{t_{32}} & \textcircled{t_{33}} \end{bmatrix} \end{array}. \tag{13.13}$$

We use the structure in Figure 13.2 to demonstrate Step 5 to Step 7. First, the unweighted supermatrix is described by the following equation.

$$\boldsymbol{W} = \begin{array}{c} \\ C_1 \\ C_2 \\ C_3 \end{array} \begin{array}{c} \begin{array}{ccc} C_1 & C_2 & C_3 \end{array} \\ \begin{bmatrix} W_{11} & 0 & W_{13} \\ W_{21} & W_{22} & W_{23} \\ 0 & W_{32} & W_{33} \end{bmatrix} \end{array}. \tag{13.14}$$

Then the α-cut total-influence matrix $\boldsymbol{T}_\alpha$, as in Equation 13.8, is

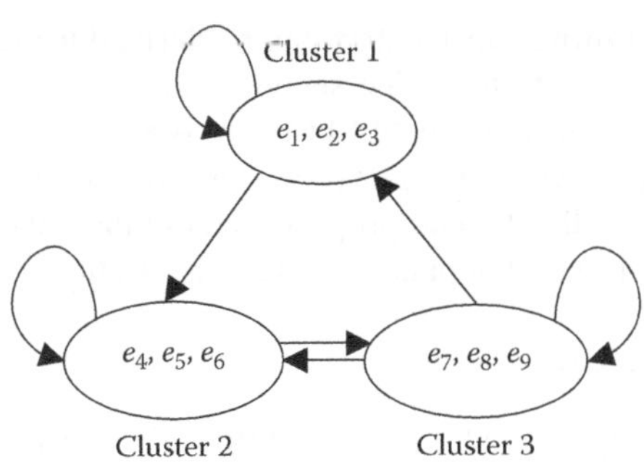

FIGURE 13.2 The structure of Case 1.

$$T_\alpha = \begin{matrix} & C_1 & C_2 & C_3 & \\ C_1 & t_{11} & t_{12} & 0 & \to d_1 \\ C_2 & 0 & t_{22} & t_{23} & \to d_2 \\ C_3 & t_{31} & t_{32} & t_{33} & \to d_3 \end{matrix}. \tag{13.15}$$

Then $d_i = \sum_{j=1}^{3} t_{ij}$ is used to divide its columns, as in the following matrix (the normalized matrix $\boldsymbol{T}_s$)

$$\boldsymbol{T}_s = \begin{matrix} & C_1 & C_2 & C_3 \\ C_1 & t_{11}/d_1 & t_{12}/d_1 & 0 \\ C_2 & 0 & t_{22}/d_2 & t_{23}/d_2 \\ C_3 & t_{31}/d_3 & t_{32}/d_3 & t_{33}/d_2 \end{matrix} = \begin{matrix} & C_1 & C_2 & C_3 \\ C_1 & t_{11}^s & t_{12}^s & 0 \\ C_2 & 0 & t_{22}^s & t_{23}^s \\ C_3 & t_{31}^s & t_{32}^s & t_{33}^s \end{matrix}$$

Next, we adopt the normalized matrix $\boldsymbol{T}_s$ and the unweighted supermatrix $\boldsymbol{W}$ and use Equation 13.11 to calculate the weighted supermatrix $\boldsymbol{W}_w$, as in Equation 13.16.

$$\boldsymbol{W}_w = \begin{matrix} & C_1 & C_2 & C_3 \\ C_1 & t_{11}^s W_{11} & 0 & t_{31}^s W_{13} \\ C_2 & t_{12}^s W_{21} & t_{22}^s W_{22} & t_{32}^s W_{23} \\ C_3 & 0 & t_{23}^s W_{32} & t_{33}^s W_{33} \end{matrix} \tag{13.16}$$

Finally, the weighted supermatrix is limited until it has converged and become a long-term stable supermatrix, as in Equation 13.12. In addition, if the limiting supermatrix is not the only one, for example, if $N = 3$ and $\lim_{k\to\infty} \boldsymbol{W}_w^k = \{\boldsymbol{W}^1, \boldsymbol{W}^2, \boldsymbol{W}^3\}$, the final weighted limiting supermatrix is presented as the following matrix:

$$\boldsymbol{W}_f = \frac{1}{3}\boldsymbol{W}^1 + \frac{1}{3}\boldsymbol{W}^2 + \frac{1}{3}\boldsymbol{W}^3. \tag{13.17}$$

In short, a stable limiting supermatrix can be derived using the above steps. The overall weights are also obtained. The second aim of this chapter is to propose a feasible model that combines the DEMATEL and ANP to achieve the normalization of the unweighted supermatrix in ANP procedures and to deal with the problem of interdependence and feedback. The proposed model described above is more suitable and rational in real-world applications than the traditional method.

13.2.3 VIKOR Method

The compromise ranking method (called VIKOR method) was proposed by Opricovic and Tzeng as one applicable technique to implement within MCDM (Tzeng et al. 2002, 2005; Opricovic and Tzeng 2002, 2004, 2007). We assume the alternatives are denoted as $A_1, A_2, \ldots, A_i, \ldots, A_m$. The rating (performance score) of the jth criterion is denoted by f_{ij} for alternative A_i, w_j is the weight of the jth criterion, expressing the relative importance of the criteria, where $j = 1,2,\ldots,n$, and n is the number of criteria. The VIKOR method began with the following form of L_p-*metric*:

$$L_i^p = \left\{ \sum_{j=1}^{n} \left[w_j \left(\left| f_j^* - f_{ij} \right| \right) / \left(\left| f_j^* - f_j^- \right| \right) \right]^p \right\}^{1/p}, \tag{13.18}$$

where $1 \le p \le \infty$; $i = 1,2,\ldots,m$; and weight w_j by ANP according to NRM based on DEMATEL method. The VIKOR method also uses $L_i^{p=1}$ (as S_i) and $L_i^{p=\infty}$ (as Q_i) to formulate the ranking measure (Tzeng et al. 2002, 2005; Opricovic and Tzeng 2002, 2004, 2007).

$$S_i = L_i^{p=1} = \sum_{j=1}^{n} \left[w_j \left(\left| f_j^* - f_{ij} \right| \right) / \left(\left| f_j^* - f_j^- \right| \right) \right], \tag{13.19}$$

$$Q_i = L_i^{p=\infty} = \max_j \left\{ w_j \left(\left| f_j^* - f_{ij} \right| \right) / \left(\left| f_j^* - f_j^- \right| \right) \middle| j = 1,2,\ldots,n \right\}. \tag{13.20}$$

The compromise solution $\min_i L_i^p$ will be chosen because its value is closest to the ideal/aspired level. In addition, when p is small, the group utility is emphasized (such as $p = 1$) when p increases, the individual regrets/gaps receive more weight (Freimer and Yu 1976). Therefore, $\min_i S_i$ emphasizes the maximum group utility, whereas $\min_i Q_i$ emphasizes selecting the minimum of the maximum individual regrets. Based on the above concepts, the compromise ranking algorithm VIKOR has the following steps.

Step 8: *Determine the best* f_j^* *and the worst* f_j^- *values of all criterion functions,* $j = 1,2,\ldots,n$.

Assuming the jth function represents a benefit, $f_j^* = \max_i f_{ij}$ (or setting an aspired level) and $f_j^- = \min_i f_{ij}$ (or setting a tolerable level). Alternatively, assuming the jth function represents a cost, $f_j^* = \min_i f_{ij}$ (or setting an aspired level) and $f_j^- = \min_i f_{ij}$ (or setting a tolerable level). Moreover, an original rating matrix is transformed into a normalized weight-rating matrix with the following formula:

$$r_{ij} = \left(\left| f_j^* - f_{ij} \right| \right) / \left(\left| f_j^* - f_j^- \right| \right). \tag{13.21}$$

Step 9: *Compute the values* S_i *and* Q_i, $i = 1,2,\ldots,m$, using the relations

$$S_i = \sum_{j=1}^{n} w_j r_{ij}, \tag{13.22}$$

$$Q_i = \max_j \left\{ w_j r_{ij} \mid j = 1,2,\ldots,n \right\}. \quad (13.23)$$

Step 10: *Compute the index values R_i, $i = 1,2,\ldots,m$*, using the relation

$$R_i = v\left(S_i - S^*\right)/\left(S^- - S^*\right) + (1-v)\left(Q_i - Q^*\right)/\left(Q^- - Q^*\right), \quad (13.24)$$

where $S^* = \max_i S_i$, $S^- = \max_i S_i$, $Q^* = \max_i Q_i$, $Q^- = \max_i Q_i$ (here, we can also set the best value to 0 and the worst value to 1) and $0 \le v \le 1$, where v is introduced as a weight for the strategy of maximum group utility, whereas $1 - v$ is the weight of the individual regret. In other words, when $v > 0.5$, this represents a decision-making process that could use the strategy of maximum group utility (i.e., if v is big, group utility is emphasized), or by consensus when $v \approx 0.5$, or with veto when $v < 0.5$. We also can rewrite Equation 13.24 as $R_i = vS_i + (1 - v_i)\, Q_i$, when $S^* = 0$, $S^- = 1$, $Q^* = 0$, and $Q^- = 1$.

Step 11: *Rank the alternatives, sorting by the value of $\{S_i, Q_i,$ and $R_i \mid i = 1,2,\ldots,m\}$, in decreasing order.* Propose as a compromise the alternative ($A^{(1)}$), which is ranked first by the measure $\min\{R_i \mid i = 1,2,\ldots,m\}$ if the following two conditions are satisfied.

C1. *Acceptable advantage*: $R(A^{(2)}) - R(A^{(1)}) \ge 1/(m-1)$, where $A^{(2)}$ is the alternative with second position in the ranking list by R; m is the number of alternatives.

C2. *Acceptable stability in decision making*: Alternative $A^{(1)}$ must also be the best ranked by S_i or/and $\{Q_i \mid i = 1,2,\ldots,m\}$.

If one of the conditions is not satisfied, then a set of compromise solutions is proposed, which consists of:

1. Alternatives $A^{(1)}$ and $A^{(2)}$ if only condition **C2** is not satisfied.
2. Alternatives $A^{(1)}, A^{(2)},\ldots,A^{(M)}$ if condition **C1** is not satisfied; $A^{(M)}$ is determined by the relation $R(A^{(M)}) - R(A^{(1)}) < 1/(m-1)$ for maximum M (the positions of these alternatives are close).

The compromise solution is determined by the compromise-ranking method; the obtained compromise solution could be accepted by the decision makers because it provides maximum group utility of the majority (represented by $\min S$, Equation 13.22), and minimum individual regret of the opponent (represented by $\min Q$, Equation 13.23). The model uses the DEMATEL and ANP procedures in Sections 13.2.1 and 13.2.2 to obtain the weights of criteria with dependence and feedback. Using the VIKOR method, the compromise solution is then obtained.

13.3 NUMERICAL EXAMPLE WITH APPLICATIONS

In this section, we provide a numerical example with applications to demonstrate the proposed method. We construct the network structure using the DEMATEL procedures, i.e., from Step 1 to Step 4. Next, we calculate the limited supermatrix using

TABLE 13.1
Total-Influence Matrix *T*

	Cluster 1	Cluster 2	Cluster 3
Cluster 1	1	0.7	0
Cluster 2	0	1	0.8
Cluster 3	3.6	1.9	1

Step 5 to Step 7 to obtain the weights of the features in the network structure of the ANP. Finally, we use the weights from the ANP and the VIKOR method (Step 8 to Step 11) to obtain the ranking index.

We assume a simple example (Case 2) for Step 1 to Step 3 of DEMATEL to obtain the total-influence matrix ***T***, as in Table 13.1. Using Step 4, if a threshold value of 0.1 is chosen, then the NRM of the relations is as shown above (Figure 13.2). We assume Cluster 1 has three elements/criteria, e_1, e_2, e_3, Cluster 2 has e_4, e_5, e_6, and Cluster 3 has e_7, e_8, e_9.

Then, we normalize the total-influence matrix ***T***, as in Table 13.2.

Using the structure of Figure 13.2, we can then obtain the unweighted supermatrix as follows.

		e_1	e_2	e_3	e_4	e_5	e_6	e_7	e_8	e_9	
	e_1	1	0	0	0	0	0	0.235	0.187	0.105	
	e_2	0	1	0	0	0	0	0.178	0.257	0.637	→ W_{13}
W_{11} ←	e_3	0	0	1	0	0	0	0.587	0.556	0.258	
	e_4	0.637	0.582	0.105	1	0	0	0.621	0.474	0.235	
W_{21} ←	e_5	0.105	0.109	0.637	0	1	0	0.127	0.203	0.178	→ W_{23}
	e_6	0.258	0.309	0.258	0	0	1	0.252	0.323	0.587	
	e_7	0	0	0	0.637	0.582	0.105	1	0	0	
W_{31} ←	e_8	0	0	0	0.105	0.109	0.637	0	1	0	→ W_{33}
	e_9	0	0	0	0.258	0.309	0.258	0	0	1	

The weighted supermatrix is now obtained by Equation 13.11, as below.

TABLE 13.2
Normalized Matrix T_s of Case 2

	Cluster 1	Cluster 2	Cluster 3
Cluster 1	0.588	0.412	0.000
Cluster 2	0.000	0.556	0.444
Cluster 3	0.554	0.292	0.154

$$
\begin{array}{c|ccccccccc}
 & e_1 & e_2 & e_3 & e_4 & e_5 & e_6 & e_7 & e_8 & e_9 \\
\hline
e_1 & 0.59 & 0.00 & 0.00 & 0.00 & 0.00 & 0.00 & 0.13 & 0.10 & 0.06 \\
e_2 & 0.00 & 0.59 & 0.00 & 0.00 & 0.00 & 0.00 & 0.10 & 0.14 & 0.34 \\
e_3 & 0.00 & 0.00 & 0.59 & 0.00 & 0.00 & 0.00 & 0.32 & 0.30 & 0.14 \\
e_4 & 0.26 & 0.24 & 0.04 & 0.56 & 0.00 & 0.00 & 0.19 & 0.15 & 0.07 \\
e_5 & 0.04 & 0.04 & 0.26 & 0.00 & 0.56 & 0.00 & 0.04 & 0.06 & 0.06 \\
e_6 & 0.11 & 0.13 & 0.11 & 0.00 & 0.00 & 0.56 & 0.08 & 0.10 & 0.18 \\
e_7 & 0.00 & 0.00 & 0.00 & 0.28 & 0.26 & 0.05 & 0.15 & 0.00 & 0.00 \\
e_8 & 0.00 & 0.00 & 0.00 & 0.05 & 0.05 & 0.28 & 0.00 & 0.15 & 0.00 \\
e_9 & 0.00 & 0.00 & 0.00 & 0.11 & 0.14 & 0.11 & 0.00 & 0.00 & 0.15
\end{array}
$$

Next, using Equation 13.12 to obtain the limiting supermatrix $\mathbf{W}_f$, the weights are as follows.

$$
\mathbf{W}_f = \begin{array}{c|ccccccccc}
 & e_1 & e_2 & e_3 & e_4 & e_5 & e_6 & e_7 & e_8 & e_9 \\
\hline
e_1 & 0.059 & 0.059 & 0.059 & 0.059 & 0.059 & 0.059 & 0.059 & 0.059 & 0.059 \\
e_2 & 0.103 & 0.103 & 0.103 & 0.103 & 0.103 & 0.103 & 0.103 & 0.103 & 0.103 \\
e_3 & 0.155 & 0.155 & 0.155 & 0.155 & 0.155 & 0.155 & 0.155 & 0.155 & 0.155 \\
e_4 & 0.179 & 0.179 & 0.179 & 0.179 & 0.179 & 0.179 & 0.179 & 0.179 & 0.179 \\
e_5 & 0.132 & 0.132 & 0.132 & 0.132 & 0.132 & 0.132 & 0.132 & 0.132 & 0.132 \\
e_6 & 0.137 & 0.137 & 0.137 & 0.137 & 0.137 & 0.137 & 0.137 & 0.137 & 0.137 \\
e_7 & 0.108 & 0.108 & 0.108 & 0.108 & 0.108 & 0.108 & 0.108 & 0.108 & 0.108 \\
e_8 & 0.063 & 0.063 & 0.063 & 0.063 & 0.063 & 0.063 & 0.063 & 0.063 & 0.063 \\
e_9 & 0.064 & 0.064 & 0.064 & 0.064 & 0.064 & 0.064 & 0.064 & 0.064 & 0.064
\end{array}. \tag{13.25}
$$

Finally, we assume the range of rating for each criterion is 1 (the worst value) to 10 (the best value); the rating of alternatives and the weights of criteria (from Equation 13.25) are listed in Table 13.3. Then we use the VIKOR method (Section 13.2.3) to obtain the ranking index R_i of alternatives, as in Table 13.4. (Here, we assume the concern is maximum group utility and minimum individual regret simultaneously, so we should select $v = 0.5$. We also set $S^* = Q^* = 0$ and $S^- = Q^- = 1$.)

The results show $S_{A_1} > S_{A_5} > S_{A_4} > S_{A_2} > S_{A_3}$, $Q_{A_1} > Q_{A_5} > Q_{A_4} > Q_{A_2} > Q_{A_3}$, and $R_{A_1} > R_{A_5} > R_{A_4} > R_{A_2} > R_{A_3}$, and the ranks of alternatives are $A_3 \succ A_2 \succ A_4 \succ A_5 \succ A_1$. Here, these alternatives satisfy condition **C2**, which represents acceptable stability. However, alternatives A_3 and A_2, A_2 and A_4, A_3 and A_4 do not satisfy condition **C1**, which represents

TABLE 13.3
Performance and Weight for Each Criterion (Case 2)

		Alternatives				
Criteria	**Weight**	A_1	A_2	A_3	A_4	A_5
e_1	0.059	4	8	7	9	5
e_2	0.103	3	7	8	8	3
e_3	0.155	3	6	8	5	3
e_4	0.179	2	8	7	9	3
e_5	0.132	3	7	8	7	2
e_6	0.137	3	7	9	7	3
e_7	0.108	4	7	8	8	4
e_8	0.063	5	8	6	7	4
e_9	0.064	5	8	7	6	5

that A_3 is not more obviously advantageous than A_2, A_2 is not more obviously advantageous than A_4, and A_3 is not more obviously advantageous than A_4. Only A_4 and A_5 satisfy condition **C1**. Therefore, a set of compromise solutions is $\{A_3, A_2, A_4\}$.

13.4 DISCUSSIONS AND COMPARISONS

In order to compare the traditional methods and this research, we also calculate the limiting supermatrix using the traditional normalization method. The result is as follows.

$$
W_f = \begin{array}{c} \\ e_1 \\ e_2 \\ e_3 \\ e_4 \\ e_5 \\ e_6 \\ e_7 \\ e_8 \\ e_9 \end{array}
\begin{array}{c} \begin{array}{ccccccccc} e_1 & e_2 & e_3 & e_4 & e_5 & e_6 & e_7 & e_8 & e_9 \end{array} \\
\begin{bmatrix}
0.041 & 0.041 & 0.041 & 0.041 & 0.041 & 0.041 & 0.041 & 0.041 & 0.041 \\
0.072 & 0.072 & 0.072 & 0.072 & 0.072 & 0.072 & 0.072 & 0.072 & 0.072 \\
0.109 & 0.109 & 0.109 & 0.109 & 0.109 & 0.109 & 0.109 & 0.109 & 0.109 \\
0.185 & 0.185 & 0.185 & 0.185 & 0.185 & 0.185 & 0.185 & 0.185 & 0.185 \\
0.117 & 0.117 & 0.117 & 0.117 & 0.117 & 0.117 & 0.117 & 0.117 & 0.117 \\
0.142 & 0.142 & 0.142 & 0.142 & 0.142 & 0.142 & 0.142 & 0.142 & 0.142 \\
0.151 & 0.151 & 0.151 & 0.151 & 0.151 & 0.151 & 0.151 & 0.151 & 0.151 \\
0.092 & 0.092 & 0.092 & 0.092 & 0.092 & 0.092 & 0.092 & 0.092 & 0.092 \\
0.090 & 0.090 & 0.090 & 0.090 & 0.090 & 0.090 & 0.090 & 0.090 & 0.090
\end{bmatrix} \end{array}. \quad (13.26)
$$

According to Equations 13.25 and 13.26, we find the ranks of weights for the two matrices are different. In Equation 13.25, using the DEMATEL method to normalize the unweighted supermatrix (our proposed method), the ranks of weights (the limiting supermatrix) are $e_4 > e_3 > e_6 > e_5 > e_7 > e_2 > e_9 > e_8 > e_1$. On the

TABLE 13.4
Ranking Index for Case 2

The Ranking Index	A_1	A_2	A_3	A_4	A_5
S_{A_i}	0.751 (5)	0.298 (2)	0.255 (1)	0.299 (3)	0.746 (4)
Q_{A_i}	0.159 (5)	0.069 (2)	0.060 (1)	0.086 (3)	0.139 (4)
R_{A_i} ($v = 0.5$)	0.455 (5)	0.183 (2)	0.157 (1)	0.192 (3)	0.443 (4)

other hand, in Equation 13.26, using the traditional normalized method, the ranks of weights are $e_4 > e_7 > e_6 > e_5 > e_3 > e_8 > e_9 > e_2 > e_1$. This study further analyzes the obtained weights between the two different methods and shows them in Table 13.5 and Figure 13.3, respectively.

Table 13.1 reveals several facts: (a) each cluster has feedback and dependence; (b) the effect of Cluster 3 on Cluster 1 is 3.6, the effect of Cluster 1 on Cluster 2 is 0.7, the effect of Cluster 3 on Cluster 2 is 1.9, and the effect of Cluster 2 on Cluster 3 is 0.8. In other words, the value for the degree to which Cluster 1 is affected is high (3.6), for Cluster 2 this value is 0.7 and 1.9, respectively, and for Cluster 3 it is low (0.8). Therefore, Cluster 1 would then be paid more attention than the other clusters in the real world, i.e., it should carry much weight, whereas Cluster 3 should have reduced weight. We thus find that the weights of criteria e_7, e_8, and e_9 in the traditional method are higher than in the proposed method, but the weights of criteria e_1, e_2, and e_3 in the traditional method are lower than in the proposed method (Table 13.5 and Figure 13.3). If we use the assumption of equal weight for each cluster to normalize the unweighted supermatrix to gain the weighted supermatrix, the results of

TABLE 13.5
Comparisons of Weights of Each Criterion between the Traditional Hybrid Method and Our Proposed Method

Criteria	Traditional Hybrid Method	Proposed Method	Difference
e_1	0.041	0.059	(0.018)[a]
e_2	0.072	0.103	(0.031)[a]
e_3	0.109	0.155	(0.046)[a]
e_4	0.185	0.179	0.006
e_5	0.117	0.132	(0.015)[a]
e_6	0.142	0.137	0.005
e_7	0.151	0.108	0.043
e_8	0.092	0.063	0.029
e_9	0.090	0.064	0.026

[a] Parentheses represent negative values.

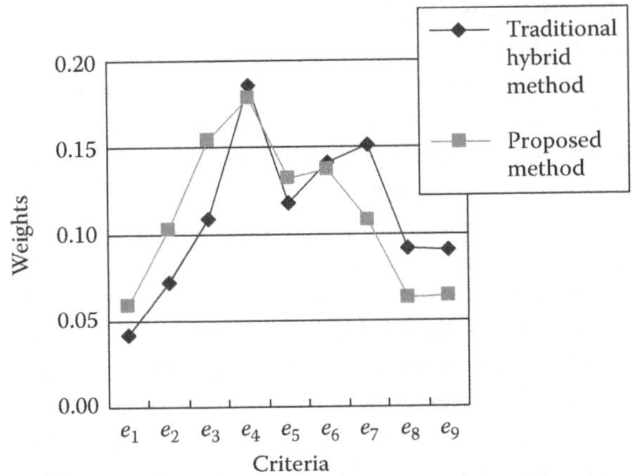

FIGURE 13.3 Comparisons of weights of each criterion between the traditional hybrid method and our proposed method.

the assessed weights would be higher or lower than the real situation. Figure 13.3 shows the criteria of Cluster 1 (e_1, e_2, e_3) are underestimated, whereas the criteria of Cluster 3 (e_7, e_8, e_9) are overestimated if we adopt the traditional method. Therefore, we use the DEMATEL method combined with the ANP to obtain better and more accurate results in real-world applications.

This study finally uses the VIKOR method to aggregate the criteria that have dependent and feedback characteristics to obtain the ranking index, as in Table 13.4. If we are concerned with maximum group utility and minimum individual regret simultaneously ($v = 0.5$), then the results are $A_3 \succ A_2 \succ A_4 \succ A_5 \succ A_1$, and a set of compromise solutions is $\{A_3, A_2, A_4\}$. Thus, A_3 is the closest to the ideal/aspired level among these alternatives, whereas A_1 is the farthest from the ideal/aspired level. If we consider improving these alternatives, then A_1 should be given priority. In addition, $\{A_3, A_2, A_4\}$ is a set of compromise solutions; the decision maker can select one from among these solutions according to his/her preference.

To sum up, the hybrid model combining the ANP and DEMATEL has been widely used in MCDM. The DEMATEL method is used to construct interrelations between criteria/factors and the ANP can overcome the problems of dependence and feedback. This study shows that using DEMTATEL to normalize the unweighted supermatrix in the ANP procedure is more reasonable than using the assumption of equal weight in each cluster. In addition, the weights obtained from the ANP and VIKOR methods are used to derive the ranking index. We also demonstrated two examples to illustrate this proposed method, and the results show this method is both suitable and effective.

13.5 CONCLUSIONS

Many papers have proposed analytical models to resolve the questions in conflict-management situations. Among the numerous approaches available for conflict management, one of the most prevalent is MCDM. The VIKOR method is one applicable technique to implement within MCDM; it is based on an aggregating function representing closeness to the ideal, which originated in the compromise programming method. However, most decision-making methods assume independence between the criteria of a decision and the alternatives of that decision, or simply among the criteria or among the alternatives themselves. Assuming independence among criteria/variables is, however, too strict to overcome the problem of dependent criteria in the real world. Therefore, many papers have suggested the ANP to overcome this problem. The ANP is used to deal with problems involving dependence and feedback. Besides, a hybrid model combining the ANP and DEMATEL methods has been widely and successfully used in various fields. The DEMATEL method is used to detect complex relationships and build the NRM of relations among criteria. The method adopted to overcome normalization for the weighted supermatrix in the ANP procedure assumes equal weight for each cluster; however, it ignores the different effects among clusters. This research proposes a new concept to overcome this irrational situation. We adopt the DEMATEL method to transform the unweighted supermatrix to a weighted supermatrix. The novel combined model is more suitable than the traditional method to solve problems with different degrees of effects among clusters. We also provide the ANP and VIKOR method to obtain the compromise ranking index. In addition, we demonstrate two cases to illustrate the effectiveness and feasibility of the proposed method to suit real-world applications. Consequently, using the method proposed in this research is an appropriate approach to overcome the compromise solution method and the problem of interdependence and feedback among criteria.

14 TOPSIS and VIKOR: An Application

In the last two decades of the twentieth century, there was growing concern about pollution in major cities, and in particular about the large contribution made by road transportation sources to this problem (McNicol, Rand, and Williams 2001). Government legislation on internal combustion engine (ICE) emissions and fuel quality substantially improved the air quality in cities through a reduction of regulated pollutants. For example, in the United States, California introduced the so-called zero-emission vehicles (ZEV) mandate, which called for 2% of all new vehicles offered for sale in California in model years 1998–2000 to be ZEVs. Initially, it was intended that such vehicles would be battery-powered electric vehicles (EVs). Owing to the limitation of EV development, the regulations were relaxed to allow additional time for the technology to develop. During the development period, alternative-fuel vehicles were also considered. The advantages of hybrid electric vehicles (HEVs) are regeneration of braking energy, engine shutdown instead of idling, and engine driving under high-load conditions; these advantages are more noticeable in city driving. The key weakness of EVs, on the other hand, is that time is needed to recharge the batteries (Morita 2003). Bus systems possess features such as stable depots, routes, groups of commuters, times of operation and frequencies, so that research on finding alternative-fuel modes for public transportation is of high interest. Therefore, the purpose of this research is to evaluate the best alternative-fuel buses suitable for the urban area and to explore the potential direction of development in the future.

The trends of the latest worldwide technological developments of a bus with new alternative fuel are considered in this chapter. Morita (2003) thought that the leading types of automobiles in the twenty-first century would probably be the following four: internal combustion engine vehicles (ICEVs), HEVs, EVs, and fuel cell vehicles (FCVs). McNicol et al. (2001) pointed out that the principal competitors of FCVs are EVs, HEVs, and advanced conventional ICEVs. Based on the literature mentioned above, several types of fuel are considered as alternative-fuel modes, i.e., EVs, HEVs, fuel cell (hydrogen), methanol, and natural gas (Morita 2003; McNicol et al. 2001; Sperling 1995).

Current research on alternative-fuel vehicles is well grounded. The scope of research includes the direction of development (Morita 2003; Harding 1999), comparison of alternative-fuel vehicles (Maggetto and Van Mierlo 2001; Johnsson and Ahman 2002), impact evaluation (Kazimi 1997; Matheny et al. 2002; Brodrick et al. 2002; Zhou and Sperling 2001; Kempton and Kubo 2000), batteries (Moseley 1999), policy (Nesbitt and Sperling 1998), costs (DeLucchi, Murphy, and McCubbin 2002; DeLucchi and Lipman 2001), markets (Sperling, Setiawan, and Hungerford 1995), etc. Most of the research focuses on comparing and describing the performance of single or several types of alternative-fuel vehicles.

In addition, some research is related to the evaluation of alternative fuel. Poh and Ang (1999) applied forward and backward analytic hierarchy processes (AHP) to analyze transportation fuels and policy for Singapore. Winebrake and Creswick (2003) also applied the AHP to evaluate the future of hydrogen fueling systems for transportation. Both of these teams utilize scenario analysis to build their evaluation model.

A similar approach is applied in this current research. In this chapter, alternative fuels are considered for their potential to displace oil as the main and only source of transport fuel. The characteristics of buses make them suitable for using such fuels in populated modern cities. Therefore, evaluating a moderate fuel mode for buses in an urban area is the purpose of this research.

The evaluation of alternative-fuel buses should be considered from various perspectives, for example, energy efficiency, emissions, technologies, costs, facilities, and so on. The multiattribute evaluation process is thus used in this chapter. The AHP is used to determine the weights of evaluation criteria. The AHP, introduced by Saaty (1980), is known as a pairwise comparison method and a popular method in evaluation problems. There is a lot of research on the application of decision analysis techniques to transportation, energy, and environmental planning, such as the research of Tzeng et al. (1992, 1994), and Tzeng and Tsaur (1993, 1997). Technique for Order Preference by Similarity to Ideal Solution (TOPSIS) and VIKOR are compared and used to rank the alternative-fuel buses. The details of these two methods are shown in Section 14.4. A multiattribute evaluation of alternative buses was performed by experts from different decision-making groups, such as bus users, the social community, and the operators. The best fuel mode has to be selected according to several competing (conflicting) criteria. This decision-making problem has no solution satisfying all criteria simultaneously. The compromise solution of the problem of conflicting criteria should be determined and the criteria could help the decision makers to reach their final decision. The compromise ranking method is applied to determine the best compromise alternative-fuel bus. For testing and verifying the usability of this methodology, we illustrate the evaluation of alternative-fuel buses of Taiwan urban areas as an empirical example. The results can prove the effectiveness of this method, and illustrate directions for future development and the weakness of the best alternative, which make it easy to implement in the future.

14.1 ALTERNATIVE SOLUTIONS

The main parameter in defining alternative solutions is the fuel mode. According to the data collected in this study, the alternatives are classified into four groups: the conventional diesel engine, new mode of alternative fuel, EV, and HEV. Worldwide, much effort is being put into developing a transportation means utilizing new alternative fuels, including methanol, fuel cell (hydrogen), and compressed natural gas (CNG). Vehicles operating on electricity are of high interest, but the appropriate technology is still being developed. The advantages of EVs are that they perform efficiently under low-load conditions and do not discharge any pollutants during use (Morita 2003). Their key weakness is that time is needed to recharge the batteries. In addition, disadvantages such as a short cruising distance (usually less than 200 km) and lack of support infrastructure significantly reduce their convenience (Morita 2003). The HEV, which retains both the electric motor and ICE, has been widely accepted by users (Griffith and Gleason 1996; Harding 1999; McNicol 2001; Maggetto and Van Mierlo 2001). Morita (2003) pointed out that HEVs have the potential to rank alongside conventional vehicles in terms of cost and convenience. The advantages of HEVs are regeneration of braking energy, engine shutdown instead of idling, and engine driving under high-load conditions; these advantages are more noticeable in city driving. The advantages of HEVs are that they can incorporate any type of ICE, or fuel cells, and show good efficiency, no matter what type of fuel the engine uses. In this chapter the following alternatives are considered: gasoline-electric, diesel-electric, CNG-electric, and liquid propane gas (LPG)-electric. Based on global development results, 12 alternatives of fuel mode are considered, and the features of each alternative-fuel mode are described in this chapter.

1. *Conventional diesel engine:* The conventional diesel engine bus is employed by Taiwanese transportation companies. In fact, the diesel engine is the most efficient of all existing ICEs, making it one of the major contenders as a power source in the twenty-first century (Morita 2003). It is introduced in the set of alternatives in order to compare it with the new fuel modes.
2. *Compressed natural gas (CNG):* Natural gas is used in several forms as vehicle fuel, i.e., CNG, liquid natural gas, and attached natural gas. The CNG vehicle has already been commercialized around the world and is mature in its technology (there are about four million CNG vehicles in the world). The CNG vehicle is widespread in countries with their own natural gas. CNG vehicles emit only small amounts of carbon dioxide and have high-octane value; thus they are suitable for utilization as public transportation vehicles (Sperling 1995). The natural gas supply, distribution, and safety are the most urgent issues needing improvement.
3. *Liquid propane gas (LPG):* There are countries that have used this mode of fuel for public transportation. In Japan, Italy, and Canada as much as 7% of the buses are powered by LPG (Sperling 1995), and some European countries are planning to employ LPG vehicles, due to pollution considerations.
4. *Fuel cell (hydrogen):* The so-called fuel cell battery can transform hydrogen and oxygen into power for vehicles (Sperling 1995); however, hydrogen is not suitable for onboard storage (Morita 2003). Research on a fuel cell-hydrogen bus has already been concluded with success and test results with an experimental vehicle operating on hydrogen fuel indicate that this vehicle has a broad surface in the burning chamber, low burning temperature, and the fuel is easily inflammable (DeLucchi 1989). Daimler-Benz Company has already developed a prototype vehicle with a fuel cell. To date, the only vehicle offered for sale with fuel cell technology is the Zevco

London taxi, which was launched in July 1998 (Harding 1999). Due to the fact that the energy to operate this vehicle comes from the chemical reaction between hydrogen and oxygen, no detrimental substance is produced and only pure water, in the form of air, is emitted. A fully loaded fuel tank can last as far as 250 km.

5. *Methanol:* Research on methanol is related to vehicles with gasoline engines. The combination rate of methanol in the fuel is 85% (so-called M85). Engines that can use this fuel with different combination rates are termed as flexible fuel vehicles (FFVs). The FFV engine can run smoothly with any combination rate of gas with methanol, and methanol will act as an alternative fuel and help to reduce the emission of black smoke and nitrous oxides. Fuel stations providing methanol have been available in Japan since 1992 (Sperling 1995). The thermal energy of methanol is lower than that of gasoline and the capability of continuous traveling by this vehicle is inferior to that of conventional vehicles. Furthermore, the aldehyde compound that comes along with burning methanol forms a strong acid. Researchers should pay more attention to this fuel mode.
6. *Electric vehicle: Opportunity charging:* The source of power for the opportunity charging electric vehicle (OCEV) is a combination of a loaded battery and fast opportunity charging during the time the bus is idle when stopped. Whenever the bus starts from the depot, its loaded battery will be fully charged. During the 10–20 seconds when the bus is stopped, the power reception sensor on the electric bus (installed under the bus) will be lowered to the charging supply plate installed in front of the bus stop to charge the battery. Within ten seconds of a stop, the battery is charged with 0.15 kWh power (depending on the design of power supply facility), and the power supplied is adequate for it to move to the next bus stop.
7. *Direct electric charging:* This type of electric bus is in the prototype design stage. The power for this vehicle comes mainly from the loaded battery. When the battery power is insufficient, the vehicle has to return to the plant for recharging. The development of a suitable battery is critical for this mode of vehicle. If a greater amount of electricity can be stored in the battery, the cruising distance for this vehicle will increase.
8. *Electric bus with exchangeable batteries:* The objective of an electric bus with an exchangeable battery is to effect a fast battery charge and to achieve longer cruising distance. The bus is modified to create more on-board battery space and the number of on-board batteries is adjusted to meet the needs of different routes. The fast exchanging facility has to be ready to conduct a rapid battery exchange so that vehicle mobility can be maintained.
9. *Hybrid electric bus with gasoline engine:* The electric-gasoline vehicle has an electric motor as its major source of power and a small gasoline engine. When electric power fails, the gasoline engine can take over and continue the trip. The kinetic energy rendered during the drive will be turned into electric power to increase the vehicle's cruising distance.

10. *Hybrid electric bus with diesel engine:* The electric-diesel vehicle has an electric motor and small diesel engine as its major sources of power. When electric power fails, the diesel engine can take over and continue the trip, while the kinetic energy rendered during the drive will be turned into electric power to increase the vehicle's cruising distance.
11. *Hybrid electric bus with CNG engine:* The electric-CNG vehicle has an electric motor and a small CNG engine as its major sources of power. When electric power fails, the CNG engine takes over and provides the power, with the kinetic energy produced converted to electric power to permit continuous travel.
12. *Hybrid electric bus with LPG engine:* The electric-LPG vehicle has an electric motor and a small LPG engine as its major sources of power. When electric power fails, the LPG engine takes over and provides the power, with the kinetic energy produced converted to electric power to permit continuous travel.

14.2 EVALUATION CRITERIA

According to the above description, we establish the evaluation criteria in Section 14.2.1 and assess the criteria weight in Section 14.2.2.

14.2.1 Establishing the Evaluation Criteria

The evaluation of alternative fuel modes can be performed according to different aspects. Four aspects of evaluation criteria are considered in this chapter: social, economic, technological, and transportation. In order to evaluate alternatives, eleven evaluation criteria are established, as follows:

1. *Energy supply:* This criterion is based on the yearly amount of energy that can be supplied, on the reliability of energy supply, the reliability of energy storage, and on the cost of energy supply.
2. *Energy efficiency:* This criterion represents the efficiency of fuel energy.
3. *Air pollution:* This criterion refers to the extent a fuel mode contributes to air pollution, since vehicles with diverse modes of fuel impact on air differently.
4. *Noise pollution:* This criterion refers to the noise produced during the operation of the vehicle.
5. *Industrial relationship:* The conventional vehicle industry is a locomotive industry, and it is intricately related to other industrial production; the relationship of each alternative to other industrial production is taken as the criterion.
6. *Costs of implementation:* This criterion refers to the costs of production and implementation of alternative vehicles.
7. *Costs of maintenance:* The maintenance costs for alternative vehicles are the criterion.

8. *Vehicle capability:* This criterion represents the cruising distance, slope climbing, and average speed.
9. *Road facility:* This criterion refers to the road features needed for the operation of alternative vehicles (like pavement and slope).
10. *Speed of traffic flow:* This criterion refers to the comparison of the average speed of alternative vehicles for certain traffic. If the speed of traffic flow is higher than the vehicle speed, the vehicle would not be suitable to operate on certain routes.
11. *Sense of comfort:* This criterion refers to the particular issue regarding sense of comfort and to the fact that users tend to pay attention to the accessories of the vehicle (air-conditioning, automatic doors, etc.)

14.2.2 Assessment of Criteria Weights

In the assessment of criteria weights, the relevant decision-making experts participating were from the electric bus manufacturing, academic, research, and bus operations sectors. They assessed the relative importance (subjectively) for each of the criteria. The average values of weights are presented in Table 14.1. These data show that the speed of traffic is the most important factor in evaluating the alternative vehicles; second in importance is air pollution, indicating the need for new alternative-fuel modes.

Good analytical procedure requires making histograms of the data, to check the form of their distribution, before proceeding with multicriteria analysis. If the data are not normally distributed, and the standard deviation is not small, the sensitivity analysis covering the range of weights should be performed within the multicriteria decision-making procedure.

TABLE 14.1
Criteria Weights

Criterion	Manufacture	Academic Institute	Research Organization	Bus Operator	Average
Energy supply	0.0357	0.0314	0.0340	0.0249	0.0313
Energy efficiency	0.1040	0.0943	0.1020	0.0748	0.0938
Air pollution	0.1355	0.2090	0.1595	0.1605	0.1661
Noise pollution	0.0452	0.0697	0.0532	0.0535	0.0554
Industrial relationship	0.0923	0.0357	0.0480	0.0757	0.0629
Employment cost	0.0900	0.0680	0.0343	0.1393	0.0829
Maintenance cost	0.0300	0.0227	0.0114	0.0464	0.0276
Vehicle capability	0.1373	0.0953	0.1827	0.0803	0.1239
Road facility	0.0827	0.0590	0.1520	0.0283	0.0805
Speed of traffic flow	0.1520	0.2420	0.1400	0.2637	0.1994
Sense of comfort	0.0957	0.0730	0.0833	0.0523	0.0761

14.3 EVALUATION OF THE ALTERNATIVES

The evaluation approach applied in this chapter is based on the assessment by the professional experts. The average assessed value for alternative j according to criterion i is determined by the following relation

$$f_{ij} = \frac{1}{N} \sum_{l=1}^{N} u_{lij},$$

where u_{lij} is the performance value given by expert l to the alternative j according to criterion i and N is the number of experts participating in the evaluation process. The "value function" u has the following properties: $0 \leq u \leq 1$, and $u_{ij} > u_{ik}$ means that the alternative j is better than the alternative k according to criterion i.

The selection of the expert group members is of extreme importance in the evaluation process of the multiple criteria analysis (MCA)/multiple criteria decision making (MCDM) problem. The selection of alternative-fuel buses is a problem related to public affairs and credible experts for evaluating this problem are very important. In Taiwan, experts from manufacturing industries, related government departments, and academic and research institutes are acknowledged as credible experts. For this reason, the experts were invited from the Transportation Bureau of Taipei City, Environmental Protection Administration, the Transportation Institute of the Ministry of Communications, Vehicle Association, Energy Committee, and research personnel on EVs. The investigation information found in previous research (emissions of black smoke, the capability of continuous traveling) was the basic reference information and it was listed in the questionnaire prepared for the experts. Within the evaluation process (Delphi method) the evaluation results were presented to the experts for the second evaluation. They had to reconsider the performance values of each alternative-fuel mode and to reevaluate the alternatives. Seventeen valid questionnaires were retrieved from the evaluation process.

The evaluation results following the second evaluation are presented in Table 14.2. According to energy supply criterion, the average performance value is highest for the diesel bus (0.820) and lowest for the hydrogen bus (0.360). With respect to energy efficiency, the average performance values are very high for EVs. The average performance values for EVs are the highest according to air pollution and to noise pollution, but the values are very low according to capability of vehicle and road facilities needed. Analyzing the data from Table 14.2, we can conclude that EVs rate very well according to the criteria of energy, environmental impact, industrial relationship, and implementation cost; the transportation mode using conventional diesel rates high according to vehicle capability and needed (new) road features; whereas the transportation modes using natural gas, methanol, and hydrogen are associated with the "middle" values.

14.4 MULTICRITERIA OPTIMIZATION

The MCDM methods VIKOR and TOPSIS are based on an aggregating function representing closeness to the reference point(s). For the details of these two methods,

TABLE 14.2
Values of the Criterion Functions

Alternatives	Energy Supply	Energy Efficiency	Air Pollution	Noise Pollution	Industrial Relations	Employment Cost	Maintenance Cost	Capability of Vehicle	Road Facility	Speed of Traffic	Sense of Comfort
Diesel bus	0.82	0.59	0.18	0.42	0.58	0.36	0.49	0.79	0.81	0.82	0.56
CNG bus	0.77	0.70	0.73	0.55	0.55	0.52	0.53	0.73	0.78	0.66	0.67
LPG bus	0.79	0.70	0.73	0.55	0.55	0.52	0.53	0.73	0.78	0.66	0.67
Hydrogen	0.36	0.63	0.86	0.58	0.51	0.59	0.74	0.56	0.63	0.53	0.70
Methanol	0.40	0.54	0.69	0.58	0.51	0.52	0.68	0.52	0.63	0.60	0.70
Charging	0.69	0.76	0.89	0.60	0.72	0.80	0.72	0.54	0.35	0.79	0.73
Electric dir.	0.77	0.79	0.89	0.59	0.73	0.80	0.72	0.47	0.44	0.87	0.75
Electric bat.	0.77	0.79	0.89	0.59	0.73	0.80	0.72	0.51	0.48	0.87	0.75
Hybrid-gas	0.77	0.63	0.63	0.52	0.66	0.63	0.65	0.67	0.70	0.80	0.74
Hybrid-diesel	0.77	0.63	0.51	0.58	0.66	0.63	0.65	0.67	0.70	0.80	0.74
Hybrid-CNG	0.77	0.73	0.80	0.48	0.63	0.66	0.65	0.67	0.71	0.62	0.78
Hybrid-LPG	0.77	0.73	0.80	0.48	0.63	0.66	0.65	0.67	0.71	0.62	0.78

please refer to Tzeng and Opricovic (2003), which are summarized in the appendix. These two methods introduce different forms of an aggregating function (L_p-metric) for ranking. The VIKOR method introduces Q_j the function of L_1 and L_∞, whereas the TOPSIS method introduces C_j^* the function of L_2. They use different kinds of normalization to eliminate the units of criterion functions: the VIKOR method uses linear normalization and the TOPSIS method uses vector normalization. The difference between these two methods is described in Section 14.5.1. We find the compromise solution of alternative-fuel buses selection by them and the results are shown in Section 14.5.2.

14.4.1 Comparison of TOPSIS and VIKOR

MCA is appropriate to solve the problems relating to several aspects. TOPSIS and VIKOR are two methods that are easy to apply among the ranking methods of MCA. However, these two methods are different in their basic definitions. Opricovic and Tzeng (2003) have discussed the differences of these two methods. In this current research, we applied these two methods to find the comprise solution of the alternative-fuel buses selection and have shown the difference of these methods. The main features of VIKOR and TOPSIS are summarized here in order to clarify the differences between these two methods.

Procedural basis. Both methods assume that there exists a performance matrix $\|f\|_{nxJ}$ obtained by the evaluation of all the alternatives in terms of each criterion. Normalization is used to eliminate the units of criterion values. An aggregating function is formulated and it is used as a ranking index. In addition to ranking, the VIKOR method proposes a compromise solution with an advantage rate.

Normalization. The difference appears in the normalization used within these two methods. The VIKOR method uses linear normalization (Opricovic and Tzeng 2003) and the normalized value does not depend on the evaluation unit of a criterion. The TOPSIS method uses vector normalization and the normalized value can be different for different evaluation units of a particular criterion. A later version of the TOPSIS method uses linear normalization (Opricovic and Tzeng 2003).

Aggregation. The main difference appears in the aggregation approaches. The VIKOR method introduces an aggregating function, representing the distance from the ideal solution. This ranking index is an aggregation of all criteria, the relative importance of the criteria, and a balance between total and individual satisfaction. The TOPSIS method introduces the ranking index (a6), including the distances from the ideal point and from the negative-ideal point. These distances in TOPSIS are simply summed in Table A14.3, without considering their relative importance. However, the reference point could be a major concern in decision making, and to be as close as possible to the ideal is the rationale of human choice. Being far away from a nadir point could be a goal only in a particular situation and the relative importance remains an open question. The TOPSIS method uses *n*-dimensional Euclidean distance that by itself could represent some balance between total and individual satisfaction, but uses it in a different way than VIKOR, where weight v is introduced in (a3).

Solution. Both methods provide a ranking list. The highest ranked alternative by VIKOR is the closest to the ideal solution. However, the highest ranked alternative by TOPSIS is the best in terms of the ranking index, which does not mean that it is always the closest to the ideal solution. In addition to ranking, the VIKOR method proposes a compromise solution with an advantage rate.

14.4.2 Compromise Solution

The compromise ranking method was applied with data given by the expert group (average evaluation values in Table 14.2 and average weights in Table 14.1). The obtained ranking list (by VIKOR) is presented in Table 14.3.

The ranking results are obtained by applying another method, named TOPSIS, which is also a modification of compromise programming. TOPSIS was developed based on the concept that the chosen alternative should have the shortest distance from the ideal solution and the farthest from the negative-ideal solution, using Euclidean distance (Hwang and Yoon 1981). The ranking results (by TOPSIS) are presented in Table 14.3.

TABLE 14.3
Multicriteria Ranking Results

Ranking by VIKOR			Ranking by TOPSIS			
			I Evaluation		II Evaluation	
Rank	Alternative	Q	Rank	Index	Rank	Index
1	Hybrid electric bus, gasoline engine	0.168	4	0.749	9	0.756
2	Electric bus, exchangeable battery	0.172	1	0.945	1	0.975
3	Electric bus, opportunity charging	0.224	2	0.933	3	0.964
4	Electric bus, direct charging	0.253	3	0.931	2	0.967
5	Hybrid electric bus, diesel engine	0.281	7	0.700	11	0.488
6	Liquid propane gas (LPG)	0.479	11	0.345	8	0.830
7	Compressed natural gas (CNG)	0.480	10	0.399	7	0.830
8	Hybrid electric bus with CNG	0.510	5	0.700	4	0.889
9	Hybrid electric bus with LPG	0.510	6	0.700	5	0.889
10	Conventional diesel engine	0.806	12	0.301	12	0.097
11	Methanol	0.852	9	0.527	10	0.698
12	Fuel cell (hydrogen)	0.925	8	0.563	6	0.865

There are four compromise solutions obtained by VIKOR, because the top four are "close." This result shows that the hybrid electric bus is the most suitable substitute bus, followed by EVs on the ranking list (Table 14.3).

Preference stability analysis was performed (by VIKOR) and the weight stability intervals for a single criterion are obtained, as follows:

$0.021 \leq w_1 \leq 0.213$ (input $w_1 = 0.031$); $0.000 \leq w_2 \leq 0.096$ (input $w_2 = 0.094$);
$0.116 \leq w_3 \leq 0.168$ (input $w_3 = 0.166$); $0.000 \leq w_4 \leq 0.063$ (input $w_4 = 0.055$);
$0.000 \leq w_5 \leq 0.175$ (input $w_5 = 0.063$); $0.000 \leq w_6 \leq 0.099$ (input $w_6 = 0.083$);
$0.000 \leq w_7 \leq 0.040$ (input $w_7 = 0.028$); $0.123 \leq w_8 \leq 0.298$ (input $w_8 = 0.124$);
$0.073 \leq w_9 \leq 0.358$ (input $w_9 = 0.081$); $0.105 \leq w_{10} \leq 0.202$ (input $w_{10} = 0.199$);
$0.000 \leq w_{11} \leq 0.188$ (input $w_{11} = 0.076$).

The weight stability intervals show that the obtained compromise solution (by VIKOR, Table 14.3) is very sensitive to changes of criteria weights.

With different weights from Table 14.1, the following sets of compromise solutions (by VIKOR) are obtained:

- EVs (three modes) are in the set of compromise solutions with the weights given by "bus operator" and by "academic institute";
- HEVs, with gasoline and diesel engine, are the compromise solution obtained with the weights given by "manufacture";
- HEV with gasoline engine, fuel mode CNG and LPG, and HEV with diesel engine are in the set of compromise solutions obtained with the weights given by "research organization."

The ranking results obtained by the TOPSIS method indicate that the EVs may be considered as the best compromise solution and the HEVs may be considered as the second best compromise solution.

14.4.3 Discussions

According to the results from Table 14.3, the conventional diesel engine is ranked very low, reflecting the need for an alternative-fuel mode. We can conclude that the hybrid electric bus is the most suitable substitute bus for the Taiwan urban areas in the short and medium terms. But, if the cruising distance of the electric bus can be extended to an acceptable range, the pure electric bus could be the best alternative.

It seems that the experts have unanimously agreed that it is necessary to develop an alternative-fuel mode for public transportation.

In comparison with conventional vehicles, alternative-fuel vehicles would contribute significantly to the improvement of air quality in urban areas. However, EVs demand recharging and remain uncompetitive with fuel-engine vehicles because of frequent recharging needs. Because the bus system has such features as permanent terminals, routes, groups of user, times of operation, and frequencies, it is expected that the implementation of an alternative-fuel bus might become a very important option for the development of public transportation.

Since the technology of EVs remains to be matured, a hybrid electric bus would be employed as the transitional mode of transportation for the improvement of environmental quality. These vehicles will be replaced when the technological characteristics of EVs, or other new technology, are improved. In terms of a short-term implementation strategy, the Environmental Protection Administration (EPA) should devote funding from air pollution taxes to the city government to the development of an electric bus. In the medium term, the city government should stimulate purchase of EVs by every bus company and replace old buses. For long-term consideration, it is necessary to establish the appropriate industrial policy to facilitate the development of the relevant domestic industry.

In the discussion with experts on the electric bus held in the graduate institute of the Bureau of Transportation of Taipei City, the government expressed a desire to rent natural gas vehicles to operate in Taipei City, thus contributing to the improvement of air quality. It was proposed that relevant data—such as energy consumption, cost of operation, cost of maintenance, and environmental impact—should be recorded when these vehicles operate in Taiwan. The collected data will be used for future study of the alternative-fuel vehicle in Taiwan.

It is widely acknowledged that the automobile industry is a locomotive industry, as it can help upgrade many of its relevant industries; therefore, if relevant industries in the country can be upgraded because of the development of EVs, it would be beneficial for the domestic industry. At present, there are already many institutions, i.e., the Asia Pacific Investment Company, Min Kun Company, Fang Fu Company Limited, and New Journey Company, working on relevant technological developments for EVs, and they have obtained patents covering all areas throughout the world.

The definition of EV does not include HEV, and the Ministry of Transportation and Communications did not specify the EV and alternative-fuel vehicle of low pollution, thus there may be confusion regarding subsidy and reward when HEVs start to operate in Taiwan. It is then suggested that the Ministry of Transportation and Communications of Taiwan clarifies the position of HEVs.

14.5 CONCLUSIONS

The result of multicriteria optimization is that the hybrid electric bus is more suitable at present for public transportation in order to improve the environmental quality. These vehicles will be replaced when the technological characteristics of EVs, or other new technology, are improved.

The multicriteria optimization of the alternative-fuel mode is performed with data given by experts from relevant engineering fields. The assessment method is based on experts gathering data and evaluating alternatives without using a mathematical model of evaluating criteria, and this approach could be considered as a contribution of this chapter.

The results of multicriteria analysis indicate what to do first in developing the alternative-fuel mode, in order to improve environmental quality. To answer the questions of how and when to implement alternative-fuel mode improvement, further research should endeavor to solve the development problems under budgetary constraints.

According to the compromise ranking method, the compromise solution which should be accepted by the decision makers is that which provides a maximum "group utility" of the "majority," and a minimum of the individual regret of the "opponent."

15 ELECTRE: An Application

15.1 INTRODUCTION

A network design problem (NDP) is a common decision-making problem that arises in urban transportation planning when selecting improvements or additions to an existing network to decrease traffic congestion or pollution, or other appropriate objectives. NDPs, according to their characteristics, are classified into several types. If the link improvement variables are 0-1 integers or continuous variables, then either a discrete problem or a continuous problem can be formulated. Chen and Alfa (1991) further divided the problems into three groups: (a) those with linear objective functions, (b) those with non-linear objective functions, the solutions of which satisfy the system-optimal criterion, and (c) those with non-linear objective functions, the solutions of which satisfy the user-optimal equilibrium criterion.

An NDP is endowed with a linear objective function, and this is to minimize the sum of travel and investment costs, subject to all feasible link flow and all combinations of alternative improvement projects. A linear objective function indicates that the travel time on each link is a constant and does not vary with the link flow. With its objective function being linear, the system optimal problem and user-optimal equilibrium problem become identical. Although the assumption of linear cost can be a simplified solution, the outcome rendered is rather impractical as the purpose of network improvement is to decrease congestion.

The objective of a solution for an NDP of the second type is to minimize the sum of travel and investment costs, subject to the same constraints as those for problems in the first category. What differs is that the objective function is non-linear, which shows that the travel time that the driver experiences on each link is a function of that link flow. The link flow rendered from the problem of the sort becomes the optimal flow of the whole system, rather than the flow of user-optimal equilibrium.

An NDP of the third type encompasses a bilevel non-linear objective function; its objective is to minimize the total cost and the link flow has to satisfy the condition of user-optimal equilibrium. As the flow of user-optimal equilibrium is a network structure, a problem of this kind is difficult to solve. Besides, the total user cost is not necessarily a decreasing function of decision variables (network improvements); thus, this explains the occurrence of Braess's paradox (Murchland 1970; LeBlanc 1975). In theory, expanding the capacity of certain links might result in an increased total travel cost. Therefore, to work on improvement of the network, a planner needs to predict the accumulated reaction of users in advance to avoid this situation.

Travel time cost was given and was considered as the sole objective in early transportation planning, and mathematical programming was devised to find the optimal solution. To evaluate transportation planning, the multiobjective technique started to be used in the middle of the 1960s; then authors increasingly revealed their investigations of multiobjective design problems in a transportation network (Current and Min 1986).

Being on a large scale, a transportation system is therefore confronted with varied needs from every perspective. As the investment in a transportation system is a sunk cost, it is natural that diversified evaluation should be made from different points (needs) of view when network construction or improvement is to be achieved; ultimately the optimal project can then be selected.

The NDP proposed in this chapter has the following characteristics:

a. Because most roads in the urban planning of the Taiwan metropolis are already constructed, the intention to widen entire car lanes to improve traffic conditions is rather impractical. Thus, what is discussed here is a continuous network design model.
b. In this chapter we discuss issues of a non-linear objective function; the improvements affect the equilibrium flow assignment (i.e., the user-optimum rather than the system-optimum assignment).
c. The matrix of a trip demand is assumed to be fixed, not influenced by increased capacity.
d. A diversified evaluation considering different points (needs) of view is made to select projects.

15.2 MULTIPLE OBJECTIVE EQUILIBRIUM NETWORK DESIGN PROBLEM

Most NDPs are conventionally formulated as a single objective problem for management. LeBlanc (1975) first used a branch-and-bound algorithm to solve the discrete network optimal design problem of a fixed investment budget. The single overall design objective is to minimize the total travel cost incurred by users, while the total budget serves as a constraint. Abdulaal and LeBlanc (1979) formulated a network design model with continuous decision variables. The budget constraint was put into the objective function after it had been converted into travel time units. A fixed budget may exclude many potentially good designs that exceed the budget only slightly. If such a design is placed in budgetary constraints or is given with a parameter and then placed into the objective function after being converted into a time unit, it will be difficult for one to interpret the design as its parameter value is arbitrary. The number of objectives should be as large as needed to represent the total behavior value of the system. As each objective has played a particular role in the decision-making process, any attempt to transform these variously measured and scaled objectives into comparable units is inappropriate. The best way to treat this problem is to consider each objective independently and to give each objective a relative importance (weight) throughout the process of management.

As for the solution to the NDP, Steenbrink (1974) initially proposed the concept of iterative optimization assignment (IOA) to solve the continuous NDP. This algorithm consists of iterating between a user-optimized equilibrium with fixed improvements and a system-optimized design with fixed flows. The IOA algorithm is efficient in computation as it is used to solve a network problem of realistic size. The defect is that the iterative process may not converge to an optimal solution.

Abdulaal and LeBlanc (1979) used the Hooke-Jeeves method to solve the continuous NDP. Because this algorithm does not employ derivatives, one is able to consider the user equilibrium constraints and to find the true local minimum. Because of the existing non-convexity of the network equilibrium design problem, no global optimal solution has yet been found. Regarding the Hooke-Jeeves algorithm for solution, as substantial calculation resources are needed to handle the practical network problem, its application is thus very handicapped. Suwansirikul, Friesz, and Tobin (1987) proposed an equilibrium-decomposed optimization (EDO) that decomposes the original NDP into interacting subproblems. Each one is simultaneously solved by using a one-dimensional search routine. Under the condition that the variables of all link improvements are fixed, the equilibrium assignment will proceed. Its approximate solution is obtained with the iterative algorithm. Such a method proves more efficient than the Hooke-Jeeves algorithm.

Choi (1984, 1985, 1986) proposed the Land Use Transport Optimization (LUTO) model to solve the problem of joint optimization of a land use plan and a transportation plan. The LUTO model is a computerized system that enables the planners to simultaneously choose between the land development area and new transport links by optimizing an objective function consisting of both land development cost and transportation costs. The model is successfully used to derive the physical development strategy in Hong Kong and is being applied to devise an implementation plan of the strategy.

As the theory of multiple criteria decision making has developed during the past twenty years, its application has gradually appeared in various fields. Considering the utility of various community groups, Li (1982) conceived the hierarchical multiobjective network design model, and the utility function per household for groups includes the objectives of disposable income and leisure time so that the optimal solution is obtained according to a heuristic algorithm under budgetary constraints.

The flow pattern on the links is the user-optimal equilibrium flow. Friesz and Harker (1983) established a multiobjective spatial-price equilibrium network design model for freight transportation. Two objectives are the maximization of total economic surplus and the minimization of transportation costs. The exact solution of the objective function cannot be found, because it involves linear integral calculus and the flow pattern on the links is constrained to be a spatial price equilibrium. Current, Revelle, and Cohon (1987) considered minimization of total travel time and the minimization of total path length from demand point to network to establish two objectives of the shortest path problem. Based on these, he established the median shortest path problem. Li (1988) designed the framework of an expert system and intended to use multicriteria decision making to evaluate and to select an improvement project for a transportation network. Tzeng and Chen (1993) took into account three objectives—the total travel time for road users, air pollution for non-users, and total travel distance for government, which were employed to formulate an effective multiobjective model for traffic assignment.

The concept and method to solve bilevel programming were successively published in academic papers after the 1980s (Fortuny-Amat and McCarl 1981; Candler and Townsley 1982). As the concept of bilevel programming is adequate to explain the decision-making operation of the NDP, LeBlanc and Boyce (1986) took advantage of a piecewise linear bilevel programming model to devise the NDP with user-optimal flows. In the next section we used the ideas of bilevel programming to explain and to discuss the network improvement problem with multiobjective decision making.

15.3 MODELING THE NETWORK IMPROVEMENT PROBLEM WITH MULTIOBJECTIVE DECISION MAKING

The purpose of examining the NDP in this chapter is to seek feasible alternatives at a bottleneck link under an existing network structure and travel demands, including the enlargement of link capacity and each link flow under the designated alternative. Then, the multicriteria decision making of ELECTRE III developed by Roy (1989, 1990) and the group decision making by Cook and Seiford (1978) are exploited to evaluate and to select a compromise alternative from feasible projects. In the design phase, multiobjective mathematical programming is adopted to devise a continuous network design model. In the phase of evaluation, multicriteria evaluation decision making is used to solve the discrete NDP. The stage of project searching is solved through the concept of bilevel programming. After the viewpoints of government and users are taken into account, the preferences of users are tentatively influenced when a link improvement is about to embark so as to minimize the total system costs. As for the travel time, the decision is determined by the route choice behavior of users (Tzeng et al. 1989). After criteria weights and project performance are set, project evaluation and selection are conducted with multiple criteria and group decision making to obtain a compromise alternative. The model of the framework is shown in Figure 15.1.

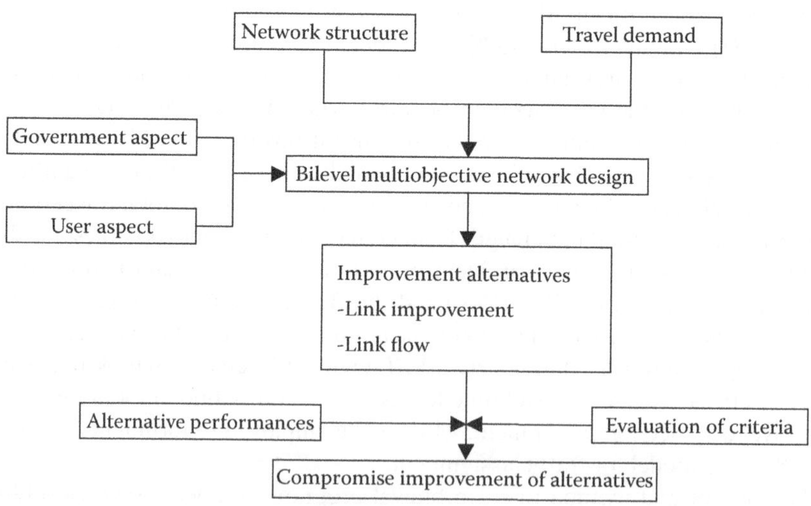

FIGURE 15.1 Framework of the network improvement model with multiobjective decision making.

15.4 MODEL AND THE SOLUTION OF THE BILEVEL MULTIOBJECTIVE NETWORK DESIGN

In this chapter, we attempted to use the concept of the preceding bilevel programming, then to devise a continuous network design model under the given trip demand matrix. The model is shown as follows:

$$(\text{P1}) \quad \min Z_1 = \sum_a Z_a = \sum_a C_a(f_a, y_a) f_a \tag{15.1}$$

$$\min Z_2 = \sum_{a \in I} G_a(y_a) \tag{15.2}$$

$$\text{s.t. } y_a \geq 0, a \in I. \tag{15.3}$$

$$(\text{E1}) \quad \min \sum_a \int_0^{f_a} C_a(x_a, y_a) dx_a \tag{15.4}$$

$$\text{s.t. } f_a = \sum_{r \in R} h_r \delta_{ar}, \quad \forall a \tag{15.5}$$

$$T_{ij} = \sum_{r \in R_{ij}} h_r, \quad \forall i, j \tag{15.6}$$

$$h_r \geq 0, \quad \forall r, \tag{15.7}$$

where

- a: link a in the network
- r: path r between origin-destination pair in the network
- i,j: nodes in the network
- R: the set of all paths of the network
- R_{ij}: the set of all paths from origin i to destination j
- C_a: the average travel time on link a as a function of flow and capacity
- y_a: the capacity improvement for link a
- y: $(\ldots, y_a, \ldots)$ denotes the vector of improvement capacity of all links
- G_a: the improvement cost for link a
- I: the set of links considered for improvement to the network
- f_a: the flow on link a
- f: $(\ldots, f_a, \ldots)$ denotes the vector of all link flows
- h_r: the flow on path r
- d_{ar}: the link-path incidence matrix element; if link a is on path r, then $d_{ar} = 1$, otherwise $d_{ar} = 0$
- T_{ij}: the travel demand from origin i to destination j.

In the above mathematical equations, Equations 15.1 through 15.3 constitute a high-level decision-making problem; Equation 15.1 represents the objective of minimization of a user's total travel time; Equation 15.2 represents the objective of minimization of the government's total investment cost. Equations 15.4 through 15.7 constitute a low-level decision-making problem, which is actually the network assignment problem of user equilibrium, and the equilibrium flow on the link can be obtained only through the link improvement variables. Equations 15.5 and 15.6 indicate, respectively, the definition and conservation of flow constraints. Equation 15.7 indicates that the flow on each link should be greater than or equal to zero. The integrated mathematical model is constituted from the high-level problem (P1) and low-level problem (E1). The travel time function is assumed to be that of the BPR (the foundation used by the U.S. Bureau of Public Roads, BPR). The cost function is based on the recommendation of Abdulaal and LeBlanc (1979). Therefore, if Equations 15.1 and 15.4 are non-linear objective functions, the integrated model becomes a bilevel non-linear programming problem.

This bilevel network design model as constructed is a typical price-control problem (Bialas and Karwan 1984). The decision variables controlled by high level are link improvement variables y, whereas the decision variables controlled by low level are the link equilibrium flows f. On the whole, the low-level decision variables generally affect the performance of the high-level objective and vice versa. The bilevel decision-making operation forms the Stackelberg game, with its high-level decision maker as leader and low-level decision makers as followers.

Our heuristic algorithm combines the ideas of the constraint method (Marglin 1967) with the IOA algorithm in order to find alternatives in the non-inferior solution set. According to the improvable performance values of each improvement link, the total budget is allocated. The procedure of the algorithm is as follows:

a. Based on the objective of investment cost minimization and the objective of travel time minimization, set the greatest value M_2 and the smallest value N_2 of the allowable budget.
b. Transform the original multiobjective programming problem of the high level into:

$$\min Z_1(y) \tag{15.8}$$

$$\text{s.t. } y \in Fd \tag{15.9}$$

$$Z_2(y) \le L_2, \tag{15.10}$$

where $L_2 = N_2 + (t/(s-1)) \times (M_2 - N_2)$, $t = 0,1,2,\ldots, s-1$; s is the number of cutting points for the section; and Fd is the feasible region.
c. Focus on various L_2 values and acquire a non-inferior solution set under varied Z_2 objective constraint values.

The following steps are repeated whenever non-inferior solutions of link improvements are to be found:

Step 0: Select the initial vector $I^0 = (0,0,\ldots,0,0)$ to be the initial value of the link improvement variables, and solve the user equilibrium problem with $y = I^0$ to obtain $f(y^0)$. Set $j = 1$ and return to step J.

Step *J*:

a. Under the assumption that is fixed, the multiplied value of each link and the value of the improvable capacity of unit cost are normalized; then the constant budget is allocated according to the normalized value of each improvement link. If the travel time function is the BPR type, then

$$C_a(f_a, y_a) = A_a + B_a(f_a/k_a + y_a)^4, \tag{15.11}$$

$$Z'_a(y) = \frac{\partial Z_a}{\partial y_a} = -4B_a f_a^5(y)(k_a + y_a)^{-5}. \tag{15.12}$$

where C_a: the average travel time on link a; A_a: the travel time of free flow; B_a: the congestion parameter for link a; k_a: the original capacity of link a.

b. The allocated budget for each improvement link is transformed into the capacity improvement value and the user equilibrium problem is solved with $y = y^J$ to obtain $f(y^J)$.
c. If $I_a^j - I_a^{j-1} \leq \varepsilon$ for link a, set $y_a^* = (I_a^j - I_a^{j-1})/2$ and link a is not improved thereafter; otherwise, set $j = j + 1$ and repeat Step J ($\varepsilon = 0.1$).
d. If all links need no further improvement, solve the user equilibrium problem with $y = y^*$ to obtain $f(y^*)$.

In the above algorithm, the idea to solve steps 0 and J is similar to marginal analysis. At first, the original problem is decomposed into many subproblems to consider each improvement link (Suwansirikul et al. 1987) and the objective performance in each link is defined as the decrease of congestion cost in the investment of per unit cost. The objective performance is also defined as the product value of the improvement capacity for per unit investment cost and the travel cost decrease for per unit improvement capacity. The product value of each link shows the link performance improved in each link per unit improvement cost. According to the degree of that performance value, the budget allocation can be made. The budget to be obtained from an allocation in each link can be transferred into a capacity improvement value. This algorithm uses the idea of the IOA algorithm; hence, in all algorithms the equilibrium network flow is obtained from the previous stage, then put into for finding solution, when the variables of the link improvement are obtained each time. As the bilevel programming model is NP-hard (non-deterministic polynomial hard), it is impossible to use a polynomial algorithm to solve the bilevel programming problem. For this reason, only an approximation approach is usable to solve a large-scale network problem. For a non-linear bilevel problem, it suffices to indicate the complementarity condition to show non-convexity. Non-convexity implies that even if the solution to the problem is identified, the solution may be only a local solution rather than a global one. Therefore, the solution from this algorithm cannot be guaranteed to be the global optimum, but this result can be regarded as an approximately optimal solution.

The techniques developed by LeBlanc, Morlok, and Pierskalla (1975) can be, without constraint, applied to solve the equilibrium assignment problem with link improvement variable to be fixed and no discussion appears here. The program written in C language in this study can be experimented on a microcomputer; the results are favorable.

15.5 EVALUATION AND GROUP DECISION MAKING FOR NETWORK IMPROVEMENT PROJECT

Various non-inferior solution alternatives are obtained under various budget constraints perceived from the results of the preceding network design model; we used the multicriteria evaluation method of ELECTRE III developed by Roy (1986, 1989, 1990) to evaluate and to select the compromise alternative from feasible alternatives with multiple evaluation criteria. ELECTRE III provides abundant information during the decision-making process. The uncertainty is taken into account throughout the decision-making process. The solution that a certain criteria performance is the best and other criteria performances are all worse can be avoided. Then the best compromise alternative is obtained. After a pseudo-criterion is introduced, distinguishing itself from other conventional models, the judgment of projects is diverted to become more coherent to reality. In this chapter, we employed the group decision making of Cook and Seiford (1978) to integrate preferences of all decision makers.

15.5.1 Consensus Ranking of Cook and Seiford

Cook and Seiford (1978) proposed a consensus ranking that uses the concept of minimal distance to integrate the preferences of decision makers. Armstrong, Cook, and Seiford (1982) suggested its applicability to the consensus ranking of alternatives in ties. Because similar ranking of non-inferior solutions is attained by ELECTRE III, it is thus suitable for use in this method.

If the ranking is $R = (A_1, A_2, \ldots, A_m)$ of non-inferior solutions, then A_m indicates the ranking of m non-inferior solutions; the average value is used to manifest if equivalent ranking occurs. If there are n members, the ranking of member i towards an alternative is r_{ij}; the consensus ranking of all members is towards alternative j, and the definition of consensus is the minimal distance of all members towards the preferences of all alternatives and the consensus preferences, which are shown as follows:

$$d = \sum_{i=1}^{n} \sum_{j=1}^{m} \left| r_{ij} - r_j^c \right|.$$

In the above equation, r_j^c can only provide a ranking number k ($k = 1,2,\ldots,m$) if we let $r_j^c = k$. Then the definition d_{jk} is the total cognition difference of n decision makers when the consensus ranking of alternative j is k,

$$d_{jk} = \sum_{i=1}^{n} \left| r_{ij} - d_k \right|,$$

thus,

$$d_k = \sum_{j=1}^{m} d_{ij}, \quad k = 1,2,\ldots,m.$$

Hence, the efforts to solve the consensus ranking problem with the minimal cognition difference are indicated by the 0-1 linear programming assignment problem as follows:

$$\min d = \sum_{j=1}^{m} \sum_{k=1}^{m} d_{ij} x_{jk},$$

$$\text{s.t.} \sum_{j=1}^{m} x_{jk} = 1, \quad k = 1,2,\ldots,m,$$

$$\sum_{k=1}^{m} x_{jk} = 1, \quad j = 1,2,\ldots,m,$$

where

$$x_{jk} = \begin{cases} 1, & \text{if the consensus ranking of alternative } j \text{ is } k; \\ 0, & \text{otherwise} \end{cases}.$$

This method is used for managing the problems of many decision makers. Although a decision maker expresses only his/her preference of the rank of each alternative in a practical application, the problems need to let the decision maker clearly understand and cleverly use this operating procedure for consensus rankings.

15.5.2 Case Study of Metropolitan Taipei

Due to the rapid growth of traffic flow and the high concentration of transportation in metropolitan Taipei, serious traffic congestion has occurred. In this chapter the Taipei metropolitan area is our case study area. The network and traffic data are extracted from materials established by the Bureau of Taipei Rapid

Transit System. Altogether, there are 995 nodes and 2727 links. The total 330 traffic zones are used for modeling the travel demands. We have selected two major roadways of Taipei for formulation of possible improvement (feasible alternatives). The selected east-west-bound one is Chung Hsiao East and West Roads (having 36 links), and the north-south-bound one is Fu Hsing North and South Roads (having 28 links). Furthermore, the travel time function is assumed to be a BPR cost function and the investment function is assumed to be a linear function. To calculate the unit cost for the capacity at peak hours, the improvement unit capacity (PCU) per kilometer requires NT$10,000 (the width of the car lane is reckoned to be 3.5 m) and if the furnished road capacity for a full day is taken into the calculation, each kilometer would require NT$1800/PCU. In this case study the major purposes are to test the proposed network improvement plan for operational procedures in usable ways and to demonstrate the applicability of the proposed method for practical planning.

15.5.3 Non-inferior Solutions of Network Improvement Alternatives

If data on network and travel demands are inserted into the bilevel multiobjective network design model, the non-inferior solutions can be obtained while minimizing the two objectives. The objectives of improvement performance of the non-inferior solution of network design alternative and computation time are indicated in Table 15.1. Accordingly, the total travel time gradually decreases, after investment cost increases, because of the trade-off between these two objectives. The computation time required to locate the solution is approximately 30 minutes for a large-scale network. Based on the outcomes of the network assignment, the capacity expansion of those improved links would change the travel pattern in which some links have comparatively higher traffic flow than before. Hence, the

TABLE 15.1
Improved Performance of Non-inferior Solutions of Network Improvement Alternatives and Computation Time

	Total Travel Time (10,000 pcu.hour/day)	Total Investment Cost (10 Million NT$)	No. of Frank-Wolfe Iterations	Computation Time (Seconds)
Original	18191			
Alt. 1	18164	5	43	1528
Alt. 2	18150	10	43	1506
Alt. 3	18142	15	44	1543
Alt. 4	18130	20	42	1461
Alt. 5	18116	25	43	1500
Alt. 6	18093	30	50	1732

Note: The computation results have been experimented on 80486 PC; pcu: passenger car units.

service level of improvement is evidently less than the improved degree of adjacent roads. The effects or the capacity expansion stay mainly within the bounds of old urban areas, rippling insignificant reaction beyond the bounds.

15.5.4 Evaluation Criteria of the Network Improvement Alternatives

In view of the mutually conflicting criteria involved in the evaluation of transportation networks, four evaluation criteria are selected: the total travel time saved, the investment cost, the improvement of air pollution, and the complexity of underground cables. Among these criteria, the total travel time saved is accomplished upon consideration of the user (driver) aspect, the investment cost and the complexity of underground cables are conducted from the viewpoint of government, whereas the improvement of air pollution is conducted from the interest of a non-user (the general public). The hierarchical structure of evaluation criteria formulated is shown in Figure 15.2. The complexity of underground cables requires both the practical experience and judgments of the construction division, and the other performance values of criteria are derived from calculated results of the network design model. They are explained as follows:

15.5.4.1 Total Travel Time Saved

The purpose of the network optimal assignment is to assign the travel demands for all origin-distination (O-D) pairs onto each link of the network. When the equilibrium condition is attained, the travel time of any used routes for each O-D pair would be equal. The travel time function of the link is as follows:

$$C_a\left(f_a, y_a\right) = A_a + B_a\left(f_a/\left(k_a + y_a\right)\right)^4,$$

where f_a is the traffic flow of link a; A_a is the travel time of free flow on link a; k_a is the original capacity of link a; and y_a is the improved capacity of link a.

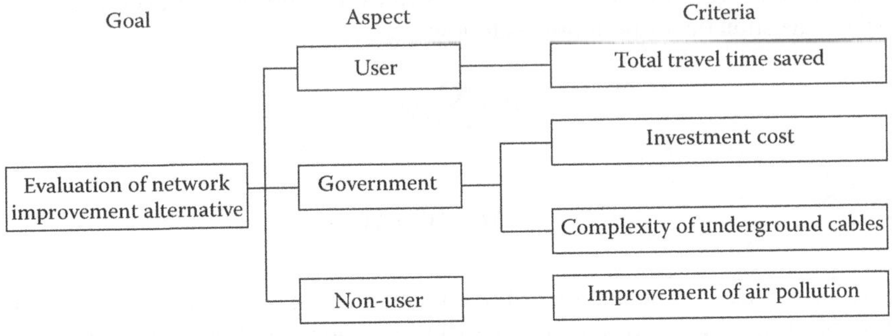

FIGURE 15.2 Hierarchical structure of evaluation criteria on network improvement alternatives.

For the whole network, the total travel time (TT) is the aggregate of travel time of the vehicle flow on each link:

$$\mathrm{TT} = \sum_a f_a C_a.$$

The total time saving refers to the differential value between the total travel time required in the original network and that of the improvement alternatives.

15.5.4.2 Total Investment Cost

Total investment cost can be obtained as

$$\sum_{a \in I} d_a y_a,$$

where d_a is the investment cost of the unit capacity of link a and I is the set of recommended links that require improvement in the network.

15.5.4.3 Improvement of Air Pollution

Improvement of air pollution is conducted from the viewpoint of a non-user. According to investigation of the environmental quality cognition of the Taipei metropolitan area at present, air quality is the environmental attribute that concerns the metropolitan residents most and it is evaluated to be the most unsatisfactory. Government experiments have already indicated that of the air pollution compounds in Taipei about 99% of carbon monoxide (CO) is from emission of motor vehicles. If CO is used to represent air pollutant, the total amount of air pollution (TP) is the aggregate emission of all the traffic flow on each link for the whole network, whereas the improvement of air pollution refers to the differential value between the total amount of pollution of the original network and that of improvement alternatives. Of these factors, the amount of pollution emission is associated with the driving distance and the coefficient of pollution emission of the unit driving distance is related to driving speed. Hence, the coefficient of the pollution emission decreases as driving speed increases. Such evidence indicates the impacts of travel distance and traffic congestion on air pollution as follow:

$$\mathrm{TP} = \sum_a p_a d_a f_a,$$

$$P_a = \alpha + \beta V_a + \gamma V_a^2,$$

where p_a is the coefficient of pollution emission of the unit driving distance (gram per kilometer, g/km); d_a is the distance (kilometer, km) of link a; V_a is the driving speed of link a; and α, β, and γ are the parameters of the relation between P_a and V_a.

15.5.4.4 Complexity of Underground Cables

The installment of the communal pipe culvert in the Taipei metropolitan area is still on the construction calendar, but there are complications because the underground cables of many divisions are involved. We intend to consider the complexity of underground cables within the excavation bounds of the car lane width as its evaluation criteria for network movement alternatives. With practical experience of management from the construction divisions, the manner of a rating scale (0–10) is thus established as a measure; the smaller the rating is, the smaller the commensurable complexity is, which facilitates the construction work.

The evaluation matrix established according to the four evaluation criteria on six alternatives is shown in Table 15.2.

The ELECTRE III method and application of the method of Cook and Seiford.

In this chapter fourteen scholars were invited from the transportation bureau, environmental protection bureau, and academic institutes to establish a decision-making group to evaluate improvement alternatives of six networks; then those evaluation criteria with the assistance of the AHP method (see Appendix) are introduced into a pairwise comparative questionnaire, after which the weight of each criterion is given by the decision makers provided with consistent confirmation. The results of the preference investigation (weights of evaluation criteria) are shown in Table 15.3. The processes are concluded as follows. The methods of ELECTRE III and that of Cook and Seiford are exploited to evaluate alternatives:

a. As in our study the concept of threshold values is not clear to the decision makers, we therefore decided to determine each threshold value according to the following equations: (i) calculate the differential values of each alternative under the same criterion, (ii) select those smaller differential values from the leading 1/5, 1/3, 1/2 differential values and calculate their average values to form the indifference threshold value, preference threshold value, and veto threshold value.
b. Based on these threshold values, the concord index and the disconcord index derived from the evaluation value of the alternatives, the concept of fuzzy theory is exercised to locate the credibility degree.

TABLE 15.2
Evaluation Matrix of Network Improvement Alternatives

	Total Travel Time (10000 pcu.hour/day)	Total Investment Cost (10 Million NT$)	Improvement of Air Pollution	Complexity of Underground Cables
Alt. 1	27	50	0.412	1
Alt. 2	41	100	0.510	3
Alt. 3	49	150	0.600	4
Alt. 4	61	200	0.752	6
Alt. 5	75	250	1.064	8
Alt. 6	98	300	1.529	10

Note: The ELECTRE III method and application of the method of Cook and Seiford method

TABLE 15.3
Weights of Evaluation Criteria of Network Improvement Alternatives

Criteria Evaluators	Total Travel Time Saved	Total Investment Cost	Improvement of Air Pollution	Complexity of Underground Cables
P 1	0.286	0.130	0.156	0.428
P 2	0.333	0.128	0.205	0.334
P 3	0.278	0.107	0.171	0.444
P 4	0.385	0.154	0.154	0.307
P 5	0.364	0.124	0.19	0.363
P 6	0.250	0.250	0.167	0.333
P 7	0.286	0.208	0.149	0.357
P 8	0.436	0.114	0.136	0.314
P 9	0.400	0.080	0.117	0.403
P 10	0.385	0.154	0.066	0.395
P 11	0.400	0.167	0.167	0.266
P 12	0.267	0.107	0.160	0.466
P 13	0.333	0.125	0.125	0.417
P 14	0.318	0.156	0.117	0.409

c. The ranking order of alternatives is conducted according to the credibility degree, and the process of management is a combination of both downward distillation and upward distillation toward the final one.
d. The preferences of decision makers towards network improvement alternatives are acquired after the preceding processes; then the method of Cook and Seiford is employed to integrate the preferences of all decision makers, resulting in a final consensus ranking shown in Table 15.4.

The rankings of the second and sixth alternatives are preferable to the others among the decision makers (see Table 15.4), whereas the sixth alternative manifests itself as the most preferable alternative on consensus ranking. Hence, the preferred alternative would be that of minimizing either total travel time or total investment cost subject to certain minimum performance standards relative to other performance criteria. Because of the effects that the decisions of a threshold value might incur on the evaluation results, we attempted to replace a threshold value by sensitivity analysis. As a result, a higher degree of comparison is revealed among the alternatives; changes are witnessed between them as the threshold value is decreased, and only the rankings of the foremost and trailing alternatives remain intact.

15.6 CONCLUSIONS AND RECOMMENDATIONS

In this chapter, a multiobjective decision-making process is proposed for a metropolitan network improvement problem. From the aspect of design, multiobjective mathematical programming is used to establish a continuous network design model. From the aspect of evaluation, multicriteria decision making is employed to solve a discrete

TABLE 15.4
Evaluation Results of Network Improvement Alternatives

Criteria Evaluators	Alternatives 1	2	3	4	5	6
P 1	2	2	6	5	2	1
P 2	2	2	2	6	2	1
P 3	2	2	6	5	1	1
P 4	3	1	6	5	3	2
P 5	2	2	6	5	2	1
P 6	3	1	6	5	3	2
P 7	3	1	6	5	3	2
P 8	2	2	6	5	2	1
P 9	5	1	5	4	3	1
P 10	3	1	6	5	3	2
P 11	3	1	6	5	3	2
P 12	2	2	6	5	2	1
P 13	2	2	6	5	2	1
P 14	3	1	6	5	3	2
Consensus rank	4	2	6	5	3	1

Note: P1–P5 represent the academic institute, P6–P8 represent the construction affairs bureau, P9–P11 represent transportation bureau, and P12–P14 represent the environmental protection bureau.

network improvement problem. We also propose an effective heuristic algorithm so that applications to the practical networks can be more efficient. In the meantime, group decision making is also utilized to evaluate and to select the compromise consensus alternative from feasible alternatives. The application of multiobjective decision making provides more reasonable consideration for the network improvement problem. Recommendations of improvement presented in this chapter are as follows:

1. Because of the non-convexity property in the network design model, the solution is difficult to find. The consideration of multiobjectives complicates the problem. We attempt to divide the entire network improvement problem into two stages: design and evaluation. Although the problem is simplified, the complete decision-making process usually considers the objective in each stage should be dynamic and changeable. Thus, It is important to consider goals/objectives to capture the problems and avoid the unreasonable conditions in the begin of the problem.
2. In our society, many items of needs from different points of view show important factors, such as road safety, public opinion, and social benefit. If more thought is given to these, the decisions made would relate better to the practical issues.
3. The communication and interaction between planners and decision makers are important and can affect the quality of decision making. Thus the

combination of effective auxiliary aids of decision making has become necessary so that the decision maker can better control the alterations in regard to planning measures and the techniques of interactive multiobjective decision making can be employed to solve problems.

4. It is assumed in this chapter that an O-D matrix is fixed and does not correspond to a practical situation. Further research on NDPs will be required for varied O-D travel demand.

16 PROMETEE: An Application

16.1 INTRODUCTION

Over the past two decades, Taiwan has experienced significant changes in its economic structure and rapid industrial development. Energy consumption has increased from 8.5 million kiloliters of oil equivalent (KLOE) in 1968 to 44.9 million KLOE in 1988, at an average annual growth rate of 8.6%. Meanwhile, the proportion of domestically produced energy in the total energy supply has dropped from 54% in 1968 to 8% in 1988. This increased dependence on imported energy increases vulnerability to unstable energy supplies, especially when the dependence on imported oil has reached almost 100%.

Scarcity, uneven geographical distribution, and necessity have subjected the oil supply to cartelization and politicization, i.e., its use as a political power source by oil-rich countries. This situation was evident during the past two oil crises. To continue economic growth in Taiwan, the stability of the oil supply has to be ensured. In addition to energy conservation, new energy-supply sources are needed.

New energy systems that are being researched include solar thermal, solar photovoltaics, fuel cells, wind, geothermal, tidal power, biomass, and hydrogen. Development of these potential energy sources is promising since most of them generate less environmental pollution than fossil fuels and some show good potential for commercialization.

The establishment of the Energy Foundation in Taiwan in 1980 marked the initiation of the development of new energy systems there. The allocation of limited financial resources to diverse new energy developments is the challenging task of the Energy Committee of the Ministry of Economic Affairs. Setting priorities every year for each candidate development project is essential given the uncertainties of future developments. The purpose of this chapter is to apply the multiple criteria decision making (MCDM) method to this priority-setting task. Using this method, expertise is integrated to set priorities for possible development. The results have been forwarded to the Energy Committee to assist its decision making.

New energy developments and prospects are reviewed as a basis for setting evaluation criteria and developing alternatives (Tzeng et al. 1992).

16.1.1 Solar Thermal Energy

The solar thermal energy absorbed in the Taiwan area is about 4.46×10^{16} KC (kilocalorie) or 5 billion KLOE (kiloiter of oil equivalent), which is equivalent to about 111 times the total energy consumption in Taiwan in 1977. The development potential for solar thermal energy is high.

About a decade ago, the Departments of Mechanical Engineering at the Tatung Institute of Technology and the National Taiwan University, and the Refinery

Research Center of the China Oil Corporation began to research solar energy. Private firms have also begun the promotion of solar heating equipment. Since the establishment of the Energy & Mining Research Organization (EMRO) at the Industrial Technology Research Institute (ITRI) in 1981, small solar heating system and heat collector designs have begun to be developed and tested. Their uses cover both homes and factories. With their technological advance and market expansion their cost is expected to decrease and their payback time should eventually reach 2–3 years. Compared to traditional water and natural gas heaters, solar heaters have some important advantages. The future research direction for these heaters is to reduce production costs, improve system design, and promote their applications.

16.1.2 Solar Photovoltaics

The Nuclear Research Institute of the Atomic Energy Committee began research on solar batteries in 1975 on a small scale. With the establishment of the Energy Research Institute at ITRI, investigations on non-silicon solar batteries began in earnest. The research projects in this area were transferred to the Department of Materials in ITRI in 1987. Some research products have already been transferred to the private sector.

The advantages of solar photovoltaics are (1) automatic production, (2) no pollution, (3) equipment that can be easily expanded, and (4) absence of transmission lines, since the equipment can be set up at the energy-demand site. Almost every country in the world has shown an interest in this potential technology.

At present, due to the high costs of solar-battery components, the development of solar photovoltaics has been limited to special applications. Over the short term, the price of a solar battery is greater than that of oil. For this reason, the development of a non-crystal solar battery should be focused on consumer electronic products. Over the long run, development will shift toward high-efficiency batteries.

16.1.3 Fuel Cells

A fuel cell operates, in a sense, as a slowed-down combustion reaction. The cell structure is similar to that of a battery. It is a device to create electric current using fuel and oxygen. Its advantages include high efficiency, low environmental pollution, short establishment time, waste heat that can be used for cogeneration, and good replacement potential for oil. There are four commonly used types of fuel cells, namely the phosphoric acid fuel cell (PAFC), molten carbonate fuel cell (MCFC), solid oxide fuel cell (SOFC), and the alkaline fuel cell (AFC). Among these, PAFC technology has been used to produce the first generation of fuel cells, which are the most promising for commercialization, but their electric efficiency is the lowest. MCFC, the second-generation fuel cell, is expected to be on the market by 1995. Its electric efficiency is 45%, and its fuel use is more extensive and flexible. The third generation of fuel cell, SOFC, is expected to be on the market by the year 2000. Its electric efficiency may reach 50% and the usable fuels are far more widely available. The AFC yields the highest efficiency at 60%; however, its use is limited to special applications.

Fuel cells have been recognized as one of the most promising electricity-generation methods. Given the policy that Taiwan will import large quantities of LNG (Liquefied

Natural Gas) after 1990, the EMRO at ITRI has recently conducted a feasibility study for a fuel-cell power plant in Taiwan. The authors of this study concluded that, for fuel-cell development, the PAFC should be assigned first priority and the MCFC second priority. To foster the development of a fuel-cell power plant in Taiwan to replace traditional power plants, the Energy Committee of the Ministry of Economic Affairs has contracted the ITRI to prepare a study on the development of fuel-cell technology in Taiwan. The study was conducted over a three-year period beginning in 1989 to 1992.

16.1.4 Wind Energy

Research on wind energy has been conducted in Taiwan since 1965. In the early stages, the Taiwan Power Corporation, Tamkang University, the Academia Sinica, the Agricultural Engineering Research Institute, and Tsing-Hua University were the agencies active in this research. Currently, the EMRO at ITRI is the key energy agency for developing wind-energy devices. Wind-energy devices for 4 and 40 kW have been transferred to industry, and 150 kW wind-energy equipment is being tested.

Wind energy is a non-depleting natural resource. The most significant advantage of its use is the absence of atmospheric pollution. However, the electricity-generation cost is greater than for traditional power-generating methods. In order to lower capital and production costs and improve operational efficiencies, future research should be focused on the development of a low-cost and practical wind-energy conversion system. The generating equipment may also be exportable to other countries.

16.1.5 Biofuels

Biomass materials produced with land, sunlight, water, and carbon dioxide offer many advantages as sources of energy. They may be burned directly as solid fuels or else converted to highly-prized gaseous or liquid fuels. Rather than relying on manufactured collectors and converters, biomass collectors may be deployed by spreading seeds. These collectors align themselves toward the sun and also solve the storage problem posed by the intermittent nature of solar energy. Biomass can provide a renewable energy source that requires only periodic harvesting.

The development of biofuels in Taiwan has emphasized the utilization of waste materials. Currently, significant progress is being made in this field. The production of agricultural waste is quite large in Taiwan. As the result of the production of industrial waste water and urban wastes, solid-waste and water-pollution problems in Taiwan are very serious. Therefore, determining how to make use of the wastes while solving the pollution problems is the main direction for bioenergy research in Taiwan. Future developments should be focused on the direct burning of solid wastes and on the disposal of oxidized wastes.

16.1.6 Geothermal Energy

The development of geothermal energy in Taiwan began in 1962. At this time, the Chin-Sui and Tu-Tsong power plants provide geothermal energy, but their economic value is low. To improve the development feasibility, multipurpose development for power generation, water heating, and sightseeing is being pursued. Potential sites for geothermal development are at Tu-Tsong, Gen-Cho, and Chin-Ren.

It is estimated that both Tu-Tsong and Gen-Cho will have 10 MW of power-generating equipment by 1995. When Chin-Sui and Chin-Ren, which will have 15 and 12 MW of generating equipment, respectively, are added, the total geothermal electricity-generating capacity will reach 37 MW or 57.6 KLOE.

Current geothermal development problems are associated with thermal transmission, geothermal resources, and working fluids. Another problem is the identification and development of hot-water technology.

16.1.7 Ocean Energy

Ocean energy includes mainly tidal power, wave energy, and ocean thermal energy. Ocean thermal energy systems may be classified into closed and open systems. One closed system is going to be commercialized and an open system is undergoing prototype testing and scientific investigation. The application technologies for tidal energy and wave energy have been relatively successful and examples of practical applications are available.

Among the ocean energy methods, ocean thermal energy and wave energy are the most promising power-generating systems for Taiwan.

16.1.8 Hydrogen Energy

High-yield hydrogen is an ideal fuel for the future. In Japan and the U.S., hydrogen manufacturing technology has been developed with high efficiencies (90%) and

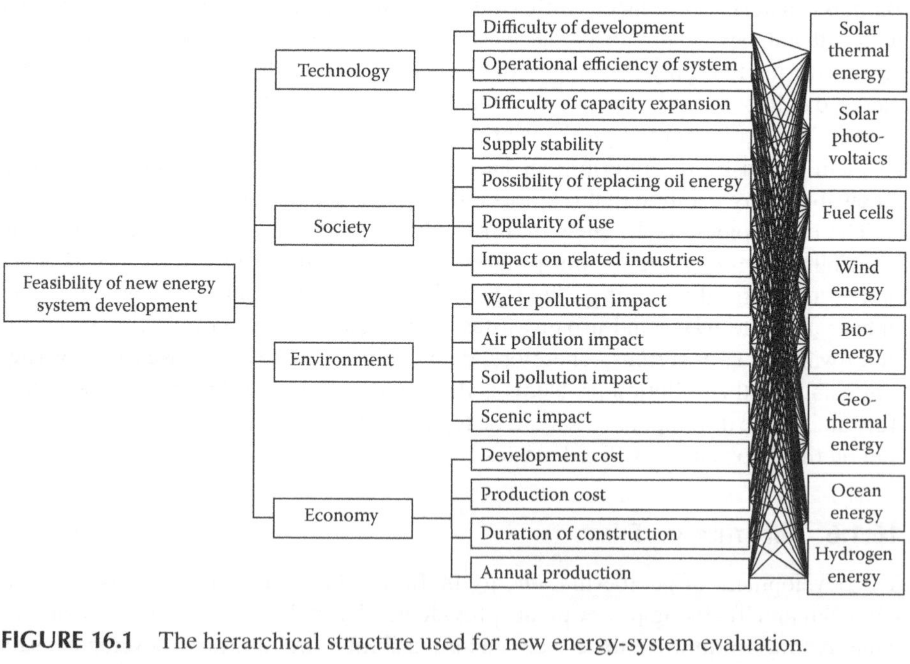

FIGURE 16.1 The hierarchical structure used for new energy-system evaluation.

TABLE 16.1
Evaluation Scale and Performance of the Criteria

Criteria	5	4	3	2	1
Difficulty of development	VE	E	M	D	VD
Operational efficiency of system	VH	H	M	L	VL
Difficulty of capacity expansion	VE	E	M	D	VD
Supply stability	VH	H	M	L	VL
Possibility of replacing oil energy	VH	H	M	L	VL
Popularity of use	VH	H	M	L	VL
Impact on related industries	HP	LP	NO	LN	HN
Water pollution impact	NO	B	M	S	VS
Air pollution impact	NO	B	M	S	VS
Soil pollution impact	NO	B	M	S	VS
Scenic impact	HP	LP	NO	L	HN
Development cost	VL	L	M	H	VH
Production cost	VL	L	M	H	VH
Duration of construction	VSH	SH	M	LG	VLG
Annual production	VH	H	M	L	VL

V, very; *M*, medium; *NO*, no; *D*, difficulty; *E*, easy; *L*, low; *H*, high; *P*, positive; *N*, negative; *B*, bit; *S*, serious; *SH*, short; *LG*, long.

hydrogen has been applied in cars on an experimental basis. These cars may reach 100 km/h for a distance of 200 km. Difficulties of hydrogen-energy development include its transmission and storage. Also, development costs are high and opportunities for commercialization are a long way off (Figure 16.1).

For technology, criteria are generated to reflect difficulty of development, system-operation efficiency, and difficulty of expanding the capacity. Social considerations are supply stability, possibility of replacing oil energy, popularity of use, and impact on related industries. Environmental criteria include the levels of water pollution, air pollution, soil pollution, and scenic impact. Economic criteria are generated, namely the development cost, production cost, construction time, and annual output.

Due to uncertainties associated with new energy developments, the performance evaluation is difficult to quantify. To deal with this difficulty, each criterion is measured on a scale from one to five. The measurement scale for each criterion is shown in Table 16.1.

16.2 EVALUATION OF NEW ENERGY-DEVELOPMENT ALTERNATIVES

To evaluate priorities for alternative new energy developments, a group decision method has been adopted. We invited 14 experts and classified them into 4 groups of different expertise and background for evaluators in new energy development from the Energy Committee, the Taipower company, the Chinese Petroleum Corporation, the Energy Research Institute of the ITRI, and the university to evaluate the performance of each alternative for a special area of Taiwan and used an evaluation scale as an example in Table 16.1 for convenient explanation of our method. The evaluation method is based on the Analytical Hierarchy Process (AHP) and the Preference Ranking Organization METHods for Enrichment Evaluations (PROMETHEE). The evaluation process is described in the following paragraphs.

16.2.1 Application of AHP

The AHP method is applied to derive weights for each criterion. Pairwise comparison is used in the evaluation for easy comparison of each item by experts. In the consistency test, we follow Saaty's suggestion that the consistency ratio be no more than 0.1. The evaluation results are shown in Tables 16.2 and 16.3.

16.2.2 Application of PROMETHEE

The PROMETHEE evaluation methods (Brans et al. 1984) consist of four variations. PROMETHEE II provides a complete order for the evaluation that will help decision makers realize the evaluation results easily. PROMETHEE II is the version used in this chapter.

The PROMETHEE process encompasses the following three steps:

1. *Establishment of a preference function for generalized criteria:*
   Generalized criteria are defined by

$$H(d) = \begin{cases} P(a,b), & d = f(a) - f(b) \geq 0; \\ P(b,a), & d = f(a) - f(b) \leq 0, \end{cases} \tag{16.1}$$

   where $P(a, b)$ represents the preference advantage of alternative a over alternative b (i.e., the measurable extent to which a is preferred to b), $P(b, a)$ represents the preference advantage of alternative b over alternative a, and $f(a)$ and $f(b)$ represent the assessed values for alternatives a and b, respectively. Greater values of $f(a)$ or $f(b)$ are better. Since the information desired by decision makers is difficult to obtain, we have defined the $H(d)$ function as

TABLE 16.2
Evaluation Weights

Sectors	Evaluators			
	1	2	3	4
Technology	0.349	0.329	0.293	0.201
Society	0.138	0.122	0.261	0.210
Environment	0.111	0.456	0.240	0.388
Economy	0.402	0.093	0.206	0.201

TABLE 16.3
Criteria Weights

Criteria	1	2	3	4
Difficulty of development	0.118	0.135	0.103	0.064
Operational efficiency of system	0.147	0.089	0.090	0.078
Difficulty of capacity expansion	0.084	0.105	0.100	0.059
Supply stability	0.039	0.026	0.068	0.065
Possibility of replacing oil energy	0.039	0.012	0.070	0.065
Popularity of use	0.033	0.037	0.065	0.040
Impact on related industries	0.027	0.047	0.058	0.040
Water pollution impact	0.033	0.119	0.062	0.105
Air pollution impact	0.033	0.128	0.062	0.105
Soil pollution impact	0.026	0.119	0.062	0.105
Scenic impact	0.019	0.090	0.054	0.073
Development cost	0.116	0.029	0.065	0.054
Production cost	0.105	0.029	0.047	0.054
Duration of construction	0.085	0.019	0.047	0.050
Annual production	0.096	0.016	0.047	0.043

$$H(d) = \begin{cases} 0, & d = 0; \\ 1, & |d| > 0. \end{cases} \tag{16.2}$$

2. *Calculation of a multicriteria preference index:*
A multicriteria preference index $\pi(a,b)$ indicating the preference advantage of alternative a over alternative b may be defined as

$$\pi(a,b) = \frac{1}{\sum_h w_h} \sum_h w_h P_h(a,b), \tag{16.3}$$

where w_h is the weight of criterion h and $P_h(a,b)$ indicates the superiority of alternative a over alternative b under criterion h. Introducing the assessed values into an evaluation matrix (Table 16.4) and also into Eqations 16.1 and 16.2, we find $P_h(a,b)$. After introducing criteria weights (Table 16.3) into Equation 16.3, we obtain $\pi(a,b)$. The results are shown in Tables 16.5 through 16.8.

From the resulting multiple-criteria preference-index values and according to the network flow concept, we may determine which alternatives are superior. The flow is defined as

$$\phi^+(a) = \sum_{b \in A} \pi(a,b);$$

$$\phi^-(a) = \sum_{b \in A} \pi(b,a);$$

TABLE 16.4
Evaluation Matrix for New Energy-Development Alternatives

Criteria	Solar Thermal Energy	Solar Photovoltaics	Fuel Cells	Wind Energy	Bioenergy	Geothermal Energy	Ocean Energy	Hydrogen Energy
Difficulty of development	4	3	2	4	3	3	2	2
Operational efficiency of system	3	2	4	2	2	3	1	3
Difficulty of capacity expansion	4	3	4	3	2	3	2	2
Supply stability	3	4	3	2	2	3	2	2
Possibility of replacing oil energy	3	3	2	2	2	2	1	2
Popularity of use	4	3	2	2	2	2	2	2
Impact on related industries	4	4	4	4	4	3	4	4
Water pollution impact	5	5	4	5	3	4	4	4
Air pollution impact	5	5	4	5	4	5	5	4
Soil pollution impact	5	5	5	5	3	3	5	5
Scenic impact	2	2	3	2	4	3	2	2
Development cost	4	2	2	3	4	3	1	1
Production cost	5	3	2	4	4	5	3	1
Duration of construction	3	3	2	4	3	3	2	1
Annual production	4	3	4	3	2	4	2	1

TABLE 16.5
Multicriteria Preference Indices and Superiority Indices for Evaluator 1

Alternative	Solar Thermal Energy	Solar Photovoltaics	Fuel Cells	Wind Energy	Bioenergy	Geothermal Energy	Ocean Energy	Hydrogen Energy	$\phi^+(a)$
Solar thermal energy	–	0.699	0.562	0.659	0.753	0.476	0.895	0.781	4.825
Solar photovoltaics	0.039	–	0.485	0.111	0.383	0.197	0.790	0.781	2.786
Fuel cells	0.166	0.346	–	0.385	0.425	0.284	0.540	0.691	2.837
Wind energy	0.085	0.424	0.490	–	0.475	0.289	0.823	0.670	3.256
Bioenergy	0.019	0.240	0.443	0.135	–	0.162	0.629	0.539	2.167
Geothermal energy	0.019	0.483	0.457	0.406	0.537	–	0.848	0.695	3.445
Ocean energy	0.000	0.000	0.138	0.000	0.092	0.053	–	0.319	0.602
Hydrogen energy	0.000	0.147	0.000	0.147	0.206	0.053	0.186	–	0.739
$\phi^-(a)$	0.328	2.339	2.575	1.843	2.871	1.514	4.711	4.476	–
$\phi(a)$	4.497	0.447	0.262	1.413	–0.704	1.931	–4.109	–3.737	–

TABLE 16.6
Multicriteria Preference Indices and Superiority Indices for Evaluator 2

Alternative	Solar Thermal Energy	Solar Photovoltaics	Fuel Cells	Wind Energy	Bioenergy	Geothermal Energy	Ocean Energy	Hydrogen Energy	$\phi^+(a)$
Solar thermal energy	–	0.420	0.380	0.343	0.815	0.603	0.616	0.655	3.852
Solar photovoltaics	0.026	–	0.505	0.075	0.562	0.360	0.587	0.655	2.770
Fuel cells	0.179	0.300	–	0.326	0.474	0.360	0.367	0.403	2.409
Wind energy	0.019	0.212	0.459	–	0.641	0.439	0.553	0.580	2.903
Bioenergy	0.090	0.090	0.302	0.119	–	0.166	0.403	0.318	1.488
Geothermal energy	0.090	0.253	0.340	0.250	0.512	–	0.550	0.577	2.572
Ocean energy	0.000	0.000	0.157	0.000	0.366	0.166	–	0.192	0.881
Hydrogen energy	0.000	0.089	0.000	0.089	0.327	0.166	0.101	–	0.772
$\phi^-(a)$	0.402	1.384	2.143	1.202	3.697	2.260	3.177	3.387	–
$\phi(a)$	3.448	1.386	0.266	1.701	−2.209	0.312	−2.296	−2.608	–

TABLE 16.7
Multicriteria Preference Indices and Superiority Indices for Evaluator 3

Alternative	Solar Thermal Energy	Solar Photovoltaics	Fuel Cells	Wind Energy	Bioenergy	Geothermal Energy	Ocean Energy	Hydrogen Energy	$\phi^+(a)$
Solar thermal energy	–	0.517	0.521	0.552	0.776	0.585	0.764	0.736	4.451
Solar photovoltaics	0.068	–	0.524	0.203	0.536	0.385	0.717	0.736	3.169
Fuel cells	0.124	0.291	–	0.359	0.429	0.310	0.494	0.518	2.545
Wind energy	0.047	0.262	0.386	–	0.483	0.332	0.631	0.533	2.674
Bioenergy	0.054	0.166	0.316	0.119	–	0.177	0.476	0.363	1.671
Geothermal energy	0.054	0.137	0.324	0.306	0.476	–	0.691	0.593	2.581
Ocean energy	0.000	0.000	0.109	0.000	0.186	0.120	–	0.203	0.618
Hydrogen energy	0.000	0.090	0.000	0.090	0.214	0.120	0.160	–	0.674
$\phi^-(a)$	0.367	1.463	2.180	1.629	3.100	2.029	3.933	3.682	–
$\phi(a)$	4.084	1.706	0.365	1.045	–1.429	0.552	–3.315	–3.008	–

TABLE 16.8
Multicriteria Preference Indices and Superiority Indices for Evaluator 4

Alternative	Solar Thermal Energy	Solar Photovoltaics	Fuel Cells	Wind Energy	Bioenergy	Geothermal Energy	Ocean Energy	Hydrogen Energy	$\phi^+(a)$
Solar thermal energy	–	0.392	0.537	0.458	0.783	0.532	0.677	0.704	4.083
Solar photovoltaics	0.065	–	0.548	0.170	0.587	0.420	0.623	0.704	3.117
Fuel cells	0.151	0.253	–	0.318	0.455	0.282	0.437	0.476	2.372
Wind energy	0.050	0.222	0.432	–	0.531	0.364	0.572	0.534	2.705
Bioenergy	0.073	0.181	0.295	0.127	–	0.167	0.438	0.338	1.619
Geothermal energy	0.073	0.302	0.327	0.313	0.509	–	0.605	0.567	2.696
Ocean energy	0.000	0.000	0.159	0.000	0.315	0.145	–	0.252	0.871
Hydrogen energy	0.000	0.078	0.000	0.078	0.288	0.145	0.143	–	0.732
$\phi^-(a)$	0.412	1.428	2.298	1.464	3.468	2.055	3.495	3.575	–
$\phi(a)$	3.671	1.689	0.074	1.241	−1.849	0.641	−2.624	−2.843	–

TABLE 16.9
Multicriteria Preference Indices and Superiority Indices for Evaluator 2

Evaluators	Solar Thermal Energy	Solar Photovoltaics	Fuel Cells	Wind Energy	Bioenergy	Geothermal Energy	Ocean Energy	Hydrogen Energy
1	1	4	5	3	6	2	8	7
2	1	3	5	2	6	4	7	8
3	1	2	5	3	6	4	8	7
4	1	2	5	3	6	4	7	8

$$\phi(a) = \phi^{+}(a) - \phi^{-}(a),$$

where A is the set of all alternatives and $a,b \in A$, $\phi^{+}(a)$ is the superiority of alternative a over all other alternatives, $\phi^{-}(a)$ is the inferiority of alternative a compared to all other alternatives, and $\phi(a)$ is the final score of alternative a. Introducing $\pi(a,b)$ into Eqations 16.4 through 16.6, we obtain $\phi^{+}(a)$, $\phi^{-}(a)$, and $\phi(a)$ as given in Tables 16.5 through 16.8.

3. *Ranking of alternatives:*
PROMETHEE II ranks the alternatives according to the following relation:

$$aPb \text{ if and only if (iff) } \phi(a) > \phi(b), aIb \text{ (iff) } \phi(a) = \phi(b),$$

where aPb means a is preferred to b and aIb means no difference is perceived between alternatives a and b.

Using the values in Tables 16.5 through 16.8 and Equation 16.7, we obtain the rankings of the four evaluators for the alternative forms of energy development shown in Table 16.9.

Our evaluation shows a consistent preference among the four expert groups for solar thermal energy, fuel cells, and bioenergy. The scores for solar photovoltaics, wind energy, and geothermal energy are similar (i.e., the ranking order is consistent). Ocean energy and hydrogen energy are ranked at the bottom. The resulting priorities for alternative new energy developments are as follows: (1) solar thermal energy, (2) solar photovoltaics, wind energy, and geothermal energy, (3) fuel cells, (4) bioenergy, and (5) ocean energy and hydrogen energy.

16.3 CONCLUSIONS

Energy consumption is expected to increase as the Taiwanese economy continues to grow. Determining how to ensure a stable energy supply with the least potential environmental pollution is a challenging task for Taiwan. However, the development of a new energy system entails many uncertainties and requires an abundance of resources to support research and development. In this chapter, we have applied multicriteria evaluation to set priorities for alternative new energy systems. The results ranked solar thermal energy as the first priority for development. That was also the choice of all experts participating in the evaluation panel. Solar photovoltaics, wind energy, and geothermal energy were assigned second priorities for future developments.

17 Fuzzy Integral and Gray Relation: An Application

17.1 INTRODUCTION

During the latter half of the twentieth century, the leading industries of some nations, Germany, Japan, Korea, Singapore, Taiwan, and the United States for instance, evolved from agriculture to manufacturing, service industries, and then to knowledge-based hi-tech information industries. In each phase of this evolution, the technological and vocational education system of Taiwan coped well with national developmental goals by readjusting its configuration and redesigning the training programs. It thus supplied the workforce with the appropriate quantity and quality of workers, with the right timing, to fulfill those human resource demands. This background provides a solid foundation for national development.

The improvement of both the learning environment and academic performance are two major conventional requests by the public for their schools. Due to the introduction of huge quantities of related information and the urgent pressure from competition in the global market, the holistic concept of "school efficiency" has been developed (DeRoche 1987; Fidler and Bowles 1989; Reynolds and Cuttance 1992). The concept of school efficiency emphasizes the problem-solving competency of individual schools and the improvement of instructional efficiency (David 1989; White 1989; Cheng 1993). For example, Morris and Young (1976) observed that school management bears intense pressure to present acceptable statistics of educational productivity to satisfy the concerns of education consumers. Critical and objective evaluations do have some strengths: they empirically examine schooling production, sort out accurate feedback for further improvement, and point out clear and objective performance criteria of school efficiency. However, a commonly agreed-on definition of school efficiency has not been achieved due to different theoretical bases, research methodologies, evaluation models, and interpretations. In such circumstances, the presentation of a new evaluation model or research paper is a difficult challenge (Fitz-Gibbon 1994). Related professionals consider the design of an evaluation instrument and the formulation of its criteria to be an important field that merits research and development efforts. In order to cultivate better school efficiency, it is believed that the development of a more appropriate monitoring system to supervise educational performance is urgently needed (Fitz-Gibbon 1994).

The two purposes of this study are the enhancement of quality improvement of Taiwan's technological and vocational education system, together with proposing a more effective evaluation model to determine outstanding schools as candidates for upgrade to technological institutes. For those conventional treatments, the application of the multiple criteria decision-making method includes two major procedures: (1) the calculation of the relative weightings, which are subjective by nature, of individual evaluation criteria and (2) the sorting of alternative schools. Most previous researchers have used the analytic hierarchy process (AHP) (Saaty 1977) to calculate these specific weightings of evaluation criteria through the additive approach, which sums up the multiplicands of the performance values of an evaluation sample (an alternative school) and its respective weightings for each criterion. The application of the additive approach is based on the assumption that all of the attributes are independent. However, those attributes are actually not completely independent. To eliminate the drawbacks of the independent criterion assumption, Sugeno (1974) presented the fuzzy integral theory that introduced general fuzzy measures and thus allowed the AHP more flexibility to deal with the independent criteria assumption. In the implementation phase of an evaluation, if researchers fail to gain a clear dependence between evaluation criteria, they can do partitioning fuzzy integral for those dependent criteria so as to reconstruct a hierarchical system of evaluation criteria.

When decision makers determine the weightings of evaluation criteria, they need to gain a clear picture of the relationships between criteria, which requires a huge amount of information, together with extra manpower and higher cost. There are two possible strategies to reduce this burden: (1) Simplification of the evaluation hierarchical system by reducing the number of evaluation criteria. However, in practical situations, it would be very difficult for researchers to accomplish a satisfactory evaluation if they substantially reduced the number of evaluation criteria; (2) Reduction of the amount of needed information to a reasonable degree, as suggested by Chen, Wang, and Tzeng (2000). This approach is helpful in reducing the amount of information to be investigated, although it has some technical problems. This current study suggests a more balanced and feasible problem-solving approach, which uses factor analysis of multivariate analysis and cluster analysis to regroup school attributes, reconstruct a partial hierarchical system, and thus make the implementation of the fuzzy integral for multicriteria evaluation possible for a gray relation model. In this way, the amount of information between related attributes could be reduced by using the fuzzy integral. In order to verify the feasibility of our suggested model, this study presents a case study of Taiwan's technological and vocational institutions, analyzing the relationship between parameters of those outstanding junior colleges and the influences of individual parameters objectively and reasonably. Analysis results show that the proposed model has the following strengths:

1. It identifies factors that are more characteristic for evaluating outstanding junior colleges;
2. Parameters, i.e., attributes, that influence outstanding junior colleges are extremely complicated and not mutually independent; this makes the use of non-additive measures for handling those influential characteristics of parameters more reasonable and fitting practical behavior;

3. By means of the ranking orders produced by the gray relation model, we can gain a clearer picture of the organizational behaviors of outstanding junior colleges;
4. Our suggested model strengthens the quantitative analysis capacity of gray relation models, making them more generalized and fitting practical behavior.

17.2 NON-ADDITIVE TYPE FUZZY MULTICRITERIA EVALUATION

In this section, we present a non-additive type fuzzy multicriteria evaluation method that is specifically for handling research problems with incompletely independent attributes. Firstly, we employ factor analysis and cluster analysis to screen out independent common factors, which contain some dependent attributes. Secondly, we integrate these independent common factors by the non-additive (i.e., superadditive) type fuzzy integral to obtain the compound performance value of each single common factor. Thirdly, we employ an AHP to estimate the relative weighting of each common factor, whose source data come from a questionnaire survey of pairwise comparison, considering that each common factor may exert an affect to different degrees on the system. Finally, we employ the gray relation approach to obtain the efficiency readings. We will briefly discuss both concepts of the fuzzy integral for multicriteria evaluation of the gray relation model.

17.3 MULTICRITERIA EVALUATION THROUGH THE FUZZY INTEGRAL

This study employs the fuzzy integral to work for a compound evaluation because this approach is free from the independent attribute assumption and is applicable to non-additive or non-linear problems with incompletely independent attributes. We believe the use of the fuzzy integral for evaluation of this kind is more appropriate than conventional treatments because the evaluators may occasionally subjectively think attributes are independent, although they are objectively not.

Whether the attributes are independent or not, the conventional multicriteria compound evaluation method is theoretically based on the additive concept, which means that the implementation of a compound evaluation in a system with multiple attributes is accomplished through an additive type calculation summing simple weightings of both the contributions of individual attributes and their respective efficiency values. Actually in those practical problems, their attributes are incompletely independent, which makes the additive approach inapplicable (Ralescu and Adams 1980; Chen and Tzeng 2000). Therefore, there is a need to administer the partial type fuzzy integral to those relational criteria to construct a new hierarchical system of evaluation criteria and to employ the fuzzy integral method proposed by Sugeno (1974) and Sugeno and Kwon (1995) to calculate the new compound performance values of related evaluation criteria.

17.4 MULTIPLE CRITERIA EVALUATION OF THE GRAY RELATION MODEL

The gray system theory presented in Deng (1982) is a powerful approach for the systematic analysis of relationships and for model construction of a system with uncertainty and incomplete data. It also employs methodologies of prediction and decision making to study the relationships between attributes and to help gain a clearer picture of the relationships between characteristics of a system. The major concepts of the gray system theory include gray generating, gray relation model, gray prediction model, gray programming, and gray control. The gray relation model, which is an evaluation method, has several primary functions: determining the relationships between individual attributes of the target system, screening out important attributes that would heavily affect the operational objectives of a system, and enhancing the effective development of a system. It is an evaluation method of quantitative description and comparison, specifically studying the changing conditions of a system's development. In contrast, conventional evaluation methods have two major shortcomings. First, they employ mathematical statistical techniques that demand a huge number of samples, which must be linear, potential, and/or log probability distributions; and second, it is easy for researchers to make polar errors when they accidentally eliminate some of the statistics. In comparison with conventional evaluation methods the gray relation model has some strengths listed below (Deng 1989; Shi 1990): (1) Computations are effort-saving and demand fewer techniques; (2) The number of samples is free from special limitations; (3) Its statistics are free from the classic distribution rules of probability; (4) The quantitative results of relationships will not conflict with their qualitative counterparts; (5) The reconstructed model is a sequential model of non-function type and is an effective means of handling distributed statistics. This study used the gray relation model for evaluation (Tzeng and Tsaur 1994; Mon, Tzeng, and Lu 1995).

17.5 CONSTRUCTING THE GRAY RELATION MODEL WITH NON-ADDITIVE MEASURES

To simplify the hierarchical system by reducing the numbers of its dimensions, this study employs factor analysis and cluster analysis because of their several advantages. These two methods present the original data structure with fewer dimensions; they help to refine common factors that are by nature independent from numerous factors, and they allow the use of a conventional additive evaluation method. Because individual common factors contain some factors that are mutually reactive and affective, this study uses the fuzzy integral method to administer the non-additive computation to these common factors and obtains the compound performance values of individual common factors. By doing this we are able to describe the system characteristics with fewer characteristic factors and fully consider the mutual influences between factors.

To display how individual data items contribute to the status change in the system's development, this study uses a questionnaire, multiple criteria decision-making method,

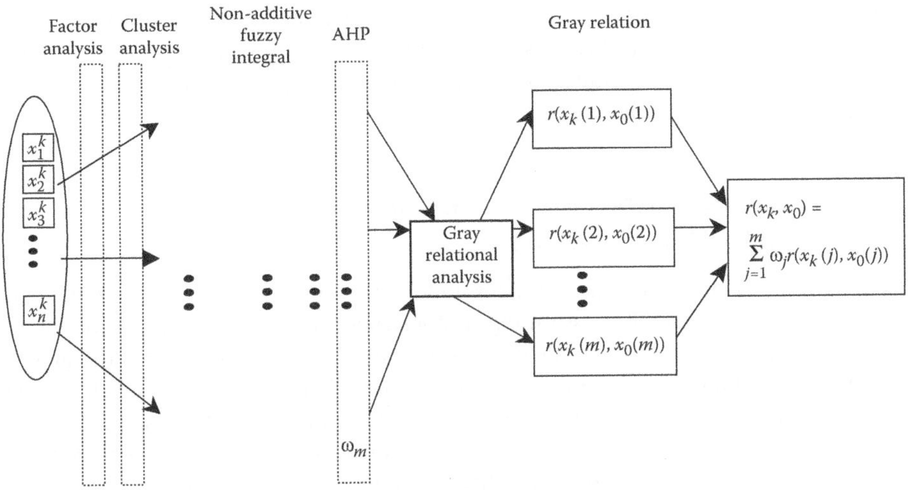

FIGURE 17.1 The concept diagram of gray relation evaluation with non-additive measures.

and AHP technique to obtain the relative weightings of individual data items. By using procedures in step 4 of the gray relation model listed above, we calculate the grades of gray relation of common factors with critical consideration of the issue of unequal weightings and with a strong willingness to show common factors' relative importance to the system. In Figure 17.1, we present the conceptual diagram of the partial fuzzy integral evaluation model.

17.6 EMPIRICAL ANALYSIS: A CASE STUDY OF TAIWAN'S JUNIOR COLLEGES

In order to offer a more detailed description of the implementation of the proposed model, the partial fuzzy integral multiple criteria evaluation model with non-additive superadditive measures, and to verify its feasibility and effectiveness, this study uses a case study to examine the performance values of individual evaluation criteria for eight of Taiwan's junior colleges. Our suggested evaluation method and its analysis outcome provide the educational authorities with innovative perspectives to do a better evaluation job of outstanding junior colleges.

17.7 PROBLEM DESCRIPTIONS

Since 1995 the Ministry of Education (MOE) of Taiwan, according to educational policies and workforce demands, has been releasing related mandates to select outstanding junior colleges that will be eligible for upgrading to the status of technological institutes. We randomly selected eight of Taiwan's junior colleges as research samples and evaluated their managerial efficiencies. Concerning the availability and the completeness of data, we objectively collected data for 12 parameters in the selection of outstanding junior colleges. These parameters are assessment results, inspector's examination results, administrative support manpower, financial support ability, school land area, campus building area, library facilities, instrument facilities, overall faculty, teaching efficiency, research efficiency, and service efficiency.

17.8 CONSTRUCTING MULTIPLE CRITERIA EVALUATION SYSTEMS

Considering the problem characteristics and the current systems of junior college in Taiwan, this study uses brainstorming to construct a holistic "evaluation hierarchy system for the selection of outstanding junior colleges," which includes the goal in first level, the dimensions in the second level, the criteria in the third level, and the measuring indices for declining the criteria. Criteria indices of the related Ministry of Education mandates include: managerial performance, practical research achievement, achievements of both school-industry collaboration and continuing education, administrative performance, faculty structure, training equipment, and school land and buildings. Based on two Ministry of Education mandates, "The 1997 Junior College Evaluation Handbook" and "The Ministry of Education Regulation for the Selection of Outstanding Junior Colleges that are to be Upgraded to Technological Institutes and Allowed to Offer Junior College Programs," these researchers used brainstorming to design a multiple criteria evaluation system that includes six dimensions: (1) school efficiency, (2) school administration, (3) school land and campus buildings, (4) facilities, (5) faculty, and (6) teachers' working efficiency. Twelve criteria were included: Assessment Results (A_1), Inspector's Examination Results (A_2), Administrative Support Manpower (A_3), Financial Support Ability (A_4), School Land Area (A_5), Campus Building Area (A_6), Library Facilities (A_7), Instrument Facilities (A_8), Overall Faculty (A_9), Teaching Efficiency (A_{10}), Research Efficiency (A_{11}), and Service Efficiency (A_{12}). Beyond this, there were a total of 36 items of assessment measures. Details are shown in Appendices 17.1 and 17.2.

17.9 ANALYSIS PROCEDURES

The analysis procedures for the implementation of the gray relational evaluation model with non-additive measures that were used by this empirical study are shown in Appendix 17.3.

Step 1: *Factor Analysis of Governing Parameters.*
This study uses conventional factor analysis, regrouping the evaluation criteria, and thus obtains factor readings of individual criteria, as shown in Table 17.1.

Step 2: *Cluster Analysis.*
After obtaining the factor readings of evaluation criteria, this study uses cluster analysis to regroup those evaluation criteria into four independent dimensions. Figure 17.2 shows the modified evaluation hierarchy system from Appendix 17.1.

TABLE 17.1
Factor Analysis Results after Varimax Rotated

Influential Parameters (Relations)	Common Factors				Communality
	1	2	3	4	
School land area	**0.833**	0.300	0.292	0.348	0.991
Campus building area	**0.538**	0.411	0.533	0.483	0.976
Overall faculty	**0.823**	0.306	0.342	0.324	0.993
Teaching efficiency	**0.793**	0.398	0.060	−0.024	0.792
Financial support ability	0.291	**0.821**	0.297	0.372	0.985
Research efficiency	0.542	**0.766**	0.274	0.034	0.956
Service efficiency	0.242	**0.937**	0.061	0.086	0.947
Assessment result	−0.134	0.358	**0.864**	−0.030	0.892
Inspector's examination results	−0.326	−0.163	**−0.872**	−0.165	0.921
Administrative support manpower	0.470	−0.133	**0.690**	0.221	0.763
Library facilities	0.108	0.098	0.217	**0.940**	0.952
Instrument facilities	0.642	0.237	−0.087	**0.669**	0.923
Eigenvalue	3.497	2.887	2.694	2.014	–
Variance interpreted (%)	29.145	24.057	22.450	16.781	–
Accumulated variance (%)	29.145	53.202	75.652	92.433	–

Note: *Extraction method*: principal component analysis; *Rotation method*: varimax with Kaiser normalization.

Step 3: *Compound Evaluation Values of Partial Fuzzy Integral of Common Factors.*

The factor loadings, which indicate the relational coefficients between factors and parameters, of parameters (attributes) of common factors are relatively high, which makes it impossible for this study to use conventional additive measures to handle the assigned research questions. Consequently, this study uses the Choquet integral with non-additive measures to calculate the compound evaluation values of the partial fuzzy integral of common factors, which are mutually dependent. In order to obtain the fuzzy measures $g(\cdot)$ of parameters and parameter combinations of common factors, this study administered a questionnaire survey to 22 interviewees who were junior college evaluation committee members, junior college management, or educational experts. The survey outcome of those 22 effective questionnaires is shown in Table 17.2. This study then uses the (C) $\int fdg$ equation to calculate the compound evaluation values of the fuzzy integral of common factors, as shown in Table 17.3.

Step 4: *Gray Relation Model for Integrating Evaluation.*

We screen out four independent common factors through the treatments of structural simplification, by using the factor analysis and the cluster analysis,

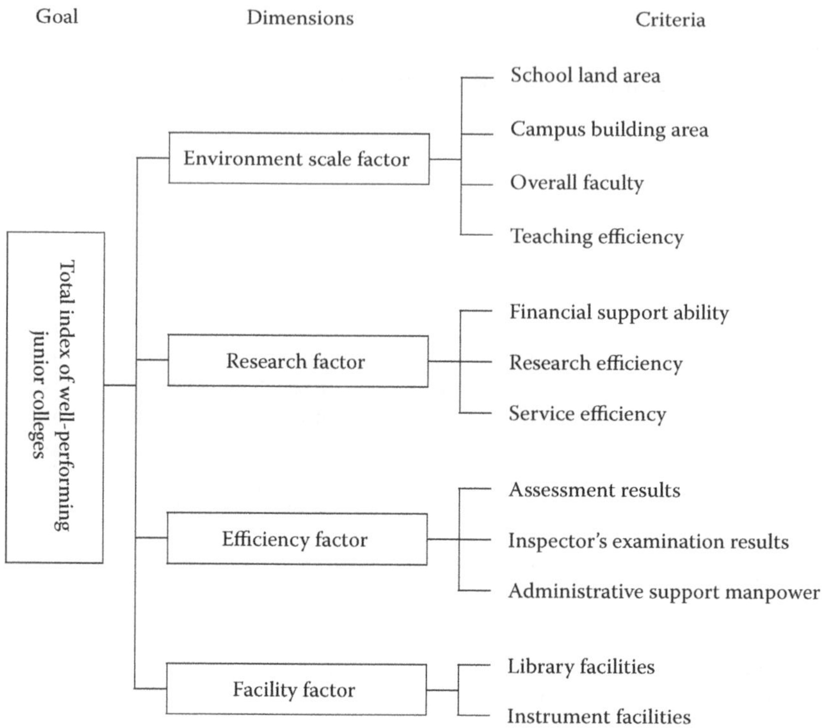

FIGURE 17.2 Revised assessment hierarchy system.

and merging, by using the Choquet integral, to those criteria within each common factor. This situation allows us to develop a non-additive gray relation model for integrating multicriteria evaluation.

The gray relation model is used to calculate the gray-relational coefficients, with a form of $r(x_i\ (k), x_0(k))$, between the comparative series and the reference series, i.e., the ideal/goal series; where $k = 1, 2, \ldots, 4$, the common factors. In this analysis we set a distinguished coefficient of 0.5, $\varsigma = 0.5$. Since individual common factors contribute differently to the empirical hypotheses of this study, we use the AHP to treat the results of the questionnaire survey and then obtain the relative weightings of common factors, as shown in Table 17.2; finally we compute the grade of gray relation by summing the weighed gray-relational coefficients $r(x_i, x_0)$, as shown in Table 17.4.

17.10 ANALYSIS RESULTS AND DISCUSSIONS

According to the above analysis, we present the following discussions in three dimensions: (1) hierarchical system of evaluation, (2) priority ranking of gray relation, and (3) fuzzy criteria evaluation.

TABLE 17.2
Fuzzy Measures $g(\cdot)$ and Relative Weightings of Each Parameter and Parameter Combination

Fuzzy Measures $g(\cdot)$				Relative Weights $\left[\sum_{k=1}^{4} w_k = 1\right]$
Environment Scale Factor				
$\{A_5\} = 0.02$	$\{A_5, A_6\} = 0.12$	$\{A_5, A_6, A_9\} = 0.5$	$\{A_5, A_6, A_9, A_{10}\} = 1$	$\omega_1 = 0.4173$
$\{A_6\} = 0.04$	$\{A_5, A_9\} = 0.3$	$\{A_5, A_6, A_{10}\} = 0.5$		
$\{A_9\} = 0.18$	$\{A_5, A_{10}\} = 0.44$	$\{A_5, A_9, A_{10}\} = 0.65$		
$\{A_{10}\} = 0.15$	$\{A_6, A_9\} = 0.48$	$\{A_6, A_9, A_{10}\} = 0.7$		
	$\{A_6, A_{10}\} = 0.45$			
	$\{A_9, A_{10}\} = 0.6$			
Research Factor				
$\{A_4\} = 0.26$		$\{A_4, A_{11}\} = 0.7$	$\{A_4, A_{11}, A_{12}\} = 1$	$\omega_2 = 0.2137$
$\{A_{11}\} = 0.27$		$\{A_4, A_{12}\} = 0.5$		
$\{A_{12}\} = 0.21$		$\{A_{11}, A_{12}\} = 0.6$		
Efficiency Factor				
$\{A_1\} = 0.33$		$\{A_1, A_2\} = 0.5$	$\{A_1, A_2, A_3\} = 1$	$\omega_3 = 0.2601$
$\{A_2\} = 0.23$		$\{A_1, A_3\} = 0.6$		
$\{A_3\} = 0.28$		$\{A_2, A_3\} = 0.4$		
Facility Factor				
$\{A_7\} = 0.46$		$\{A_7, A_8\} = 1$		$\omega_4 = 0.1089$
$\{A_8\} = 0.50$				

TABLE 17.3
Synthetic Evaluation Values of Partition Fuzzy Integral of Common Factors

Alternatives	Environment Scale Factor	Research Factor	Efficiency Factor	Facility Factor
School A	0.205	0.446	0.196	0.271
School B	0.101	0.095	0.166	0.483
School C	0.195	0.690	0.264	0.058
School D	0.756	0.822	0.763	0.882
School E	0.269	0.199	0.172	0.293
School F	0.187	0.332	0.640	0.194
School G	0.184	0.474	0.275	0.560
School H	0.246	0.549	0.217	0.236

TABLE 17.4
Comparison of Three Evaluation Models

	Gray Relational Grades				Ranking of Human Cognitive Investigation
	Simple Additive Weighting		Non-Additive Fuzzy Integral		
Alternatives	Scores	Rankings	Scores	Rankings	
School A	0.4334	5	0.4960	6	6
School B	0.3759	8	0.4520	8	8
School C	0.5008	2	0.5380	2	2
School D	0.7252	1	0.8660	1	1
School E	0.4159	7	0.4760	7	7
School F	0.4460	3	0.5280	3	3
School G	0.4397	4	0.5200	4	4
School H	0.4260	6	0.5190	5	5

1. This study suggests using factor analysis of multivariate analysis and the cluster analysis to reconstruct an evaluation hierarchy system, and to regroup those criteria into several dimensions. This saves computations and assures the independence of individual dimensions.
2. According to the grades of gray relation obtained in this empirical case study, the ranking order of sample junior colleges is listed below, where the symbol $\succ$ denotes the outranking priority:

   a. Outcomes of the conventional gray-relational evaluation with additive measures:
   school D $\succ$ school C $\succ$ school F $\succ$ school G $\succ$ school A $\succ$ school H $\succ$ school E $\succ$ school B.
   b. Outcomes of the gray-relational evaluation with non-additive measures:
   school D $\succ$ school C $\succ$ school F $\succ$ school G $\succ$ school H $\succ$ school A $\succ$ school E $\succ$ school B.
   c. Human Cognitive Investigation (HCI)

   We asked scholars who knew the eight target schools very well to do a cognitive investigation of priority ranking and obtained the following results:

   school D $\succ$ school C $\succ$ school F $\succ$ school G $\succ$ school H $\succ$ school A $\succ$ school E $\succ$ school B.

   These empirical analysis outcomes show some interesting phenomena. The conventional gray relational evaluation with additive measures differs slightly with the gray relational evaluation with non-additive measures, switching the priority rankings of the fifth and the sixth schools in these two approaches. The gray relational evaluation with non-additive measures and the human cognitive investigation (HCI) produce equivalent ranking results. The conventional gray relational evaluation with additive measures differs slightly with HCI. These findings verify our proposed models meet practical and cognitive results better because we cover some governing issues, like uneven weightings of factors and interaction between parameters.

3. The multiple criteria fuzzy evaluation model critically considers the problem of the disagreed recognition and provides objective space for expansion of the factor loadings. The multiple criteria fuzzy evaluation model that was suggested by this empirical study also seemingly produces better evaluation results. This indicates that the fuzzy integral for multiple criteria evaluation is superior to the conventional multiple criteria evaluation in terms of feasibility and reasonability. Therefore, our suggested model would significantly aid in the selection of the governing factors for outstanding junior colleges.

17.11 CONCLUSIONS

In a practical system, governing factors are mutually reactive and influential, which makes the use of the conventional additive approach inapplicable. To resolve this problem of inapplicability mentioned above, this study suggests employing factor analysis of the multivariate analysis and the cluster analysis, making use of the powerful ability of the fuzzy integral to handle non-additive measures, and assigning adequate weightings to individual data items according to their contributions to the system, thus constructing the fuzzy integral model with non-additive measures. The implementation of our suggested model would eliminate those potential troubles that are brought about by the conventional additive measures and would confirm the independence between attributes since we regrouped the dimensions by the multivariate approach.

We then tested our modified fuzzy integral evaluation method with multiple criteria through an empirical case study of eight Taiwanese junior colleges, which had upgraded successfully into technological institutes. Study outcomes confirmed the feasibility of our suggested model, which cultivates a better regrouping of factor types, in comparison with the conventional approach, and whose evaluation results coincided well with the subjective cognition of the public. This verifies our suggested model surpasses the conventional multiple criteria evaluation in terms of feasibility and reasonability. We could conclude confidently that the fuzzy integral evaluation method with multiple criteria would significantly help select the governing factors of the outstanding junior colleges.

The proposed fuzzy integral evaluation method with multiple criteria critically considers the mutual reactions between governing parameters of outstanding junior colleges, screens out common factors, and thus confirms that the attributes are completely independent. It evaluates outstanding junior colleges through less common factors that are more representative, administers a reasonable and objective evaluation job, produces accurate outcomes that cope well with the real results, and effectively evaluates the selection of outstanding junior colleges.

APPENDICES

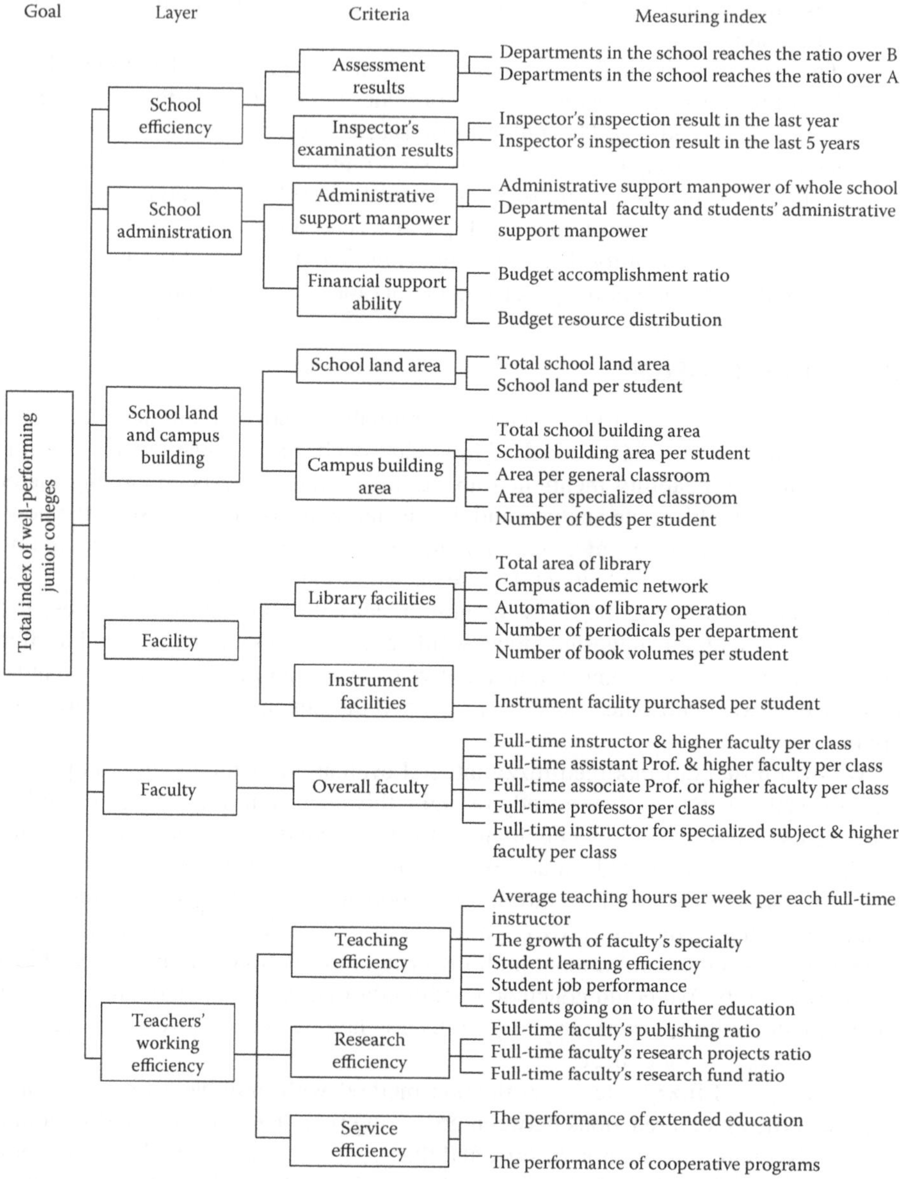

APPENDIX 17.1 Assessing hierarchy system for well-performing junior colleges.

APPENDIX 17.2

The Illustration of Assessment/Measuring Method for Well-Performing Junior College Selection Model

Criterion	Measuring Index	Formula or Illustration	Index Code
Criteria and Measurement Relating to School Efficiency			
Assessment results	Departments in the school reaches the ratio over B	Departments in the school reaches the ratio over B/Departments were assessed	1
	Departments in the school reaches the ratio over A	Departments in the school reaches the ratio over A/Departments were assessed	2
Inspector's examination results	Inspector's inspection result in the last year	Inspector's inspection result in the recent 1 year reached A (85%) or better	3
	Inspector's inspection result in the last 5 years	The rank of inspector's inspection result in the last 5 years	4
Criteria and Measurement Relating to School Administration			
Administrative support manpower	Administrative support manpower of whole school	Administrative support manpower in the whole school includes clerk, assistant teacher, technical personnel, technician, servant, security guard, driver, and other personnel	5
	Departmental faculty and students' Administrative support manpower	Administrative manpower/total number of students and faculty	6
Financial support ability	Budget accomplishment ratio	Actual total expenses/total budgeted expenses	7
	Budget resource distribution	Actual total income/daytime students	8
Criteria and Measurement Relating to School Land and Campus Building			
School land area	Total school land area	By university law, junior college law, and relevant regulations, campus land that can be developed has to meet the minimum requirement (5 hectare)	9
	School land per student	Total utilized campus land/number of daytime students	10
Campus building area	Total school building area	By university law, junior college law, and relevant regulations, school building area has to meet the minimum requirement (12,000 square meters)	11

(*continued*)

APPENDIX 17.2 (Continued)
The Illustration of Assessment/Measuring Method for Well-Performing Junior College Selection Model

Criterion	Measuring Index	Formula or Illustration	Index Code
	School building area per student	Total area of school buildings/number of daytime classes	12
	Area per general classroom	Number of general classrooms/ number of daytime classes	13
	Area per specialized class room	Number of specialized classrooms/number of daytime classes	14
	Number of beds per student	Number of beds/number of daytime students	15
Criteria and Measurement Relating to Facility			
Library facilities	Total area of library	Library is required	16
	Campus academic network	Fund for campus academic network/number of daytime students	17
	Automation of library operation	Fund for automation of library operation/number of daytime students	18
	Number of periodicals per department	Number of periodicals/total number of departments	19
	Book volumes per student	Book volumes/number of daytime students	20
Instrument facilities	Instrument facility purchased per student	Annual fund for instrument facility/number of daytime students	21
Criteria and Measurement Relating to Faculty			
Overall Faculty	Full-time instructor and higher faculty per class	Number of full-time instructor & higher faculty/number of classes	22
	Full-time assistant prof. and higher faculty per class	Number of full-time assistant prof. & higher faculty/number of classes	23
	Full-time associate prof. or higher faculty per class	Number of full-time associate prof. or higher faculty/number of classes	24
	Full-time professors per class	Number of full-time professors/ number of classes	25
	Full-time instructors for specialized subjects and higher	Number of full-time instructors for specialized subjects & higher/number of classes	26

APPENDIX 17.2 (Continued)
The Illustration of Assessment/Measuring Method for Well-Performing Junior College Selection Model

Criterion	Measuring Index	Formula or Illustration	Index Code
Criteria and Measurement Relating to Teachers' Working Efficiency			
Teaching efficiency	Average teaching hours per week per full-time instructor	Total teaching hours/number of teachers	27
	The growth of faculty's specialty	Number of teachers that go on seminars and for further education/number of teachers	28
	Student learning efficiency	Actual number of graduates/ number of daytime graduates	29
	Student job performance	Number of students going for jobs/number of daytime graduates	30
	Students go for further education	Number of students going for further education/number of daytime graduates	31
Research efficiency	Full-time faculty's publishing ratio	Number of full-time faculty publications/number of teachers	32
	Full-time faculty's research projects ratio	Number of full-time faculty's research projects/number of teachers	33
	Full-time faculty's research fund ratio	Full-time faculty's research fund/ number of teachers	34
Service efficiency	The performance of cooperative education	Income from cooperative education/number of teachers	35
	The performance of extension education	Income from extension education/number of teachers	36

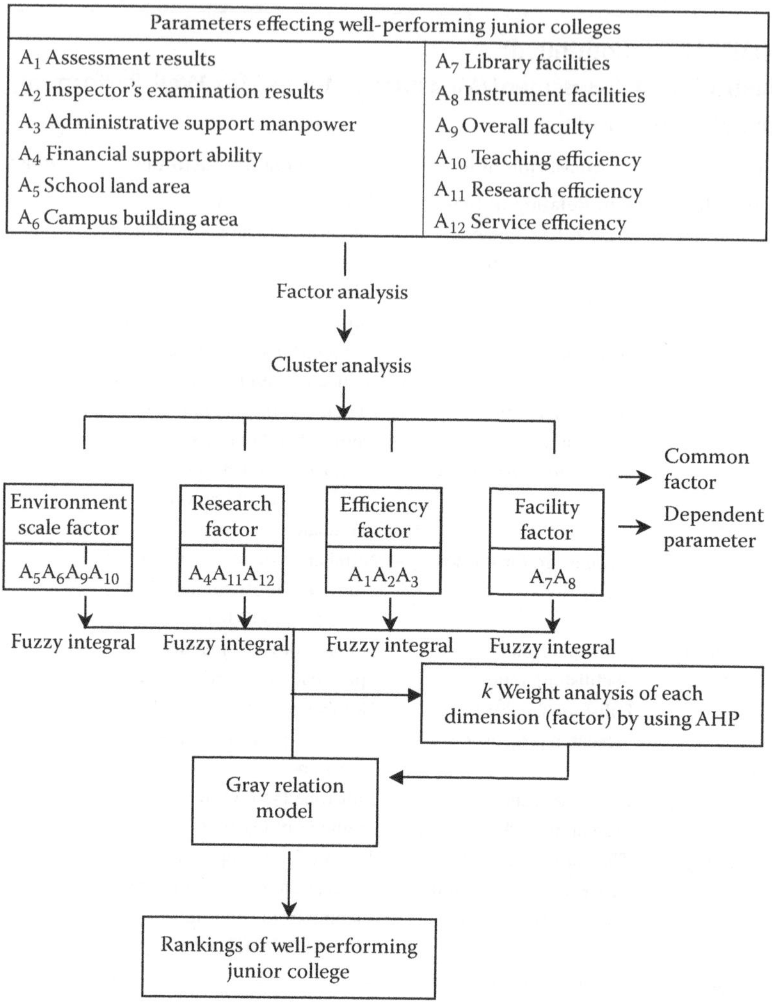

APPENDIX 17.3 Analysis procedures.

18 Fuzzy Integral: An Application

18.1 INTRODUCTION

One of the traditional tools for information aggregation is the weighted average method, for example, a linear integral or the Lebesgue integral (Lebeggue 1966). These methods assume that the information sources involved are non-interactive/independent and, hence, their weighted effects are viewed as additive type. However, this assumption is not realistic in many real-world applications. Due to some inherent interaction/interdependencies among diverse information sources, the weighted average method does not work well in many real problems. Instead of the weighted average method, the Choquet integral can be used. The Choquet integral can be applied to multiattribute evaluation such as Grabisch (1995, 1996), Lee, Liu, and Tzeng (2001), Chen and Tzeng (2001), etc. Fuzzy measures and fuzzy integrals can analyze the human evaluation process and specify decision-makers' preference structures.

The Choquet fuzzy integral is a fuzzy integral based on any fuzzy measure that provides an alternative computational scheme for aggregating information (Chiang 1999). Sugeno (1974, 1977) introduced the concepts of fuzzy measure and fuzzy integral. Fuzzy measures, according to Sugeno, are obtained by replacing the additivity requirement of classical measures with weaker requirements of monotonicity (with respect to set inclusion) and continuity. The requirement of continuity was later found to be still too restrictive and was replaced with a weaker requirement of semicontinuity. Since the specification of general fuzzy measures is extremely cumbersome, Sugeno (1974) and Sugeno and Terano (1977) proposed a λ-fuzzy measure satisfying the λ-additive axiom to reduce the difficulty of fuzzy measure identification. The λ-fuzzy measure is constrained by a parameter, λ, which describes the degree of additivity between elements. Compared with other fuzzy measure patterns, the λ-fuzzy measure is easier and is widely used in determining measure values (Chen and Wang 2001; Lee and Leekwang 1995). However, when the number of elements is sufficiently large, the identification of λ-fuzzy measure is still troublesome for users. Lee and Leekwang (1995) developed an identification method of λ-fuzzy measure based on genetic algorithms, although the information for a fuzzy measure value of an element from the data set was not complete. Chen (1998) and Chen and Wang (2001) developed a partial information sampling procedure in order to reduce the information demand, also employing genetic algorithms as the solution strategy. Their methods overcame the difficulty of data collection for subjective importance identification. Although their methods work well, their questionnaire data requires fuzzy densities and partial information about performance values. However, it is easiest to investigate only fuzzy densities to determine the λ-value (Wang et al.'s algorithm) (Wang, Chen, and Shen 2001). According to Wang et al. (2001) this research will propose an effective algorithm to determine the λ-value using the input data of fuzzy densities. Therefore, the main objectives of this research are as follows: Firstly, it uses fuzzy measures and fuzzy integrals to determine the overall performance of human subjective decision making. Secondly, it develops a hierarchical structure for evaluating the enterprise intranet websites and uses the methods of λ-fuzzy measure and Choquet integral to assess overall evaluation. Finally, it has developed a simpler and easier algorithm to determine the λ-value.

The remainder of this chapter is organized as follows. In Section 18.2, the λ-fuzzy measures and fuzzy integral for a multiattribute decision-making (MADM) process are presented. In Section 18.3, modeling the hierarchical structure of Choquet integral and its algorithm for identifying λ are presented. In Section 18.4, evaluation of enterprise intranet websites as a case and the results of analysis are discussed in detail. Finally, conclusions are given in Section 18.5.

18.2 MODELING THE HIERARCHICAL STRUCTURE OF THE CHOQUET INTEGRAL

Since criteria interact and affect each other in the real world, a fuzzy integral is employed to conduct non-additive operations for these dependent aspects, criteria, and subcriteria. Furthermore, this research constructs the hierarchical structure of a Choquet integral assessment model, as shown in Figure 18.1. In addition, this research uses an effective algorithm to determine the λ-value.

In Figure 18.1, if we see each circle as a node, we can use the evaluation values f and the grades of importance g on the lower-level objects/elements calculated on Choquet integral's Equation 9.2 to obtain the upper-level objects/elements evaluation values. For example, $f_1^{11}, f_2^{11}, \ldots, f_{s_1}^{11}$ are the evaluation values of the bottom-level objects/elements; and $g_1^{11}, g_2^{11}, \ldots, g_{s_1}^{11}$ are the grades of importance. By using

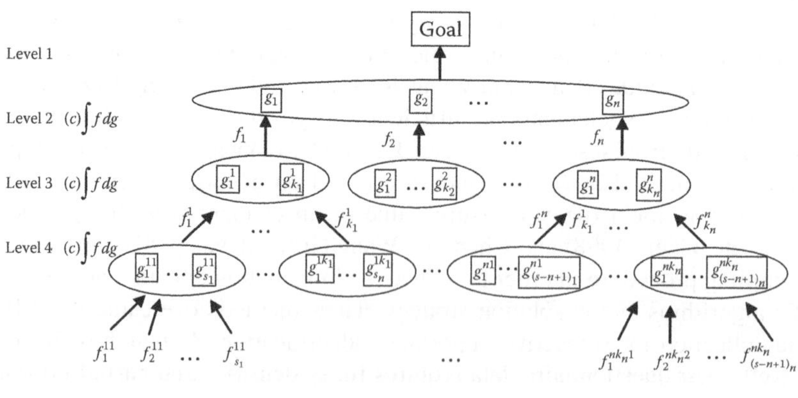

FIGURE 18.1 Concepts of Choquet integral assessment model.

Choquet integral's Equation 9.2 to compute the subtotal evaluation values of the first node on Level 4, we can get the result f_1^1. The other subtotal evaluation values can be calculated in the same way. Then all results are $(f_1^1, f_2^1, \ldots, f_{k_1}^1), \ldots, (f_1^n, f_2^n, \ldots, f_{k_n}^n)$, respectively. Likewise, there are n nodes on Level 3. Here $f_1^1, f_2^1, \ldots, f_{k_1}^1$ are the evaluation values, $g_1^1, g_2^1, \ldots, g_{k_1}^1$ are the grades of importance, and the result is f_1. The other subtotal evaluation values are $f_2, \ldots, f_n$. Finally, there is only one node on level 2. Here $f_1, f_2, \ldots, f_n$ are the evaluation values and $g_1, g_2, \ldots, g_n$ are the grades of importance. By using Equation 8 to compute the overall evaluation value, we get the final result on Level 1.

18.3 ALGORITHM FOR IDENTIFYING λ

The algorithm for identifying λ is adopted from Wang et al. (2001). According to Equation 9.4 the computing algorithm of Wang et al. is listed as follows: Step 1: if $F'(0) = 0$, then $\lambda = 0$, stop; Step 2: if $F'(0) > 0$, then let $p^* = -1$, $m^\dagger = 0$, and go to Step 5 to perform a bisection search; Step 3: if $F'(0) < 0$, then let $p = +1$, $m = 0$, and go to Step 4 to find a range of λ; Step 4: if $F(p) < 0$, let $m = p$, $p = p*2$ and continue Step 4 (repeat double p until $F(p) > 0$); Step 5: if $F((p+m)/2) = 0$, then $\lambda = (p+m)/2$, and stop; Step 6: if $F((p+m)/2) > 0$, then let $p = (p+m)/2$, else let $m = (p+m)/2$, and continue from Step 5.

Following Wang et al. (2001) we propose three steps as follows, based on the properties of λ above: Step 1: if $\sum_{i=1}^n g_i = 1$, then $\lambda = 0$, stop; Step 2: if $\sum_{i=1}^n g_i > 1$, then let $p = -1$, $m = 0$, go to Step 5 to perform a bisection search; Step 3: if $\sum_{i=1}^n g_i < 1$, then let $p = +1$, $m = 0$, go to Step 4 to find a range of λ.

The results of our modified algorithm are compared with the algorithm of Wang et al., indicating that the λ-value of this research is very close to that of Wang et al. It very clearly shows the properties of λ-fuzzy measure with respect to using $\sum_{i=1}^n g_i$ (Leszczyński, Penczek, and Grochulski 1985), and it is easy to obtain a solution by the algorithm as modified in this study.

18.4 EMPIRICAL CASE: FUZZY INTEGRAL FOR ENTERPRISE INTRANET WEBSITES

The previous section discussed the advantages of using the λ-fuzzy measure and Choquet integral. In this section, a case of intranet website evaluation is used to illustrate the feasibility of the proposed approach. The empirical case background, the problem statement, and the multiattribute assessment model are discussed below.

18.4.1 Background and Problem Statement

Websites are widely employed throughout industry, education, government, and other institutions. In addition, electronic commerce (EC) activities have been discussed widely (D'Ambra and Rice 2001). EC can help business organizations cut costs, interact

* p is pointer.
† m is initial value, $\lambda \in (p, m)$.

directly with customers, and operate more efficiently, thus helping an organization to be more competitive. Intranet websites are discussed less frequently than EC. However, it should be noted that effective enterprise intranet websites can help decision makers obtain important information or knowledge to stimulate innovation, promote exchange of knowledge, and promote company working efforts. Accordingly, some enterprises have annual contests for their intranet website performance. Therefore, assessing factors associated with website success is needed. Generally speaking, the evaluation of intranet website performance is an MADM problem, and it also involves human subjective decision making. Consequently, it can be properly characterized using the λ-fuzzy measure and Choquet integral. This research uses this empirical case to illustrate the approach using the λ-fuzzy measure and Choquet integral, and modeling a hierarchical evaluation system for assessing the enterprise's intranet websites.

18.4.2 Constructing a Hierarchical Multiattribute Evaluation System

There are three stages in this system, as follows: (1) a hierarchical system of enterprise intranet websites is established, the aspect information is collected and the ways of collecting information are stated; (2) the criteria grade of importance is determined and its performance scores of bottom criteria are obtained from questionnaire investigation; and (3) overall evaluation is obtained by the Choquet integral.

18.4.2.1 Generating Aspects and Performance Scores for Criteria

The aspect information being assessed for the case enterprise intranet websites was adopted from the evaluation plan for Taiwan government websites and relational needs of the case enterprise. However, the enterprise intranet websites provide services for employees, so they are different from the service objects of government websites, which are ordinary people. Therefore, information is updated according to the case enterprise needs. In order to promote content quality, the selected criteria and subcriteria were also discussed with workers in the case enterprise who have experience constructing websites. The proposed hierarchical evaluation system for case enterprise intranet websites is listed in Figure 18.2. All detailed aspects/criteria/subcriteria are listed in Appendix 18.3. Finally, according to the subcriteria of the bottom level, we can obtain the performance scores (f) in the case of enterprise intranet websites (E1–E7) listed on Level 5. That is, from C111–C114, C121–C123, C131–C136, C211–C212, C221–C222, and C231–C236, every item has an answered question and their performance scores are f_1^{11}–f_4^{11}, f_1^{12}–f_3^{12}, f_1^{13}–f_6^{13}, f_1^{21}–f_2^{21}, f_1^{22}–f_2^{22}, and f_1^{23}–f_6^{23} respectively. Likewise, C31–C39 have nine answered questions and their performance scores are f_1^3–f_9^3. C41–C39 also have nine answered questions and their performance scores are f_1^4–f_9^4.

18.4.2.2 Determining the Grade of Criteria Importance

The grade of criteria importance is determined from the questionnaire investigation and the mapping membership function of the seven scales of linguistic variables. Every evaluator obtains his/her own λ_i^h-value by our modified algorithms, as introduced in Section 18.3.1. Here, λ_i^h is the λ-value of the hth evaluator toward the ith group of criterion (see Figure 18.2). In Figure 18.2, the aspects C1, C2, C3,

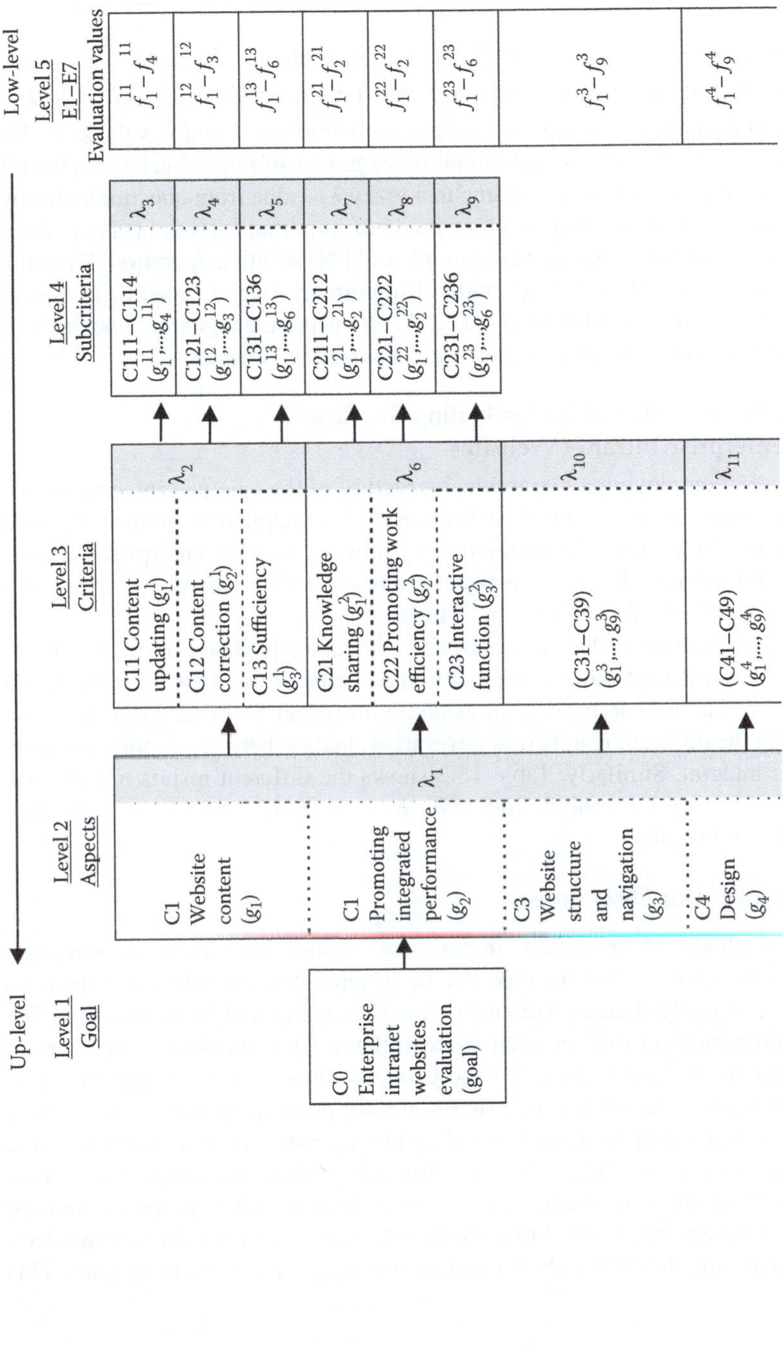

FIGURE 18.2 Hierarchical evaluation systems for the case enterprise intranet websites.

TABLE 18.1
Results of λ-values for the Enterprise Intranet Websites

λ_1	λ_2	λ_3	λ_4	λ_5	λ_6	λ_7	λ_8	λ_9	λ_{10}	λ_{11}
−0.6865	0.0574	−0.7854	−0.1036	−0.8838	−0.4584	1.1924	2.9619	−0.8273	−0.9693	−0.9797

and C4 generate a λ_1^h-value in Level 2 for the hth evaluator. In Level 3 there are $\lambda_2^h, \lambda_6^h, \lambda_{10}^h$, and λ_{11}^h for the hth evaluator. In Level 4, there are λ_3^h, λ_4^h, λ_5^h, λ_7^h, λ_8^h, λ_9^h, for the hth evaluator. Consequently, every evaluator has eleven λ_i values. In this research, every evaluator obtains individual fuzzy grades of importance (weight) for criteria and criteria interactions according to a fuzzy λ_i^h-value from questionnaires by using Equation 9.4, and the integral fuzzy grade of importance (weight) is obtained through the mean of the triangular fuzzy number (TFN) additive operation. By using the defuzzifying approach, the crisp grade of importance (weight) can be obtained. In addition, the λ_i-value is obtained ($i = 1,\ldots,11$) by using the algebraic operation of the TFN and using the defuzzifying approach, as listed in Table 18.1.

18.4.2.3 Choquet Integral for Evaluating the Case Enterprise Intranet Websites

The procedure of employing a hierarchical structure of the Choquet integral for the multiattribute assessment, as stated in Section 18.3, is applied to analyze the real cases. In Figure 18.2, since the evaluation system of the case enterprise intranet websites has a tree hierarchy, we investigate evaluation values (scores of assessment) (f_1^{11}–f_6^{23}, f_1^3–f_9^3, f_1^4–f_9^4) for the bottom of the tree.

The overall evaluation of the case enterprise is obtained by determining the grades of importance, assessment scores, and hierarchical structure of the Choquet integral in Section 18.3. This research gives an example for intranet websites (E1–E7), and for illustrating the different results of an overall evaluation between arithmetic mean and Choquet integral. Similarly, Table 18.2 shows the different results of the overall evaluation between the analytic hierarchy process (AHP) and Choquet integral, together with their results.

18.4.3 Results and Discussions

Traditional methods of assessing importance cannot effectively approximate the human subjective evaluation process. In general, human subjective decision making can be properly characterized using fuzzy measures and fuzzy integrals. The expected contributions of this research are as follows: (i) a hierarchical structure of human subjective decision making for the Choquet integral is developed to assess the overall evaluation for which the criteria are not completely independent; (ii) a modified algorithm is proposed, corresponding to the perspective of λ-fuzzy measure adapted from Wang et al. (2001) (these modified algorithms are simpler and easier to understand than those in Wang); (iii) an assessment model is proposed and the assessment results can be provided to relevant authorities for their reference in selecting better-performing intranet websites and improving poorly performing ones. This evaluation can improve the quality of the case enterprise intranet websites and the efficiency of e-works, knowledge share, etc. In order to demonstrate that the Choquet integral is more suitable than a traditional multicriteria evaluation method for human subjective evaluation, or when criteria are dependent, we have provided some simple examples in Appendix 18.2. These examples demonstrate that the fuzzy hierarchical analytic approach can cope better with the non-independent situations that frequently occur in real-world decision-making problems. Consequently, we have successfully demonstrated that the non-additive fuzzy integral technique can overcome the non-independent criteria cases, and the results of the hierarchical MADM of the Choquet integral for evaluating the enterprise intranet websites are discussed as follows.

1. From the viewpoint of the grade of importance, Table 18.1 represents the interactions in each criterion, since $\lambda \neq 0$. In contrast, if the λ-value is positive, that implies the criteria relations have a multiplicative effect, and also that criteria would be enhanced simultaneously. If the λ-value is negative, that implies the existence of the substitutive effect. However, authorities can analyze the improved methods according to the results, which is the major contribution of our paper. However, the λ-value can adjust the underestimation or overestimation of the grades of importance, according to whether they are positive or negative, respectively.
2. From the results based on these methods, this research uses average values to compare with those of AHP, as listed in Table 18.2, and to compare with those of the arithmetic mean, as listed in Table 18.2.

In Table 18.2, the AHP ranking order is E6 $\succ$ E7 $\succ$ E5 $\succ$ E3 $\succ$ E1 $\succ$ E4 $\succ$ E2, and the order using the Choquet integral is E6 $\succ$ E7 $\succ$ E1 $\succ$ E3 $\succ$ E5 $\succ$ E4 $\succ$ E2. Thus, there are different ranking orders when these two methods are used to obtain overall scores. Strategy E1 evaluates more highly, E5 is lower, and the other ranking order is the same as using the Choquet integral. According to the change of ranking order, this research discusses E1 and E5. By using the same approach as Examples 18.1

TABLE 18.2
Ranking Order of AHP and Arithmetic Mean and Choquet Integral: Practical Case

Alternatives	AHP Scores	Ranking Order	Choquet Integral Scores	Ranking Order	Arithmetic Mean	Ranking Order
E1	0.6363	5	0.7191	3	0.6504	3
E2	0.5890	7	0.6542	7	0.5480	7
E3	0.6363	4	0.6996	4	0.5821	6
E4	0.6336	6	0.6817	6	0.6138	5
E5	0.6677	3	0.6855	5	0.6309	4
E6	0.7101	1	0.7453	1	0.7016	1
E7	0.6713	2	0.7397	2	0.6837	2

and 18.2 (Appendix 18.2), we can see that E1 is actually higher than E5. Since they have interaction ($\lambda \neq 0$), E5 is overestimated by AHP. Additionally, the arithmetic mean ranking order is E6 $\succ$ E7 $\succ$ E1 $\succ$ E5 $\succ$ E4 $\succ$ E3 $\succ$ E2. Comparing the different ranking orders from Choquet integral and from arithmetic mean, we see that their evaluations of E3 are different. Similarly, using the same analytic approach as above, we can find that E3 really is better than E4 and E5 in practice. In sum, when assessable criteria of MADM interact in the real world, using the Choquet integral is more accurate than either the AHP or arithmetic mean.

From the above results and the illustration in Appendix 18.2, employing the Choquet integral to obtain an overall evaluation is more suitable, because interactions of criteria are considered, whereas with traditional methods criteria must be independent. In addition, this method does not overestimate or underestimate for the additive model (AHP and arithmetic mean) when criteria are dependent. Thus, it is easier to select better websites in this case.

18.5 CONCLUSIONS

In traditional multiattribute evaluation approaches, each attribute must be independent from the others. Therefore, the characteristics that have interactions and mutual influence among attributes or criteria in a real system cannot be handled by the concept of traditional additive measures alone. However, non-additive fuzzy measures and a fuzzy integral model are extremely effective for analyzing the relations between criteria in a real system. Currently, the most widely used fuzzy measure is the λ-fuzzy measure. However, when the number of elements is sufficiently large, the identification of the λ-fuzzy measure is troublesome. This research modifies an algorithm from Wang et al. (2001) to determine the λ-value using the input data of fuzzy densities. The modified algorithm shows the properties of the λ-fuzzy measure more clearly and simply than the algorithms of Wang et al. (2001). This research employs a practical case for the hierarchical enterprise intranet websites assessment model to assess scores for achieving overall evaluations. The results show that the fuzzy integral multiattribute evaluation process in this research is effective and applicable. In addition, the fuzzy integral results are better and more reasonable than those obtained from the traditional multiattribute assessment process. Consequently, the hierarchical structure evaluation system of human subjective decision making using λ-fuzzy measure and fuzzy integral as proposed in this research is an appropriate approach to the evaluation of case enterprise intranet websites, especially when criteria are not mutually independent.

APPENDIX 18.1

Figure A18.1.1 is illustrated as follows.

i. If $F'(0) > 0$, then one solution of $F(\lambda)$ is $\lambda = 0$, and another solution scope is $-1 < \lambda < 0$ (i.e., $\sum g_i > 1, -1 < \lambda < 0$). This is line (1).
ii. If $F'(0) > 0$, then its solution is $\lambda = 0$ (i.e., $\sum g_i = 1$, $\lambda = 0$). This is line (2).
iii. If $F'(0) < 0$, then one solution of $F(\lambda)$ is $\lambda = 0$, and another solution scope is $\lambda > 0$ (i.e., $\sum g_i < 1$, $\lambda > 0$). This is line (3).

The ideas of the algorithm for identifying λ are from the concepts above. We present the algorithm in Section 18.3.1.

APPENDIX 18.2

Example 18.1

Comparing the qualities and capabilities of computer products. Example, x_1 = easy use, and x_2 = function capabilities. If $g_\lambda(\{x_1\}) = u(x_1^*, x_2^0) = 0.5$, $g_\lambda(\{x_2\}) = u(x_1^0, x_2^*) = 0.3$, $g_\lambda(\{x_1, x_2\}) = 1$; score of computer product P: $f(x_1) = 90$, $f(x_2) = 20$; score of computer product Q: $f(x_1) = 60$, $f(x_2) = 60$, then the results according to the additive model and the Choquet integral model are as follows:

Choquet integral

P: $(c)\int f dg = 20 \times 1.0 + (90 - 20) \times 0.5 = 55$, Q: $(c)\int f dg = 60 \times 1.0 = 60$, then Q $\succ$ P.

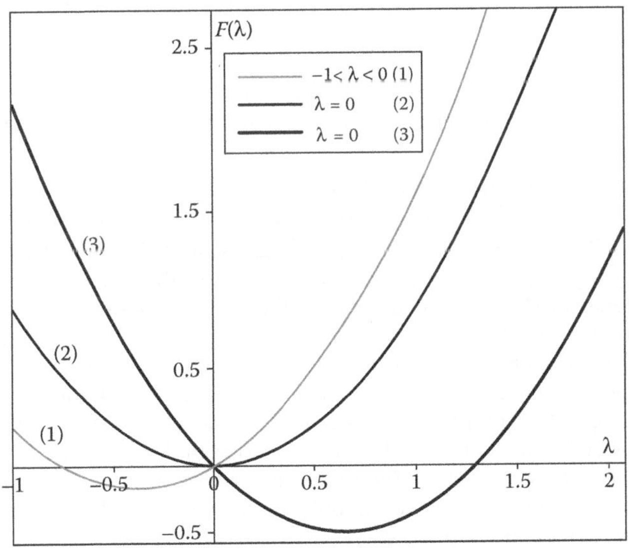

FIGURE A18.1.1 Graphing the function $F(\lambda)$.

Additive model

P: 90 × 0.5/(0.5 + 0.3) + 20 × 0.3/(0.5 + 0.3) = 63.75, Q: 60 × 1 = 60, then P ≻ Q.

1. According to mathematical reasoning, if $\sum_{i=1}^{2} g_\lambda(\{x_i\}) < 1$ (i.e., $\lambda > 0$, it implies an underestimation situation in the grades of importance, if we use the additive model), then $g(\{x_1\})/\sum_{i=1}^{2} g_\lambda(\{x_i\}) > g(\{x_1\})/g_\lambda(\{x_1, x_2\})$, and $g(\{x_2\})/\sum_{i=1}^{2} g_\lambda(\{x_i\}) > g(\{x_2\})/g_\lambda(\{x_1, x_2\})$, and thus we get an overestimated overall evaluation if we used the additive model.
2. From these results, we find that if $g_\lambda(\{x_1\}) + g_\lambda(\{x_2\}) = 0.8 < 1$, then $\lambda > 0$. This implies that their criteria relations have a multiplicative effect. In other words, it can increase overall performance if criteria are enhanced simultaneously. In practice, we hope the results of the evaluation can all reach a certain satisfying level, so Q ≻ P is more reasonable. Therefore, the Choquet integral is more suitable than a traditional evaluation method when criteria are dependent.

Example 18.2

Employment evaluation items (from interview and test). Example, x_1 = salesman's ability and x_2 = technological ability. If $g_\lambda(\{x_1\}) = u(x_1^*, x_2^0) = 0.9$, $g_\lambda(\{x_2\}) = u(x_1^0, x_2^*) = 0.9$, $g_\lambda(\{x_1, x_2\}) = 1$; score of interview and test for Mr. P: $f(x_1) = 90$, $f(x_2) = 20$; score of interview and test for Mr. Q: $f(x_1) = 60$, $f(x_2) = 60$, score of interview and test for Mr. R: $f(x_1) = 30$, $f(x_2) = 80$; then the results according to the additive model and the Choquet integral model is as follows:

Choquet integral

$$\text{P: } (c)\int fdg = 20 \times 1.0 + (90 - 20) \times 0.9 = 83, \text{ Q: } (c)\int fdg = 60 \times 1.0 = 60.$$

$$\text{R: } (c)\int fdg = 30 \times 1.0 + (85 - 30) \times 0.9 = 79.5, \text{ then P} \succ \text{R} \succ \text{Q}.$$

Additive model

P: 90 × 0.9/(0.9 + 0.9) + 20 × 0.9/(0.9 + 0.9) = 55, Q: 60 × 1 = 60.
R: 30 × 0.9/(0.9 + 0.9) + 85 × 0.9/(0.9 + 0.9) = 57.5, then Q ≻ R ≻ P.

1. According to mathematical reasoning, if $\sum_{i=1}^{2} g_\lambda(\{x_i\}) > 1$ (i.e., $\lambda < 0$ implies overestimation situation in the grades of importance, if we use the additive model), then $g(\{x_1\})/\sum_{i=1}^{2} g_\lambda(\{x_i\}) < g(\{x_1\})/g_\lambda(\{x_1, x_2\})$, and $g(\{x_2\})/\sum_{i=1}^{2} g_\lambda(\{x_i\}) < g(\{x_2\})/g_\lambda(\{x_1, x_2\})$, and thus we get an underestimated overall evaluation if we used the additive model.
2. From these results, we find that if $g_\lambda(\{x_1\}) + g_\lambda(\{x_2\}) = 1.8 > 1$, then $\lambda > 0$. This implies that criteria relations have a substitutive effect. We can enhance some criteria, if we want to increase overall performance. In practice, we hope that both the required abilities meet our needs. But if we can't have both together, we choose professional skills as the higher priority, and P ≻ R ≻ Q is better. Therefore, according to the above examples, the Choquet integral is more suitable than a traditional evaluation method when criteria are dependent.

APPENDIX 18.3

TABLE A18.3.1
Definition of Aspects/Criteria for the Enterprise Websites

Aspects/Criteria	Subcriteria
C1 Website Content	
C11 Content updating	C111 Content updated with the latest version and up-to-date data
	C112 Data updated at least every three days
	C113 Regularly update the web pages and content (check randomly at different times)
	C114 Having indications of the updated places or dates
C12 Content correction	C121 Having consistency of the title and content
	C122 Having precision of all the hyperlinks
	C123 Having indications of the sources and copyrights
C13 Sufficiency	C131 Providing an introduction to the departments, staff, and positions of the employees
	C132 Providing complete business introduction
	C133 Providing the latest news, notifications, or activities
	C124 Providing frequently asked questions (FAQ)
	C135 Providing contact-us information including service time, location, phone, fax, or e-mail, etc
	C136 Providing related links to other websites
C2 Promoting Integrated Performance	
C21 Knowledge sharing	C211 Organized with documentation and information sharing zone
	C212 Organized with the discussion zone
C22 Promoting work efficiency	C221 E-data processing
	C222 On-line e-processing of application, registration, inquiry, or work procedures
C23 Interactive function	C231 Providing service mailbox and chief director's mailbox
	C232 Providing download of documents/forms
	C233 Providing network authentication and encryption
	C234 Providing on-line calculating/simulating functions
	C235 User-friendly interfaces of interactive functions
	C236 Providing on-line help or guidance of interactive functions
C3 Website Structure and Navigation	
	C31 Website's content is classified by a convenient, understandable methodology
	C32 Distinct classification without repetition and ambiguity
	C33 All items in each catalog are easy and understandable
	C34 Moderate amount (about 4–15) of classified data in each layer
	C35 Moderate layers of classification (less than six layers)
	C36 Providing hyperlinks to the homepage, the last page, and the parent directory in all web pages

(*continued*)

TABLE A18.3.1 (Continued)
Definition of Aspects/Criteria for the Enterprise Websites

Aspects/Criteria	Subcriteria
	C37 Quick links to all catalogs on the homepage of all web pages
	C38 Hints for the latest and frequently used information on the homepage
	C39 Providing a sitemap or auxiliary help tools.
C4 Design	
	C41 Consistent editorial features and style for all web pages
	C42 Easy reading colors and fonts
	C43 Providing the resolution and browser recommendation
	C44 Complete the download of homepage and static information within fifteen seconds in normal situations
	C45 Acceptable download time of the multimedia on the website
	C46 Capably stressing the features of a department
	C47 Innovative website functions
	C48 Innovative website presentations
	C49 Animation and multimedia presentations

19 Rough Sets: An Application

19.1 INTRODUCTION

Knowing customers' consumer trends, buying behavior, and product purchase acceptability are very important in the field of sales marketing. It is well known that customer satisfaction with products or service provided by industries is deemed as key to achieving successful business operation and sustainable competitiveness. Customer satisfaction is a critical issue in keeping customers continuously purchasing and it is stimulated by the comparison of the post-purchase experience of a product or service with pre-purchase expectations (Kristensen, Martensen, and Gronholdt 1999). The intention of customer repurchases and retention of customers can be increased by customer satisfaction (Kim, Ferrin, and Rao 2003). Fulfilling customer needs is related to satisfying customer expectations, which achieves customer satisfaction.

The insurance industry is a business that needs face-to-face contact with customers and proactive motivation to provide services for satisfying the customers needs, in order to encourage them to continue insurance and/or re-buy products, which is the main revenue source of the industry. Due to its characteristics, the insurance industry may need up-to-date information to modify the products or services that may attract the attention of potential customers. Therefore, the best data source is market surveys. The results of surveys may provide inherent information on customers, such as needs and the acceptance of products and service.

A set of questionnaires have been designed about insurance products, purchase purpose, purchase expectation, acceptable premium, and participants' basic data, which may serve as a basis to understand the customers needs. Meanwhile, the influence of purchasing intention with the involvement of consumers' purchasing cognition and motivation is discussed and analyzed. Next, the consumers' purchase decisions and processes are discussed, and proper marketing strategies and management operations are proposed. The analysis results may be fully applied by enterprise management to make decisions on the strategies and processes related to consumer purchasing.

Most presented papers deal with insurance audits, product acceptability, purchase channel studies, and the methodology of investigating customer purchasing intention and customer satisfaction (Hennig-Thurau and Klee 1997). Researchers used to quantify insurance questions in order to simplify the discussed parameters, which were social parameters, and statistical tools were the common measure used to analyze data. That approach is better only for crisp types of data sets and certainty of data sets. A fuzzy theory is applied, as continuous data sets and uncertain data sets are included (Zadeh 1965).

Rough set theory is applied in this study to analyze data contents and features. Rough set theory was developed by Pawlak (1982) and became a rule-based decision-making technique that could handle crisp data sets and fuzzy data sets, without the need for preassumption membership functions, as required by fuzzy theory. Rough set theory also can deal with uncertainty, vagueness, and perceptible data sets. Perceptible data recognition has various combination choices for a subject. Until now, there has been little discussion of combination choices using rough set theory. In this study, a questionnaire with single-choice and multichoice questions is designed to apply rough set theory to investigate the relationship between them. Using expert knowledge, the value class of the multichoice questions is reclassified in order to simplify the value complexity, which is useful in the decision-making procedure.

The objective of this study is to discuss the effect of approximation accuracy by applying the combination values that result from features/attributes for satisfying needs in the decision making of insurance marketing. The results demonstrate that match the requirements of the insurance market, and the anticipation of customer is premium refund, and that most of those who purchase insurance products are women.

19.2 ILLUSTRATIVE EMPIRICAL STUDY: A CASE FOR MAKING INSURANCE MARKETING DECISIONS

This section applies rough set theory to explore the classification problem via the insurance questionnaire with single-choice and multi-choice questions. Then, we propose the reclasmethod by expert's knowledge to increase the approximation accuracy and improve decision rules.

19.2.1 Problem Descriptions

In 2002, Taiwan's life insurance market share and average people being insured reached 135 and 158%, respectively. Due to changes in administrative codes, non-life insurance companies are allowed to compete in the medical insurance market. Under highly competitive conditions, the best way to access the market and enlarge market share is to acquire necessary information from potential customers, which relies on well-designed surveys. The features of the relationship between the customer and the insurance company can be concluded from information on customer satisfaction and company service. The service provided by industries is no longer deemed as an additional value toward business promotion; a successful business should fulfill the customer's real needs but also combine with other business strategies and/or measures to improve performance. This is critical research not only for Taiwan but also globally. In this study, a series of questions, such as purpose of insurance, purchased products, acceptable premium, purchase anticipation, and reasons for not purchasing products, etc. are designed, and the results of the questionnaire are combined with participants' personal attribute

TABLE 19.1
Response Number of Questionnaire Data

Qualified Data			
Have Purchased Products		Have Not Purchased Products	
Male	Female	Male	Female
133	147	27	17
280 (86%)		44 (14%)	
Total: 324			

data to investigate purchasing trends, motivation, and reasons for not purchasing products.

19.2.2 Empirical Process

The questionnaire was used in the North and Northeast districts of Taiwan. From a total of 420 survey samples, 324 samples completed questionnaires were received. Of those 324 qualified replies, 280 people had purchased insurance products and 44 persons had not (as seen in Table 19.1). Those questions include the demographic attributes (e.g. Age, gender, etc.) and the data of respondent's feeling. The major advantage of rough set theory, unlike traditional statistics or fuzzy sets needs to assume the distribution of data or defines the membership function of a variable, is that it does not postulate any assumption for the data. In order to demonstrate this empirical study, 50 validation sample data are added to test the accuracy of the decision rules.

19.3 STUDY OF CUSTOMER NEEDS AND REASON FOR NO PURCHASE

Expert knowledge is used to process the attribute extraction, which contains eight attributes, of which seven are condition attributes and one is a decision attributes. The multichoice attributes include:

- Purchase anticipation ($d1$)
- Purchased products ($c6$): the customer for what kind of products has purchased
- Purchase purpose ($c5$): for which purpose of purchasing products
- Single choices

In Table 19.2, following the attribute name in brackets is the substitute name, which will make the paper easier to read. To improve the classification rate, the value class is redefined by using expert knowledge (the results are shown in Table 19.2) and a nominal scale is used for the value sets for attributes. The original attribute specification is shown in Appendix 19.1.

TABLE 19.2
Attribute Specification

Attribute Name	Attribute Values	Attribute Value Sets
Condition Attributes		
Area	North district; Northeast district	{1,2}
Age (*c*1)	Single: <25; marriage, housing, business, birth: 25–34; foster children, education: 35–44; retired: 45–	{1,2,3,4}
Gender (*c*2)	Female; male	{F,M}
Profession (*c*3)	1–6 (by department of insurance profession publish)	{1,2,3,4,5,6}
Purpose (*c*5)	(Endowment; life; education; tax saving) endowment; health; (endowment; health) mix	{1,2,3}
Purchased products (*c*6)	Life insurance; social insurance; group insurance	{(1), (1,2), (1,2,3), (2), (2, 3), (3)}
Acceptable premium (*c*7)	<10,000; 10,000–19,999; 20,000–29,999; 10,000–49,999; 50,000–99,999; >100,000	{1,2,3,4,5,6}
Decision Attributes		
Purchase anticipation (*d*1)	Living; endowment; premium refund	{(1), (1,2), (1,2,3), (2), (2,3), (3)}

There are 220 sets with only one object from a total of 246 elementary sets, suggesting that 79% of total data cannot be classified and no relationship is found between them. After redefining the attribute value class, the number of total elementary sets is down to 194; the results are shown in Table 19.3.

It is clear that the reduction of condition attributes resulted in removing attributes one after another and checked the similarity between condition elementary sets and the original set number while processing the "reduct" on condition attributes (Waczak and Massart 1999; Pawlak 1982, 1984). The attribute is a superfluous attribute, as if the set number is the same as the original set number (as seen in Table 19.4).

No superfluous attribute is found, only one reduct set is yielded, and the core set is the same as the reduct set, which is {*Area*, *c*1, *c*2, *c*3, *c*5, *c*6, *c*7}.

The accuracy rate of classification is 0.73. By removing 148 sets of one object and recomputing the accuracy rate, the rate will be reduced to 0.21, as shown in Table 19.5.

As for the results of the no purchase products in experiment, seven attributes are generated, of which six are conditional attributes and one is a decision attribute, as seen in Table 19.6. The original attribute specification is shown in Appendix 19.2. The results of comparison of condition sets and decision sets are listed in Table 19.7. Removing the superfluous attribute after processing yields 32

TABLE 19.3
Condition Elementary Sets

	Original Elementary Sets		Redefined Elementary Sets	
No. of Objects	No. of Sets	Set*Object	No. Sets	Set*Object
1	220	220	148	148
2	19	38	27	54
3	6	18	9	27
4	1	4	4	16
5	0	0	4	20
6	0	0	1	6
9	0	0	1	9
Total	246	280	194	280
Individual classified rate	(220 ÷ 280) × 100% = 79%		(148 ÷ 280) × 100% = 53%	

condition sets and 4 decision sets. No superfluous attribute is generated for conditional attributes but one reduct set is, and the core set is the same as the reduct set, which is $\{Area, c1, c2, c3, c4, c5, c7\}$. Table 19.7 lists 24 sets of condition sets with one object, indicating that 56% of the total data cannot be grouped.

The classification accuracy rate for no purchase products is 0.79. By removing the sets of one object, the classification accuracy rate decreases to 0.42 (Table 19.8).

19.4 EMPIRICAL RESULT

A series of single-choice questions are designed and multichoice questions are used to illustrate the attributes of combination values to extend the complex problems of classification. The attributes of combination values will increase the number of decision rules.

This study generated 8 rules for customer needs and 4 rules for reasons for not purchasing products. In Tables 19.9 and 19.10, the decision rules for customer needs and for reasons for not purchasing products are listed, respectively. The reason for those persons who did not purchase products are as follows:

- Not interesting
- People aged below 25 with weak economic ability
- Annual premium almost under NT$10,000 (US$313)
- Purposes of purchase are endowment, life, education, and tax reduction

The following is true of those customers who purchased insurance products:

- The most bought insurance product is life insurance.
- The average annual premium fee is under $30,000 (US$938), but the premium fee increases by age.
- Most of those who purchase insurance products are women.
- The purpose of purchase is endowment.
- The anticipation of insurance is the premium refund.

TABLE 19.4
Moving Superfluous Condition Attributes after Redefining the Original Data

Attribute	None		*Area*		c1		c2		c3		c5		c6		c7	
No. of Object	No. Set	Set* Object	No. Set	Set* Object	No. Set	Set* Object	No. Set	Set* Object	No. Set	Set* Object	No. Set	Set* Object	No. Set	Set* Object	No. Set	Set* Object
1	148	148	105	105	89	89	98	98	136	136	93	93	55	55	65	65
2	27	54	28	56	29	58	29	58	31	62	19	38	27	54	23	46
3	9	27	14	42	11	33	19	57	10	30	18	547	15	45	10	30
4	4	16	4	16	5	20	2	8	4	16	4	16	12	48	10	40
5	4	20	2	10	4	20	2	10	4	20	2	10	3	15	5	25
6	1	6	2	12	2	12	4	24	0	0	4	24	1	6	4	24
7	0	0	2	14	3	21	2	14	1	7	1	7	1	7	1	7
8	0	0	2	16	1	8	0	0	0	0	1	8	1	8	1	8
9	1	9	1	9	1	9	0	0	1	9	0	0	0	0	0	0
10	0	0	0	0	1	10	0	0	0	0	3	30	0	0	0	0
11	0	0	0	0	0	0	1	11	0	0	0	0	1	11	0	0
13	0	0	0	0	0	0	0	0	0	0	0	0	1	13	0	0
15	0	0	0	0	0	0	0	0	0	0	0	0	0	0	1	15
18	0	0	0	0	0	0	0	0	0	0	0	0	1	18	0	0
20	0	0	0	0	0	0	0	0	0	0	0	0	0	0	1	20
Total	194	280	160	280	146	280	157	280	187	280	145	280	118	280	121	280

TABLE 19.5
The Classification Accuracy

Class Number	Number of Objects	Lower Approx.	Upper Approx.	Approx. Accuracy	After Moving Sets with Only One Object: Lower Approx.	Upper Approx.	Approx. Accuracy
Class 1	152	85	112	0.76	14	41	0.34
Class 2	69	44	61	0.72	2	19	0.11
Class 3	36	22	35	0.63	0	13	0.00
Class 4	11	8	11	0.73	0	3	0.00
Class 5	4	2	4	0.50	0	2	0.00
Class 6	3	2	3	0.67	0	1	0.00
Class 7	3	3	3	1.00	0	1	0.00
Class 8	2	1	1	1.00	1	1	1.00
sum-up	280	167	230	0.73	17	81	0.21

19.5 DISCUSSIONS

Rough set theory uses a mathematical method to process the classification by the same values as attributes. The elementary sets can extract the same degree relationship between them, which can induce decision rules. The limitation of rough

TABLE 19.6
Attribute Specification for No Purchase Products

Attribute Name	Attribute Values	Attribute Value Sets
Condition Attributes		
Area	North district; Northeast district	{1,2}
Age ($c1$)	Single: <25; marriage, housing, business, birth: 25–34; foster children, education: 35–44; retired: 45–	{1,2,3,4}
Gender ($c2$)	Female; male	{F,M}
Profession ($c3$)	1–6 (by department of insurance profession publish)	{1,2,3,4,5,6}
Reason for no purchase products ($c4$)	No one introduced; economic; can't trust; not interesting;	{(1), (1,2), (1,2,3), (1,2,3,4), (2), (2,3), (2,3,4), (3), (3,4), (4)}
Purpose ($c5$)	(Endowment; life; education; tax saving) endowment; health; (endowment; health) mix	{1,2,3}
Acceptable premium ($c7$)	<10,000; 10,000–19,999; 20,000–29,999; 10,000–49,999; 50,000–99,999; >100,000	{1,2,3,4,5,6}
Decision Attributes		
Purchase anticipation ($d1$)	Living; endowment; premium refund	{(1), (1,2), (1,2,3), (2), (2,3), (3)}

TABLE 19.7
Elementary Sets of No Purchase Products

	Condition Sets	
No. of Objects	**No. of Sets**	**Set* Object**
1	24	24
2	5	10
3	2	6
4	1	4
12	0	0
27	0	0
Total	32	44

set theory is that out-of-sample data are undefined by the previous finding decision rules. The new merging data will generate new rules to themselves. Thus, 50 validation sample data sets are added to the hit test to check the feasibility of the decision rules in this empirical study, and the results are shown in Table 19.11; the hit rate reaches 100%. It is clear from Table 19.11 that new merging objects still can fit classes among those decision classes. Meanwhile, the wider the explanation of each rule (or degree accuracy of each rule), the higher the hit rate will be.

Essentially, the main problem is data set classification in this experiment and the multichoice attributes with combination values may enlarge the discrete degree of the data. Several approaches can be taken. One is to redefine the attribute values to narrow down the range of value class in order to solve the problem. It should be noted that to avoid misleading the participant, the usage of special terminology of insurance theory in the question is omitted. After the attribute value class is redefined, the ungrouped data rate decreases from 79% to 53%, indicating that 194 elementary sets are generated after redefining the attribute value class, of which 148 sets have only one object. This shows that redefining the attribute value class reduces the number of elementary sets from 246 to 194.

TABLE 19.8
Classification Accuracy of No Purchase Products

					After Moving Sets with Only One Object		
Class Number	**Number of Objects**	**Lower Approx.**	**Upper Approx.**	**Approx. Accuracy**	**Lower Approx.**	**Upper Approx.**	**Approx. Accuracy**
1	27	15	18	0.83	4	7	0.57
2	12	9	11	0.82	1	3	0.33
3	3	2	3	0.67	0	1	0.00
4	2	1	2	0.50	0	1	0.00
Sum-up	44	27	34	0.79	5	12	0.42

TABLE 19.9
Rules for Customer Needs

Rule #	If *Area*	*Age*	*Gender*	*Profession*	*Purpose*	*Purchased Products*	*Premium*	Then *Anticipation*
1.	North district	<25	Male	1	Endowment	Life insurance	<10,000	Premium refund
or	North district	25–34	Female	1	Mix	Mix	20,000–29,999	Premium refund
or	Northeast district	25–44	Male	1	Mix	Mix	20,000–29,999	Premium refund
or	North district	<25	Male	1	Mix	Mix	<10,000	Premium refund
or	North district	<25	Female	1	Mix	Mix	10,000–19,999	Premium refund
or	Northeast district	<25	Male	2	Endowment	Mix	<10,000	Premium refund
					...			

The unique data encountered in this study is classified into an individual class by itself. It is difficult to understand the contents of data sets that relate to the decision rule in the condition part. On the contrary, the number of decision rules will increase as if too much unique data (generated to many elementary sets) happened at decision part. This will not be beneficial to decision maker, if too many decision rules are produced. Because a single object set is in the condition part, that will match one of the decision classes and increase the lower approximation. The lower approximation is increased as well as the upper approximation. It yields a higher classification rate, which is the wrong image. The higher classification rate indicates that the objects in the class may have a higher dependency among condition attributes of

TABLE 19.10
Rules for Reasons for No Purchase of Products

Rule #	If *Area*	*Age*	*Gender*	*Profession*	*Reasons*	*Purpose*	*Premium*	Then *Anticipation*
1.	Northeast district	<25	Male	1	No one introduced and Economic	Mix	10,000–19,999	Premium refund
or	North district	<25	Female	1	Not interesting	Mix	10,000–19,999	Premium refund
or	Northeast district	<25	Female	1	Not interesting	Endowment	<10,000	Premium refund
					...			
2.	Northeast district	<25	Male	1	No one introduced and Economic	Health	30,000–49,999	Living
					...			

TABLE 19.11
Description of the Hit Rate from New Data

Original of No Objects	Decision Rules: Class No.	Decision Rules: No of New Test Objects	Decision No	Hit Decision: Class 1	Class 2	Class 3	Class 4	Class 5	Class 6	Class 7	Class 8	Class 9	Hit Rate(%)
152	Class 1	22	Class 1	22	0	0	0	0	0	0	0	0	100%
69	Class 2	18	Class 2	0	18	0	0	0	0	0	0	0	100%
36	Class 3	6	Class 3	0	0	6	0	0	0	0	0	0	100%
11	Class 4	1	Class 4	0	0	0	1	0	0	0	0	0	100%
4	Class 5	1	Class 5	0	0	0	0	1	0	0	0	0	100%
3	Class 6	1	Class 6	0	0	0	0	0	1	0	0	0	100%
3	Class 7	0	Class 7	0	0	0	0	0	0	0	0	0	0%
2	Class 8	0	Class 8	0	0	0	0	0	0	0	0	0	0%
0	Class 9	1		0	0	0	0	0	0	0	0	1	0%
280		50											

those objects. As seen in Table 19.3, 148 sets (classes) have only one object that will be added to some of the decision classes, and the lower and upper approximation will be increased. However, this yields an incorrect approximation rate. In addition, the number of every attribute value can affect the class number, suggesting that the same size number of value set for each attribute will contribute to group the data into classes.

Another approach is called a hybrid system, where rough set theory is applied along with other methods, such as fuzzy theory, artificial neural network theory, genetic algorithms (GA), genetic programming (GP), and so on to proceed with rule extraction (Hassan and Tazaki 2001; Huang et al. 2005).

The exploration of introducing rough set theory into multichoice questions is summarized above; it significantly affects the number of elementary sets. The importance of experts' contributions on compiling data into preprocessing tasks cannot be ignored, as there is a limitation derived on rough set rules by using "experts." Different experts may have different views and preferential attitudes on decision making as over-reliance on an expert. On the other hand, reclassified value classes will reduce the rule description precision. There may be other methods for attribute (feature) extraction, such as the FUSINTER technique and VPRS, which may enhance the testability of rough set theory (Beynon and Peel 2001). For now, the question of combination values is focused and the optimally discrete attribute with its values may serve as a topic for in-depth research.

19.6 CONCLUSIONS

This study demonstrates that the results match the requirements for the insurance market in Taiwan. From the results of a survey, the following findings are drawn:

- Most insurance purchasers are female.
- Age 25–35 is the highest insured class.
- The purpose of insurance is endowment.
- The acceptable annual premium is under NT$19,999 (US$625).
- The most-purchased insurance products is life insurance.
- The anticipant is premium refund.

The multichoice question will affect the number of elementary sets. The expert's contribution in the input data preprocessing task is of importance. We demonstrate that redefined attribute values can narrow down elementary sets and that too many sets of one object will decrease the classification accuracy. In addition, every attribute value's number can affect class number. Also, the multichoice questions can have many combination values. If the characters of a question's answer is unique, then reclassification will not work well. Here we suggest a hybrid system, rough set theory combined with GP to do rule extraction, that may be able to solve the shortcomings of too many rules under the attribute combination values of using rough set theory. This will be the subject of further study.

APPENDIX 19.1

The original attribute specification for customer needs described in Table A19.1.1.

TABLE A19.1.1
Attribute Specification for Customer Needs

Attribute's Name	Attribute Values
Condition Attributes	
Area	North district; Northeast district
Age *(c1)*	19–76
Gender (*c*2)	Female; male
Profession (*c*3)	Student; worker; employee; retired…
Purpose (*c*5)	Endowment; life; education; health; tax saving; others
Purchased products (*c*6)	Life Insurance; social insurance; group insurance
Acceptable premium (*c*7)	<10,000; 10,000–19,999; 20,000–29,999; 10,000–49,999; 50,000–99,999; >100,000
Decision Attributes	
Purchase anticipation (*d*1)	Living; endowment; premium refund

APPENDIX 19.2

The original attribute specification for customers purchasing no products described in Table A19.2.1.

TABLE A19.2.1
Attribute Specification for No Purchase of Products

Attribute's Name	Attribute Values
Condition Attributes	
Area	North district; Northeast district
Age *(c1)*	19–76
Gender (*c*2)	Female; male
Profession (*c*3)	Student; worker; employee; retired…
Reason for no purchase of products (*c*4)	No one introduced; economic; can't trust; not interesting
Purpose (*c*5)	Endowment; life; education; health; tax saving; others
Acceptable premium (*c*7)	<10,000; 10,000–19,999; 20,000–29,999; 10,000–49,999; 50,000–99,999; >100,000
Decision Attributes	
Purchase anticipation (*d*1)	Living; endowment; premium refund

20 Extending the DEMATEL Method for Group Decision Making in Fuzzy Environments

20.1 INTRODUCTION

The more complicated and confused the environment gets, the more profoundly decision making is desired. Today, strategies for structuring comprehensive models are essential for problem solving. As a kind of problem-solving method, structural modeling is widely applied and befitting for the use of dealing with complex problems, which may serve as a basis for arranging involved criteria, and also provides a practical mechanism for use in the formation of decision structures. In particular, structural modeling is based on graph theory so that complex systems can be broken down into several subsystems—namely, factors of the problem—and displays interactions between those subsystems by graphs. Through the graphs, the whole structure of the system is more easily understood by intuition for capturing the nature of problems.

Structural modeling is employed to structure the set of elements into a structural model that is a collection of components and their relationships (Sharma, Gupta, and Sushil 1995), which is also able to provide the computer-based environment for conceiving, representing, and manipulating a wide variety of models (Geoffrion 1987). By using a structural model, one can illustrate the ways in which subsystems are connected to each other and show the overall shape of a system. Unlike the mathematical modeling approach, which requires clear quantitative variables, structural modeling is a qualitative approach that can capture and image conceptual characters of the system explicitly. Depending on whether or not the circuit exists within the structural graph, structural modeling may be divided into the network type and the tree type. A circuit is a path which ends at the vertex where it begins. In contrast to the tree type, the network type is more suitable for analysis of the situation in which orders of subsystems are not explicit. Furthermore, the network type may be classified into two kinds of structural graphs: directed graphs and directionless graphs. Directed graphs, known as digraphs, are more useful than directionless graphs, because digraphs can demonstrate the directed relationships of subsystems (Harary, Norman, and Cartwright 1965). The decision-making trial and evaluation laboratory (DEMATEL) method is based on digraphs, which can separate involved criteria into cause group and effect group.

In practice, to achieve effective and reasonable decision making for solving complicated problems with multiple criteria, it is usually necessary to gather group knowledge and employ multiple criteria decision-making (MCDM) methods. For handling MCDM problems, there are several methods available, such as elimination and choice translating reality (ELECTRE), the technique for order preference by similarity to ideal solution (TOPSIS), and the analytic hierarchy process (AHP), etc. (Zanakis et al. 1998). Nevertheless, those MCDM methods may not work unless the structural model of evaluation is established beforehand. The application of structural modeling has become essential for group development of structural models. Hence, as a sort of structural modeling approach, the DEMATEL is a potent method that has the following benefits: (1) gather group knowledge for capturing the interactions between subsystems; (2) form a structural model of evaluation for making decisions; and (3) visualize the causal relationship of subsystems by offering a causal diagram that promotes understanding of the character of the problem and communicates opinions within a group.

Human judgments for deciding the relationship between subsystems are usually given by crisp values for establishing a structural model. However, in many cases, crisp values are an inadequate reflection of vagueness in the real world. The fact that human judgments with preferences are often unclear and hard to estimate by exact numerical values has created the need for fuzzy logic for handling problems with vagueness and imprecision. Moreover, a more sensible approach is to use linguistic assessments instead of numerical values, in which all assessments of criteria in the problem are evaluated by means of linguistic variables (Bellman and Zadeh 1970; Zadeh 1975; Delgado, Verdegay, and Vila 1992). Therefore, enabling the DEMATEL method to be suitable for solving multiperson and MCDM problems in fuzzy environments, our purpose is to develop a methodology that extends the DEMATEL method by applying linguistic variables and a fuzzy aggregation method.

This chapter is organized as follows. In Section 20.2, some of the prior literature and definitions related to the DEMATEL method and fuzzy group decision-making are reviewed. In Section 20.3, for coping with the fuzzy group decision-making problems, a methodology based on the DEMATEL method with the fuzzy logic is proposed. In Section 20.4, an empirical study is presented to illustrate the procedure of our proposed solution and to demonstrate its usefulness and validity. Finally, based upon the findings of this research, conclusions and suggestions are presented.

20.2 DEMATEL AND FUZZY GROUP DECISION MAKING

As a kind of structural modeling approach, DEMATEL is a comprehensive method for building and analyzing a structural model involving causal relationships between criteria. To lay the foundation for extending the DEMATEL method for group decision making in fuzzy environments, the essentials of the DEMATEL method and fuzzy group decision making are discussed below.

20.2.1 DEMATEL METHOD

The DEMATEL method is based on graph theory that enables us to project and solve problems visually, so that we can divide multiple criteria into cause group and effect group in order to better capture causal relationships visibly. Graph theory has grown tremendously in recent years, largely due to the usefulness of graphs as models for computation and optimization. Applying the graph theory, we can easily visually discover things inside the complex problem, because the graph displays the mathematical results with visualization clearly and unambiguously.

The DEMATEL method is based on digraphs, which can separate involved criteria into cause group and effect group. Directed graphs, known as digraphs, are more useful than directionless graphs, because digraphs can demonstrate the directed relationships of subsystems. A digraph may typically represent a communication network, or some domination relation between individuals, etc. Suppose a system contains a set of elements $S = \{s_1, s_2, \ldots, s_n\}$ and particular pairwise relations are determined for modeling with respect to a mathematical relation R. Next, portray the relation R as a direct-relation matrix that is indexed equally on both dimensions by elements from the set S. Then, except the case is not relation where the number 0 appears in the cell (i, j), if the entry is a positive integral, this means: (1) the ordered pair (s_i, s_j) is in the relation matrix $\boldsymbol{R}$, and (2) shown element s_i causes element s_j.

Both interpretive structural modeling (ISM) and DEMATEL are based on digraphs. A digraph portrays a contextual relation between the elements of a system and can be converted into a visible structural model of a system with respect to that relation (Warfield 1974). In contrast with the ISM, which is developed using binary data, the DEMATEL is applied by ranking values. The tangible product of an ISM exercise is a structural model called a "map," which is a multilevel structure like a hierarchy (Warfield 1977). Hierarchies are fundamental in the study of many kinds of complex systems (Warfield 1973). By contrast, the tangible product of a DEMATEL exercise is a structural model appearing as a "causal diagram" which may divide subsystems into cause group and effect group. In particular, DEMATEL is able not only to demonstrate directed relationships of subsystems, but also to clarify the degree of interactions between subsystems. Thus, toward analyzing a complex system, if we wish to capture the cause-effect relationship among subsystems, DEMATEL is apparently more helpful than ISM.

The Battelle Memorial Institute conducted the DEMATEL method project through its Geneva Research Centre (Gabus and Fontela 1972, 1973). The original DEMATEL was aimed at the fragmented and antagonistic phenomena of world societies and searched for integrated solutions. In recent years, the DEMATEL method has become very popular in Japan, because it is especially pragmatic to visualize the structure of complicated causal relationships with digraphs. The digraph portrays a contextual relation among the elements of the system, in which a numeral represents the strength of influence (Figure 20.1). Hence, the DEMATEL method can convert

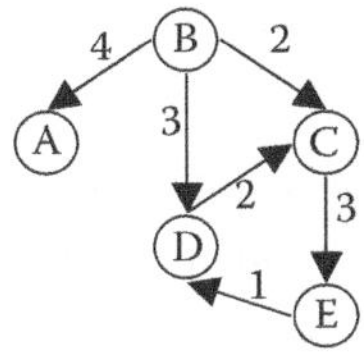

FIGURE 20.1 DEMATEL digraph.

the relationship between the causes and effects of criteria into an intelligible structural model of the system. In order to apply the DEMATEL method smoothly, we refined the version used by Hori and Shimizu (1999) and make essential definitions as below.

Definition 20.1

The pairwise comparison scale may have four designated levels, where the scores of 0, 1, 2, and 3 represent "No influence," "Low influence," "High influence," and "Very high influence," respectively.

Definition 20.2

The initial direct-relation matrix z is an $n \times n$ matrix obtained by pairwise comparisons in terms of influences and directions between criteria, in which z_{ij} is denoted as the degree to which the criterion i affects the criterion j.

Definition 20.3

The normalized direct-relation matrix $\boldsymbol{X}$ can be obtained through Equations 20.1 and 20.2, in which all principal diagonal elements are equal to zero.

$$\boldsymbol{X} = s \cdot \boldsymbol{Z} \tag{20.1}$$

$$s = \frac{1}{\max\limits_{1 \le i \le n} \sum_{j=1}^{n} z_{ij}} \tag{20.2}$$

$$s = \min \left\{ \max \sum_{j=1}^{n} z_{ij}, \max \sum_{j=1}^{n} z_{ij} \right\}, \quad i, j = 1, 2, \ldots, n$$

Definition 20.4

The total-relation matrix $\boldsymbol{T}$ can be acquired by using Equation 20.3, in which $\boldsymbol{I}$ is denoted as the identity matrix (see Appendix 20.1).

$$\boldsymbol{T} = \boldsymbol{X}(\boldsymbol{I} - \boldsymbol{X})^{-1} \tag{20.3}$$

Definition 20.5

The sum of rows and the sum of columns are separately denoted as vector $\boldsymbol{d}$ and vector $\boldsymbol{r}$ within the total-relation matrix $\boldsymbol{T}$ through the Equations 20.4 through 20.6.

$$\boldsymbol{T} = \left(t_{ij}\right)_{n \times n}, \quad i, j = 1, 2, \ldots, n \tag{20.4}$$

$$\boldsymbol{d} = \left(d_1, \ldots, d_i, \ldots, d_n\right), \quad \text{where } d_i = \sum_{j=1}^{n} t_{ij}, \tag{20.5}$$

$$\boldsymbol{r} = \left(r_1, \ldots, r_j, \ldots, r_n\right), \quad \text{where } r_j = \sum_{i=1}^{n} t_{ij}, \tag{20.6}$$

where vector $\boldsymbol{d}$ and vector $\boldsymbol{r}$ denote the sum of rows and the sum of columns from total-relation matrix $\boldsymbol{T}$, respectively.

Definition 20.6

A causal diagram can be acquired by mapping the dataset of $(d_i + r_i, d_i - r_i)$, where the horizontal axis $(d_i + r_i)$ is made by adding d_i to r_i, and the vertical axis $(d_i - r_i)$ is made by subtracting d_i from r_i.

20.2.2 Fuzzy Group Decision Making

To gain a solution for problem solving, group decision making (GDM) is important to any organization, because GDM usually impacts upon the decisions that affect organizational performance. GDM is a way to draw from varying experience, opinions, ideas, and motivations. It can facilitate learning for a broad range of informants in terms of their own perceptions and it can also be helpful in identifying variables, issues, and hypotheses. In particular, GDM is the process of arriving at a consensus based upon the reaction of multiple individuals, which has merit in that group interaction may facilitate the exchange of ideas and information whereby an acceptable judgment may be obtained (Cheng and Lin 2002).

However, in many cases, judgments for decision-making are often given by crisp values, though crisp values are an inadequate reflection of situational vagueness. In the real world, many decisions involve imprecision since goals, constraints, and possible actions are not known precisely (Bellman and Zadeh 1970). When making decisions in a fuzzy environment, the result of decision making is strongly affected by subjective judgments that are vague and imprecise. The sources of imprecision include unquantifiable information, incomplete information, non-obtainable information, and partial ignorance (Chen et al. 1992). To solve this kind of imprecise problem, fuzzy set theory was first introduced by Zadeh (1965) as a mathematical way to represent and handle vagueness in decision making. Fuzzy set theory has been developed for solving problems in which definitions of activities or expressions are imprecise. Fuzzy logic is the logic of approximate rather than exact reasoning, which is similar to human reasoning. In fuzzy logic, each number between 0 and 1 indicates a partial truth, whereas crisp sets correspond to binary logic: 0 or 1. Hence, fuzzy logic can express and handle vague or imprecise judgments mathematically (Al-Najjar and Alsyouf 2003).

Decision makers tend to give assessments based on their past experiences and knowledge, and also their estimations are often expressed in equivocal linguistic terms. To deal with the vagueness of human thought and expression in making decisions, fuzzy set theory is very helpful. In particular, to tackle the ambiguities involved in the process of linguistic estimation, it is better to convert these linguistic terms to fuzzy numbers. Based on the definition of fuzzy sets, the concept of linguistic variables is introduced to represent a language typically adopted by a human expert. A linguistic variable is a variable whose values (namely linguistic values) have the form of phrases or sentences in a natural language (von Altrock 1996). To efficiently resolve the ambiguity arising from incomplete information and the fuzziness in human judgments, it is necessary to employ a linguistic scale.

Thus, the problems of GDM in a fuzzy environment have created a need to employ fuzzy logic and require effective fuzzy aggregation methods to cope with fuzzy GDM problems. In the following, we briefly review some essential definitions of fuzzy logic.

Definition 20.7

A fuzzy set $\tilde{A}$ is a subset of a universe of discourse X, which is a set of ordered pairs and is characterized by a membership function $\mu_{\tilde{A}}(x)$ representing a mapping $\mu_{\tilde{A}}: X \rightarrow [0,1]$. The function value of $\mu_{\tilde{A}}(x)$ for the fuzzy set $\tilde{A}$ is called the membership value of x in $\tilde{A}$, which represents the degree of truth that x is an element of the fuzzy set $\tilde{A}$. It is assumed that $\mu_{\tilde{A}}(x) \in [0,1]$, where $\mu_{\tilde{A}}(x) = 1$ reveals that x completely belongs to $\tilde{A}$, while $\mu_{\tilde{A}}(x) = 0$ indicates that x does not belong to the fuzzy set $\tilde{A}$.

$$\tilde{A} = \left\{\left(x, \mu_{\tilde{A}}(x)\right)\right\}, \quad x \in X, \tag{20.7}$$

where $\mu_{\tilde{A}}(x)$ is the membership function and $X = \{x\}$ represents a collection of elements x.

Definition 20.8

A fuzzy set $\tilde{A}$ of the universe of discourse X is convex if

$$\mu_{\tilde{A}}\left(\lambda x_1 + (1-\lambda)x_2\right) \geq \min\left(\mu_{\tilde{A}}(x_1), \mu_{\tilde{A}}(x_2)\right), \quad \forall x \in [x_1, x_2], \tag{20.8}$$

where $\lambda \in [0,1]$.

Definition 20.9

A fuzzy set $\tilde{A}$ of the universe of discourse X is normal if

$$\max \mu_{\tilde{A}}(x) = 1. \quad (20.9)$$

Definition 20.10

A fuzzy number $\tilde{N}$ is a fuzzy subset in the universe of discourse X, which is both convex and normal.

Definition 20.11

The α-cut of the fuzzy set $\tilde{A}$ of the universe of discourse X is defined as

$$\tilde{A}_\alpha = \{x \in X \mid \mu_{\tilde{A}}(x) \geq \alpha\}, \quad (20.10)$$

where $\alpha \in [0,1]$.

Definition 20.12

A triangular fuzzy number $\tilde{N}$ can be defined as a triplet (a, b, c), and the membership function $\mu_{\tilde{N}}(x)$ is defined as:

$$\mu_{\tilde{N}}(x) = \begin{cases} 0, & x < a, \\ (x-a)/(b-a), & a \leq x \leq b, \\ (c-x)/(c-b), & b \leq x \leq c, \\ 0, & x > c, \end{cases} \quad (20.11)$$

where a, b, and c are real numbers and $a \leq b \leq c$.

Definition 20.13

Linguistic variables are used as variables whose values are not numbers but linguistic terms (Zadeh 1975; von Altrock 1996). The linguistic term approach is a convenient way for decision makers to express their assessments. The linguistic variable is very useful in dealing with situations that are described in quantitative expressions. Linguistic values can be represented by fuzzy numbers. In particular, the triangular fuzzy number is commonly used.

Definition 20.14

The graded mean integration representation (Chen and Hsieh 1998; Chuo 2003) of the triangular fuzzy number $\tilde{N}$ is defined as

$$\tilde{N} = \frac{1}{6}(a + 4b + c). \tag{20.12}$$

Definition 20.15

Let $\tilde{M} = (a_1, b_1, c_1)$ and $\tilde{N} = (a_2, b_2, c_2)$ be two triangular fuzzy numbers. Then the representation of the addition operation on triangular fuzzy numbers (Chuo 2003) is defined as

$$\tilde{M} \oplus \tilde{N} = \frac{1}{6}(a_1 + a_2 + 4b_1 + 4b_2 + c_1 + c_2). \tag{20.13}$$

20.3 ANALYTICAL PROCEDURE OF THE PROPOSED METHODOLOGY

The DEMATEL method is a highly pragmatic way to form a structural model of evaluation for better decision making. To further the practicality of the DEMATEL method for group decision making in a fuzzy environment, the analytical procedure of our proposed methodology is explained as follows.

Step 1: *Identifying the decision goal and forming a committee.*
Decision making is the process of defining the decision goals, gathering relevant information, generating the broadest possible range of alternatives, evaluating the alternatives for advantages and disadvantages, selecting the optimal alternative, and monitoring the results to ensure that the decision goals are achieved (Hess and Siciliano 1996; Opricovic and Tzeng 2004). Thus, the first step is to identify the decision goal. Also, it is necessary to form a committee for gathering group knowledge for problem solving.

Step 2: *Developing evaluation criteria and designing the fuzzy linguistic scale.*
In this step, it is necessary to establish sets of criteria for evaluation. However, evaluation criteria have the nature of causal relationships and are usually comprised of many complicated aspects. To gain a structural model dividing involved criteria into cause group and effect group, the DEMATEL method must be used here. For dealing with the ambiguities of human assessments, the linguistic variable "influence" is used with five linguistic terms (Li 1999), {Very high, High, Low, Very low, No}, that are expressed in positive triangular fuzzy numbers (a_{ij}, b_{ij}, c_{ij}) as shown in Table 20.1.

TABLE 20.1
The Fuzzy Linguistic Scale

Linguistic Terms	Triangular Fuzzy Numbers
Very high influence (VH)	(0.75,1.0,1.0)
High influence (H)	(0.5,0.75,1.0)
Low influence (L)	(0.25,0.5,0.75)
Very low influence (VL)	(0,0.25,0.5)
No influence (No)	(0,0,0.25)

Step 3: *Acquiring and aggregating the assessments of decision makers.*
For measuring the relationship between criteria $C = \{C_i \mid i = 1, 2,\ldots,n\}$, a decision group of P experts make sets of pairwise comparisons matrix $(\tilde{\mathbf{Z}}^1,\tilde{\mathbf{Z}}^2,\ldots,\tilde{\mathbf{Z}}^P)$ in terms of influences and directions between criteria. To aggregate the result of these assessments, we can use Equation 20.14, which is based on the representation of the addition operation (see Definitions 20.14 and 20.15). Then, the initial direct-relation matrix $\tilde{\mathbf{Z}} = [\tilde{z}_{ij}]_{n\times n}$ can be obtained, in which z_{ij} is denoted as the degree to which the criterion i affects the criterion j.

$$Z = \left[\frac{1}{P}(\tilde{\mathbf{Z}}^1 \oplus \tilde{\mathbf{Z}}^2 \oplus \cdots \oplus \tilde{\mathbf{Z}}^P)\right]_{n\times n}, \tag{20.14}$$

where

$$(\tilde{\mathbf{Z}}^1 \oplus \tilde{\mathbf{Z}}^2 \oplus \cdots \oplus \tilde{\mathbf{Z}}^P) = \left[\frac{1}{6}\left(a_{ij}^1 + 4b_{ij}^1 + c_{ij}^1 + a_{ij}^2 + 4b_{ij}^2 + c_{ij}^2 + \cdots + a_{ij}^P + 4b_{ij}^P + c_{ij}^P\right)\right]_{n\times n},$$

Step 4: *Establishing and analyzing the structural model.*
On the base of the initial direct-relation matrix $\mathbf{Z}$, the normalized direct-relation matrix $\boldsymbol{X}$ can be obtained through Equations 20.1 and 20.15. Then, the total-relation matrix $\boldsymbol{T}$ can be acquired by using Equation 20.3. According to Definitions 20.5 and 20.6, a causal diagram can be acquired through Equations 20.4 through 20.6. The causal diagram is constructed with the horizontal axis $(d_i + r_i)$ named "Prominence" and the vertical axis $(d_i - r_i)$ named "Relation". The horizontal axis "Prominence" shows how much importance the criterion has, whereas the vertical axis "Relation" can divide criteria into cause group and effect group. Generally, when the $d_i - r_i$ axis is plus, the criterion belongs to the cause group. Otherwise, if the $d_i - r_i$ axis is minus, the criterion belongs to the effect group. Hence, causal diagrams can visualize the complicated causal relationships of criteria into a visible structural model, providing valuable insight for problem solving. Further, with the help of a causal diagram, we can make proper decisions by recognizing the difference between cause and effect criteria.

$$X = \frac{Z - kI}{\max\left\{\max_{1 \le i \le n} \sum_{j=1}^{n} z_{ij}, \max_{1 \le j \le n} \sum_{i=1}^{n} z_{ij}\right\}}, \quad i, j = 1, 2, ..., n, \tag{20.15}$$

$$X = \frac{Z - kI}{\max \sum z_{ij}}$$

where k is z_{11}, and I is denoted as the identity matrix.

20.4 APPLICATION OF THE PROPOSED METHODOLOGY

When making any intellectual decision, it is necessary to grasp the significance of each criterion and to clarify the role that it plays in the problem. Our proposed methodology is able to identify and summarize relationships among specific criteria which define the problem, as well as to visualize the importance and interaction of criteria among themselves, even when the decision has to be reached using group decision making in a fuzzy environment. In order to demonstrate that the proposed methodology is useful and valid, an empirical study of high-tech industry in Taiwan's Hsin-Chu Science-Based Industry Park (HCSIP) is presented.

20.4.1 Empirical Study of an Application of the Proposed Methodology

The case Company W is a Taiwanese firm with a turnover of more than USD 120 million. The company is one of the world's leading manufacturers specializing in wafer foundry services for semiconductor products. Due to the competitive challenges of new technology and the global market, Company W set out to redesign a more forceful human resource strategy to enrich human capital, and thereby enhance their competitive advantages and market share. It is the trend that numerous companies are increasingly focusing on human capital as a source of competitive advantages. Hence, Company W intended to develop manager competency models to leverage their human capital. The following shows how Company W successfully utilized our proposed solution to establish a structural model of evaluation for making decisions.

Step 1: *Identifying the decision goal and forming a committee.*
In building a manager competency model, there are three main stages: establish a structural model for evaluation; make judgments in terms of the relative priority of criteria; and synthesize priorities in order to accomplish a favorable model. Here, the decision goal for the Company W was to establish a structural model for evaluation. Also, they set up a manager competency model committee of seven members including the general manager and several department managers.

Step 2: *Developing evaluation criteria and designing the fuzzy linguistic scale.* After the decision goals are determined, it is necessary to gather the relevant criteria in order to be able to create a structural model. The committee adopted eleven criteria from a generic manager competency model by Spencer and Spencer (1993), and decided to use a fuzzy linguistic scale (Table 20.1) for making assessments. Those eleven criteria included: Impact and Influence (C_1), Achievement Orientation (C_2), Teamwork and Cooperation (C_3), Analytical Thinking (C_4), Initiative (C_5), Developing Others (C_6), Self-confidence (C_7), Directiveness (C_8), Information-Seeking (C_9), Team Leadership (C_{10}), and Conceptual Thinking (C_{11}).

Step 3: *Acquiring and aggregating the assessments of decision makers.* Once the relationships between those criteria were measured by the committee through the use of the fuzzy linguistic scale, the data from each individual assessment can be obtained. For example, the assessment data of the general manager are shown in Table 20.2. Then, using Equation 20.14 to aggregate these assessment data, the initial direct-relation matrix (Table 20.3) was produced.

Step 4: *Establishing and analyzing the structural model.* Based on the initial direct-relation matrix, the normalized direct-relation matrix (Table 20.4) was obtained by Equations 20.1 and 20.15. Next, the total-relation matrix (Table 20.5) was acquired using Equation 20.3. Then, using Equations 20.4 through 20.6, the causal diagram (Figure 20.2) could be acquired by mapping a dataset of ($D + R$, $D - R$). Looking at the causal diagram, it is clear that evaluation criteria were visually divided into the cause group, including C_2, C_4, C_5, C_7, and C_9, while the effect group was composed of such criteria as C_1, C_3, C_6, C_8, C_{10}, and C_{11}. From the causal diagram, valuable cues are obtained for making profound decisions. For example, if Company W wished to reach a high level of performance in

TABLE 20.2
For example, the assessment data of the general manager

	C_1	C_2	C_3	C_4	C_5	C_6	C_7	C_8	C_9	C_{10}	C_{11}
C_1	No	H	No	No	H	VH	VH	VH	VL	VH	No
C_2	VL	No	No	L	VH	L	VH	VL	H	H	L
C_3	VL	VL	No	L	L	H	No	No	VL	L	No
C_4	VL	VH	VL	No	VL	H	VL	H	H	No	VH
C_5	VL	H	VL	H	No	H	No	VH	No	H	H
C_6	L	No	VL	VL	No	No	L	L	VH	No	No
C_7	H	VH	No	VH	H	L	No	VL	H	VH	VL
C_8	VH	No	VL	VL	L	VH	No	No	No	H	L
C_9	L	VH	L	VH	No	VL	L	L	No	No	VH
C_{10}	VH	L	L	H	VH	VH	H	L	No	No	L
C_{11}	No	H	VL	VH	VL	VL	No	H	H	No	No

TABLE 20.3
Initial Direct-relation Matrix of Criteria

	C_1	C_2	C_3	C_4	C_5	C_6	C_7	C_8	C_9	C_{10}	C_{11}
C_1	0.042	0.637	0.798	0.327	0.810	0.768	0.637	0.625	0.571	0.899	0.292
C_2	0.661	0.042	0.673	0.679	0.893	0.637	0.798	0.536	0.601	0.768	0.500
C_3	0.768	0.595	0.042	0.470	0.601	0.732	0.405	0.470	0.292	0.804	0.327
C_4	0.565	0.631	0.393	0.042	0.435	0.470	0.524	0.565	0.667	0.435	0.833
C_5	0.732	0.863	0.667	0.458	0.042	0.536	0.506	0.595	0.393	0.631	0.500
C_6	0.798	0.429	0.631	0.363	0.333	0.042	0.363	0.565	0.464	0.464	0.256
C_7	0.673	0.732	0.369	0.536	0.738	0.458	0.042	0.500	0.500	0.631	0.458
C_8	0.667	0.500	0.565	0.494	0.458	0.726	0.429	0.042	0.292	0.738	0.458
C_9	0.393	0.595	0.494	0.762	0.464	0.393	0.423	0.494	0.042	0.399	0.530
C_{10}	0.833	0.631	0.863	0.464	0.565	0.798	0.357	0.732	0.369	0.042	0.429
C_{11}	0.256	0.464	0.363	0.839	0.321	0.321	0.363	0.536	0.571	0.292	0.042

terms of effect group criteria, it would be required to control and pay much attention to cause group criteria. Within the cause group, the criteria of Achievement Orientation (C_2) and Self-confidence (C_7) are relatively important for leveraging manager competency. By contrast, Developing Others (C_6) is the most easily improved of the effect group criteria. Furthermore, these two cause and effect groups may be used to respectively serve as cause criteria and effect criteria clusters in an MCDM model such as the AHP method (Saaty 1980) for selecting the optimal solution.

20.4.2 Discussions

When solving any business problem, group decision making is a crucial way of gathering and synthesizing group knowledge in order to reach a reasonable decision. Moreover, when problems are complex, with multiple criteria, MCDM

TABLE 20.4
Normalized Direct-relation Matrix of Criteria

	C_1	C_2	C_3	C_4	C_5	C_6	C_7	C_8	C_9	C_{10}	C_{11}
C_1	0.000	0.094	0.119	0.045	0.121	0.115	0.094	0.092	0.084	0.135	0.040
C_2	0.098	0.000	0.100	0.101	0.135	0.094	0.119	0.078	0.088	0.115	0.072
C_3	0.115	0.087	0.000	0.068	0.088	0.109	0.057	0.068	0.040	0.120	0.045
C_4	0.083	0.093	0.056	0.000	0.062	0.068	0.076	0.083	0.099	0.062	0.125
C_5	0.109	0.130	0.099	0.066	0.000	0.078	0.073	0.087	0.056	0.093	0.072
C_6	0.119	0.061	0.093	0.051	0.046	0.000	0.051	0.083	0.067	0.067	0.034
C_7	0.100	0.109	0.052	0.078	0.110	0.066	0.000	0.072	0.072	0.093	0.066
C_8	0.099	0.072	0.083	0.071	0.066	0.108	0.061	0.000	0.040	0.110	0.066
C_9	0.056	0.087	0.071	0.114	0.067	0.056	0.060	0.071	0.000	0.056	0.077
C_{10}	0.125	0.093	0.130	0.067	0.083	0.119	0.050	0.109	0.052	0.000	0.061
C_{11}	0.034	0.067	0.051	0.126	0.044	0.044	0.051	0.078	0.084	0.040	0.000

TABLE 20.5
Total-relation Matrix of Criteria

	C_1	C_2	C_3	C_4	C_5	C_6	C_7	C_8	C_9	C_{10}	C_{11}
C_1	0.452	0.507	0.523	0.404	0.505	0.521	0.420	0.476	0.396	0.552	0.346
C_2	0.561	0.444	0.525	0.473	0.536	0.521	0.460	0.484	0.421	0.555	0.395
C_3	0.494	0.443	0.361	0.373	0.423	0.460	0.344	0.402	0.317	0.481	0.309
C_4	0.450	0.440	0.399	0.312	0.391	0.410	0.354	0.407	0.366	0.418	0.378
C_5	0.516	0.507	0.475	0.399	0.369	0.459	0.381	0.444	0.353	0.486	0.354
C_6	0.437	0.365	0.390	0.312	0.335	0.306	0.294	0.363	0.298	0.380	0.258
C_7	0.486	0.471	0.415	0.393	0.450	0.427	0.297	0.413	0.354	0.464	0.337
C_8	0.464	0.416	0.422	0.366	0.389	0.445	0.335	0.326	0.307	0.457	0.317
C_9	0.393	0.402	0.379	0.384	0.363	0.367	0.314	0.365	0.249	0.379	0.313
C_{10}	0.538	0.481	0.509	0.402	0.450	0.503	0.364	0.468	0.353	0.409	0.347
C_{11}	0.326	0.339	0.316	0.358	0.300	0.313	0.270	0.329	0.293	0.318	0.210

methods are especially necessary for making favorable decisions. However, human judgment in decision-making is not usually precise but vague. Just how to deal with such ambiguous judgments, so that the quality of decisions may be improved, is an important issue. Regarding this issue, fuzzy logic is an acceptable way to express and handle vague or imprecise judgments mathematically. Another important issue is just how to establish a structural model for evaluation, before making judgments, of the relative priority of criteria. Without this, MCDM methods may not work. In light of this concern, the DEMATEL method is an extremely useful way of forming a structural model for evaluation of decision making when orders of subsystems are not explicit.

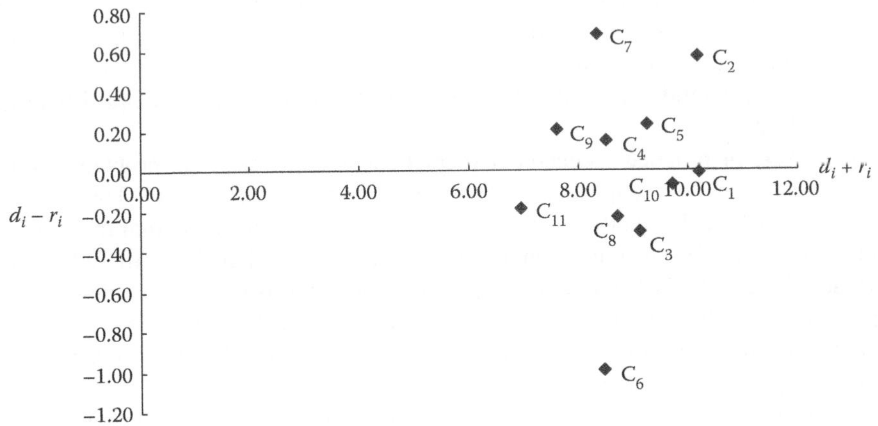

FIGURE 20.2 The causal diagram of criteria.

In particular, when tackling a problem that requires group decision making to establish a structural model in a fuzzy environment, a methodology that extends the DEMATEL method with fuzzy logic is not only valuable but indispensable. Through the empirical study, it is evident that our proposed methodology is useful and valid. The proposed methodology is comprehensive and applicable to all kinds of similar problems. In particular, we utilize an effective aggregation method based on the representation of the addition operation on triangular fuzzy numbers, so that the effort of defuzzification may be saved in the analytical procedure.

Also, our proposed methodology is well suited to situations when it is desirable to enrich the evaluation criteria by adding new ones, even if the number of criteria becomes quite large. For example, when using an MCDM method like the AHP, it is advisable to limit evaluation criteria to a number less than seven to nine within the same level or cluster (Saaty 1996; Salomon and Montevechi 2001). If this limit is not respected, decision difficulty increases, while decision quality is degraded, rendering the method unsuited to such use. Fortunately, the DEMATEL method is based on graph theory, which is able to divide involved criteria into two groups: cause and effect, and displaying causal relationships between criteria visually. That is, the DEMATEL method is useful when it is not suitable to use the AHP method directly, namely when evaluation criteria exceed seven to nine in number.

20.5 CONCLUSIONS

The purpose of structural modeling is to illustrate the ways in which subsystems affect and are affected by each other. A digraph portrays contextual relations between the elements of the system and can be converted into a visible structural model of the system with respect to those relations. As a sort of structural modeling approach, the DEMATEL method is based on digraphs, which not only works to visualize the causal relationship of criteria with a causal diagram, but also divides involved criteria into cause group and effect group. Through a causal diagram, our attention is focused on those criteria that provide the most leverage within the causal relationship. With these specialties, including understanding, conceptualizing, and representing a problem, if we aim to capture the cause-effect relationship among subsystems, it would appear that the DEMATEL method is more helpful than the ISM method.

However, in practice, decision makers hope that a favorable problem-solving method can handle group decision making in fuzzy environments. Therefore, to make the DEMATEL method suitable for solving a group decision-making problem with multiple criteria in a fuzzy environment, we have developed a methodology that extends the DEMATEL method by applying both linguistic variables and a fuzzy aggregation method. Using our proposed methodology, the interactions of criteria can be transformed into a visible structural model, making it easier to capture the complexity of a problem, whereby profound decisions can be made.

As for future research, one possible direction may be to research a more satisfying fuzzy aggregation method. Another may be to expand our proposed solution by incorporating a MCDM method, such as the AHP method, for selecting optimal alternatives.

APPENDIX 20.1 TOTAL-RELATION MATRIX *T*

Let $\boldsymbol{X}$, $\boldsymbol{H}$, $\boldsymbol{T}$ denote the normalized direct-relation matrix, the indirect-relation matrix, and the total-relation matrix, respectively. And let $\sum_{i=1}^{m}\boldsymbol{X}^{i}$ represent the total influence comprising the direct influence and the indirect influence, so that it can be shown as $\sum_{i=1}^{m}\boldsymbol{X}^{i}=\boldsymbol{X}+\boldsymbol{X}^{2}+\cdots+\boldsymbol{X}^{m}=\boldsymbol{X}+\boldsymbol{H}$. Then, with the solution given by $m\rightarrow\infty$ and $\boldsymbol{X}^{m}\rightarrow 0$, the total-relation matrix $\boldsymbol{T}$ can be obtained through $\sum_{i=1}^{\infty}\boldsymbol{X}^{i}=\boldsymbol{X}(\boldsymbol{I}-\boldsymbol{X})^{-1}$. Additionally, the indirect-relation matrix $\boldsymbol{H}$ can be obtained through $\sum_{i=2}^{\infty}\boldsymbol{X}^{i}=\boldsymbol{X}^{2}(\boldsymbol{I}-\boldsymbol{X})^{-1}$, in which $\boldsymbol{I}$ is denoted as the identity matrix.

Appendix

CHAPTER 1 INTRODUCTION

Developing the Criteria and Designing the Fuzzy Linguistic Scale

The committee followed our proposed method using the steps given. First, they defined the decision goals and developed the criteria regarding the research question.

Linguistic variables take on values defined in their term sets—its set of linguistic terms. Figure A1.1 displayed a triangular fuzzy number (TFN) in general. Linguistic terms are subjective categories for the linguistic variables. A linguistic variable is a variable whose values are words or sentences in a natural or artificial language. A TFN $x \in \tilde{A}$ and $\tilde{A} = (l, m, u)$ on $\mathbb{R}$ is to be a TFN if its membership function $\mu_{\tilde{A}}(x) : \mathbb{R} \rightarrow [0,1]$ is equal to the following equation:

$$\mu_{\tilde{A}}(x) = \begin{cases} (x-l)/(m-l), & l \leq x \leq m. \\ (u-x)/(u-m), & m \leq x \leq u. \\ 0, & \text{otherwise.} \end{cases} \tag{A1.1}$$

From Equation A1.1, l and u mean the lower and upper bounds of the fuzzy number $\tilde{A}$, and m is the modal value for $\tilde{A}$. The TFN can be denoted by $\tilde{A} = (l, m, u)$. The operational laws of TFNs $\tilde{A}_1 = (l_1, m_1, u_1)$ and $\tilde{A}_2 = (l_2, m_2, u_2)$ are displayed as Equations A1.2 through A1.5.

Addition of the fuzzy number $\oplus$

$$\tilde{A}_1 \oplus \tilde{A}_2 = (l_1, m_1, u_1) \oplus (l_2, m_2, u_2) = (l_1 + l_2, m_1 + m_2, u_1 + u_2). \tag{A1.2}$$

Multiplication of the fuzzy number $\otimes$

$$\begin{gathered} \tilde{A}_1 \otimes \tilde{A}_2 = (l_1, m_1, u_1) \otimes (l_2, m_2, u_2) = (l_1 l_2, m_1 m_2, u_1 u_2) \\ \text{for } l_1, l_2 > 0;\ m_1, m_2 > 0;\ u_1, u_2 > 0. \end{gathered} \tag{A1.3}$$

Subtraction of the fuzzy number $\ominus$

$$\tilde{A}_1 \ominus \tilde{A}_2 = (l_1, m_1, u_1) \ominus (l_2, m_2, u_2) = (l_1 - u_2, m_1 - m_2, u_1 - l_2) \tag{A1.4}$$

Division of a fuzzy number $\varnothing$

$$\begin{gathered} \tilde{A}_1 \varnothing \tilde{A}_2 = (l_1, m_1, u_1) \varnothing (l_2, m_2, u_2) = (l_1/u_2, m_1/m_2, u_1/l_2) \\ \text{for } l_1, l_2 > 0;\ m_1, m_2 > 0;\ u_1, u_2 > 0. \end{gathered}$$

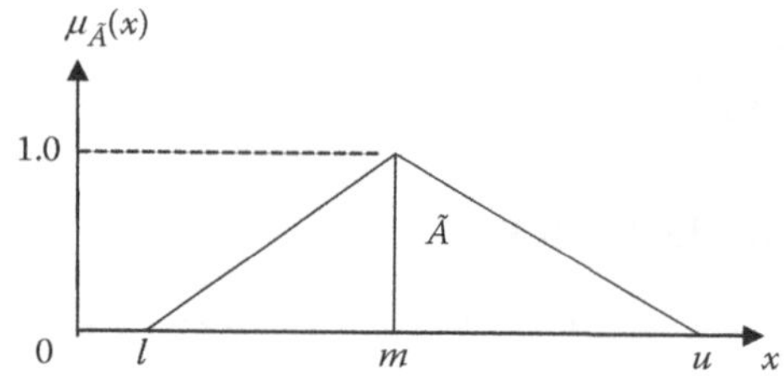

FIGURE A1.1 The membership function of the triangular fuzzy number.

Reciprocal of the fuzzy number

$$\tilde{A}^{-1} = (l_1, m_1, u_1)^{-1} = (1/u_1, 1/m_1, 1/l_1) \quad \text{for } l_1, l_2 > 0;\ m_1, m_2 > 0;\ u_1, u_2 > 0. \tag{A1.5}$$

Here, we use this kind of expression to compare two shopping websites that are evaluated by nine basic linguistic terms (natural language) for measuring influence: as "Perfect," "Very high influence," "High influence," "Low influence," "Very low influence," and "No influence," with respect to a fuzzy level scale as shown in Table A1.1 and Figure A1.2.

TABLE A1.1
Linguistic Scales for Importance (Example)

Linguistic Terms	Linguistic Values
Perfect	(1, 1, 1)
Very high influence (VH)	(0.5, 0.75, 1)
High influence (H)	(0.25, 0.5, 0.75)
Low influence (L)	(0, 0.25, 0.5)
Very low influence (VL)	(0, 0, 0.25)
No influence (No)	(0, 0, 0)

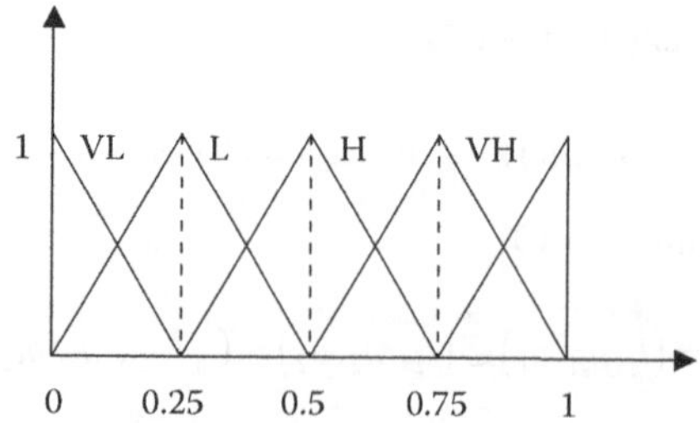

FIGURE A1.2 Triangular fuzzy numbers of linguistic variables.

CHAPTER 2 ANALYTIC HIERARCHY PROCESS

SUMMARY OF AHP AND FUZZY AHP METHODS

Concepts of Pairwise Comparison for Solving AHP

$$\boldsymbol{W}\boldsymbol{w} = \begin{array}{c} \\ w_1 \\ \vdots \\ w_i \\ \vdots \\ w_n \end{array}\begin{array}{c} \begin{array}{ccccc} w_1 & \cdots & w_j & \cdots & w_n \end{array} \\ \begin{bmatrix} w_1/w_1 & \cdots & w_1/w_j & \cdots & w_1/w_n \\ \vdots & & \vdots & & \vdots \\ w_i/w_1 & \cdots & w_i/w_j & \cdots & w_i/w_n \\ \vdots & & \vdots & & \vdots \\ w_n/w_1 & \cdots & w_n/w_j & \cdots & w_n/w_n \end{bmatrix} \end{array} \begin{bmatrix} w_1 \\ \vdots \\ w_j \\ \vdots \\ w_n \end{bmatrix} = n \begin{bmatrix} w_1 \\ \vdots \\ w_i \\ \vdots \\ w_n \end{bmatrix}$$

$$\boldsymbol{W}\boldsymbol{w} = n\boldsymbol{w} \Rightarrow (\boldsymbol{W} - n\boldsymbol{I})\boldsymbol{w} = 0,$$

where assume $w_1, \ldots, w_i, \ldots, w_n$ are known given.
In real situations, w_i/w_j is unknown, but $a_{ij} \cong w_i/w_j$ and $a_{ij} = 1/a_{ji}$ (positive reciprocal), and let $\boldsymbol{A} = [a_{ij}]_{n \times n}$.

a. $\boldsymbol{A}\boldsymbol{w} \cong n\boldsymbol{w} \Rightarrow (\boldsymbol{A} - \lambda_{\max}\boldsymbol{I})\boldsymbol{w} = 0$, find $\lambda_{\max}$ and find $\boldsymbol{w}$ with $\lambda_{\max}$, and calculate C.I. $= (\lambda_{\max} - n)/(n-1) \Rightarrow \boldsymbol{w} = (w_1, w_2, \ldots, w_n)$,

b. $$\min \sum_{i=1}^{n}\sum_{j=1}^{n}\left(a_{ij} - \frac{w_i}{w_j}\right)^2,$$

$$s.t. \sum_{i=1}^{n} w_i = 1, \qquad w_i > 0, i = 1, \ldots, n$$

c. $r_i = \left(\prod_{j=1}^{n} a_{ij}\right)^{1/n} \Rightarrow w_i = r_i / \sum_{i=1}^{n} r_i$ (normalization) $\Rightarrow \boldsymbol{w} = (w_1, w_2, \ldots, w_n)$,

d. When $\boldsymbol{A}\boldsymbol{w} = \lambda_{\max}\boldsymbol{w}$, then $\lambda_{\max}$ can be estimated by $\lambda_{\max} = \frac{1}{n}\sum_{i=1}^{n}\frac{(\boldsymbol{A}\boldsymbol{w})_i}{w_i}$.

Concepts of Pairwise Comparison for Solving Fuzzy AHP

1. Fuzzy $\tilde{\boldsymbol{A}} = [\tilde{a}_{ij}]_{n\times n} \rightarrow$ Fuzzy $\tilde{\boldsymbol{w}} = (\tilde{w}_1, \tilde{w}_2, \ldots, \tilde{w}_n)$.
   a. $\tilde{\boldsymbol{A}} \rightarrow$ solve $\tilde{\lambda}_{\max} \rightarrow$ solve $\tilde{w}_i$, i.e. $(\tilde{\boldsymbol{A}} - \tilde{\lambda}_{\max}\boldsymbol{I})\tilde{\boldsymbol{w}} = 0$
   b. $\tilde{r}_i = [\tilde{a}_{i1} \otimes \tilde{a}_{i2} \otimes \cdots \otimes \tilde{a}_{in}]^{1/n} \Rightarrow \tilde{w}_i = \tilde{r}_i \otimes [\tilde{r}_1 \oplus \tilde{r}_2 \oplus \cdots \oplus \tilde{r}_n]^{-1}$
   Inverse operation of TFN: $(a,b,c)^{-1} = (1/c, 1/b, 1/a)$
2. Fuzzy $\tilde{\boldsymbol{A}} = [\tilde{a}_{ij}]_{n\times n} \rightarrow$ Crisp $\boldsymbol{w} = (w_1, w_2, \ldots, w_n)$
   c. $\tilde{\boldsymbol{A}} = [\tilde{a}_{ij}]_{n\times n}$, $\tilde{a}_{ij} \cong w_i/w_j$, $l_{ij} \tilde{\lesssim} w_i/w_j \tilde{\lesssim} u_{ij}$, $i = 1, 2, \ldots, n-1$; $j = 1, 2, \ldots, n$; $i < j$
   $l_{ij}(\alpha) \tilde{\lesssim} w_i/w_j \tilde{\lesssim} u_{ij}(\alpha)$ in level α, then fuzzy constraints:

$$\begin{array}{l} w_i - w_j u_{ij}(\alpha) \tilde{\lesssim} 0 \\ -w_i + w_j l_{ij}(\alpha) \tilde{\lesssim} 0 \end{array} \Rightarrow \boldsymbol{R}\boldsymbol{w} \tilde{\lesssim} 0$$

where the matrix $\boldsymbol{R} \in \Re^{m\times n}$, $m \le n(n-1)/2$,

then $\mu_k(\boldsymbol{R}_k\boldsymbol{w}) = \begin{cases} 1-\dfrac{\boldsymbol{R}_k\boldsymbol{w}}{d_k}, & \boldsymbol{R}_k\boldsymbol{w} \le d_k \\ 0, & \boldsymbol{R}_k\boldsymbol{w} \le d_k \end{cases}$,

$$\lambda = \mu_D(\boldsymbol{w}) = \max_{\boldsymbol{w}}\{ \min_{k=1,2,\ldots,m} [\mu_1(\boldsymbol{R}_1\boldsymbol{w}),\ldots,\mu_m(\boldsymbol{R}_m\boldsymbol{w})] \,|\, \boldsymbol{w} \in Q^{n-1},$$

$$w_1 + w_2 + \cdots + w_n = 1\}.$$

The max-min prioritization problem:

$$\max \lambda$$

s.t.

$$\lambda \le 1 - \frac{\boldsymbol{R}_k\boldsymbol{w}}{d_k},$$

$$\sum_{i=1}^{n} w_i = 1,\ w_i > 0,\ i = 1,2,\ldots,n;\ k = 1,2,\ldots,2m.$$

3. Crisp $\boldsymbol{A} = [a_{ij}]_{n\times n} \rightarrow$ Fuzzy $\tilde{\boldsymbol{w}} = (\tilde{w}_1, \tilde{w}_2, \ldots, \tilde{w}_n)$

$$\boldsymbol{w}^k = (w_1^k, w_2^k, \ldots, w_n^k),\ k = 1,2,\ldots,K;$$

$$\tilde{\boldsymbol{w}} = (\tilde{w}_1, \ldots, \tilde{w}_j, \ldots, \tilde{w}_n);$$

$$\tilde{w}_j = (l_j, m_j, u_j);$$

where

$$l_j = \min_k\{w_j^k \,|\, k = 1,2,\ldots,K\};$$

$$m_j = \frac{1}{K}\sum_{k=1}^{K} w_j^k, \text{ or } m_j = \left[\prod_{k=1}^{K} w_j^k\right]^{1/K};$$

$$u_j = \max_k\{w_j^k \,|\, k = 1,2,\ldots,K\}.$$

we also can use α-cut by α-Level, $\tilde{W}_j^{\alpha} = (l_j(\alpha), \ldots, m_j(\alpha), \ldots, u_j(\alpha))$

CHAPTER 3 FUZZY ANALYTIC NETWORK PROCESS

Summary for ANP and Fuzzy ANP

Introduction to ANP

Step 1: Compare the ratios of weights between criteria with respect to each cluster

Step 2: Derive the local weights by solving $\boldsymbol{A}\boldsymbol{w} = \lambda_{\max}\boldsymbol{w}$ where $\lambda_{\max}$ is the largest eigenvalue of the estimated ratio matrix $\boldsymbol{A}$. Then $\boldsymbol{w} = \lim_{l\to\infty}(\boldsymbol{A}^l 1)/(1'\boldsymbol{A}^l 1)$, where $1' = (1,1,\ldots,1)$

Step 3: Forming the supermatrix

$$
\begin{array}{cc}
 & \begin{array}{cccc} C_1 & C_2 & \cdots & C_m \\ e_{11}\cdots e_{1n_1} & e_{21}\cdots e_{2n_2} & \cdots & e_{m1}\cdots e_{mn_m} \end{array} \\
\mathbf{W}= \begin{array}{cc} C_1 & \begin{array}{c} e_{12} \\ \vdots \\ e_{1n_1} \end{array} \\ C_2 & \begin{array}{c} e_{21} \\ e_{22} \\ \vdots \\ e_{2n_2} \end{array} \\ \vdots & \vdots \\ C_m & \begin{array}{c} e_{m1} \\ e_{m2} \\ \vdots \\ e_{mn_m} \end{array} \end{array} & \begin{bmatrix} \mathbf{W}_{11} & \cdots & \mathbf{W}_{12} & \cdots & \mathbf{W}_{1m} \\ \mathbf{W}_{21} & \cdots & \mathbf{W}_{22} & \cdots & \mathbf{W}_{2m} \\ \vdots & & \vdots & & \vdots \\ \vdots & & \vdots & & \vdots \\ \mathbf{W}_{m1} & \cdots & \mathbf{W}_{m2} & \cdots & \mathbf{W}_{mm} \end{bmatrix}
\end{array}
$$

Step 4: Derive the global weights by raising the weighted supermatrix to limiting power

$$\lim_{k\to\infty} \mathbf{W}^{(k)},$$

where (k) denotes the power operator.

Fuzzy ANP

Step 1: Compare the ratios of weights between criteria with respect to each cluster using the fuzzy judgments. To satisfy the condition of the fuzzy reciprocal matrix, we assume that $\tilde{a}_{ji} = 1/\tilde{a}_{ij}$ and $a_{ii} = 1$. That is, it is assumed that if $\tilde{a}_{ij} = (\underline{a}_{ij}, a_{ij}^{c}, \overline{a}_{ij})$ then $\tilde{a}_{ji} = (1/\overline{a}_{ij}, 1/a_{ij}^{c}, 1/\underline{a}_{ij})$, where $\underline{a}_{ij}$ denotes the infimum, a_{ij}^{c} denotes the center value, and $\overline{a}_{ij}$ denotes the supremum.

Step 2: *Derive the fuzzy local weight vectors*

Let $\tilde{\mathbf{\Lambda}}$ be a fuzzy positive and reciprocal matrix and choose the specific value $\alpha \in [0,1]$. In addition, let $\Gamma(\alpha)$ be a Cartesian product of intervals, i.e., $\Gamma(\alpha) = \Pi\{\tilde{\alpha}_{ij}[\alpha] \mid 1 \le i < j \le m\}$; and $v \in \Gamma(\alpha)$, where $v = (a_{12},\ldots,a_{1m}, a_{23},\ldots,a_{m-1,m})$. Then, we can define a positive and reciprocal matrix $\mathbf{\Lambda} = [e_{ij}]$ as follows: (1) $e_{ij} = a_{ij}$ if $1 \le i < j \le m$; (2) $e_{ij} = 1$, $1 \le i \le m$; and (3) $e_{ji} = a_{ij}^{-1}$ if $1 \le i < j \le m$. Let

$$w = \lim_{l\to\infty}\left(\frac{\mathbf{\Lambda}^{l}1}{1'\mathbf{\Lambda}^{l}1}\right),$$

where $1' = [1,1,\ldots,1]$ and $\mathbf{\Lambda}$ is any positive reciprocal matrix. Having described a continuous mapping $\Phi_i(v) = w_i$, $1 \le i \le m$ for each α in the range [0,1], we can obtain the following fuzzy eigenvector:

$$\tilde{w}_i[\alpha] = \left[\underline{w}_i(\alpha), \overline{w}_i(\alpha)\right], \forall 1 \le i \le m,$$

where

$$\underline{w}_i(\alpha) = \min\left\{ w_i(\alpha) \,\middle|\, \sum_{i=1}^{m} w_i = 1, v \in \Gamma(\alpha) \right\},$$

$$\overline{w}_i(\alpha) = \max\left\{ w_i(\alpha) \,\middle|\, \sum_{i=1}^{m} w_i = 1, v \in \Gamma(\alpha) \right\}.$$

Step 3: Form the fuzzy weighted supermatrix

$$\tilde{\mathbf{W}} = \begin{array}{c c} & \begin{array}{c c c c} C_1 & C_2 & \cdots & C_m \\ e_{11}\cdots e_{1n_1} & e_{21}\cdots e_{2n_2} & \cdots & e_{m1}\cdots e_{mn_m} \end{array} \\ \begin{array}{c c} C_1 & \begin{array}{c} e_{11} \\ e_{12} \\ \vdots \\ e_{1n_1} \end{array} \\ C_2 & \begin{array}{c} e_{21} \\ e_{22} \\ \vdots \\ e_{2n_2} \end{array} \\ \vdots & \vdots \\ C_m & \begin{array}{c} e_{m1} \\ e_{m2} \\ \vdots \\ e_{mn_m} \end{array} \end{array} & \begin{bmatrix} \tilde{\mathbf{W}}_{11} & \tilde{\mathbf{W}}_{12} & \cdots & \tilde{\mathbf{W}}_{1m} \\ \tilde{\mathbf{W}}_{21} & \tilde{\mathbf{W}}_{22} & \cdots & \tilde{\mathbf{W}}_{2m} \\ \vdots & \vdots & \vdots & \vdots \\ \tilde{\mathbf{W}}_{m1} & \tilde{\mathbf{W}}_{m2} & \cdots & \tilde{\mathbf{W}}_{mm} \end{bmatrix} \end{array}$$

Step 4: *Raise the fuzzy weighted supermatrix until the convergent condition is satisfied*

Let the fuzzy steady-state probabilities can be derived by a specific function of the α-cut domain:

$$\tilde{\pi}_{ij}^{(k)}[\alpha] = f_{ij}^{(k)}\left(\widetilde{\text{Dom}}[\alpha]\right),$$

where (k) denotes the limiting power that transforms the transition probabilities into steady-state probabilities.

Then, the fuzzy steady-state probabilities can be expressed using the α-cut as

$$\tilde{\pi}_i^{(k)}[\alpha] = \left[\underline{\pi}_i^{(k)}(\alpha), \overline{\pi}_i^{(k)}(\alpha)\right], \forall 1 \le i \le m,$$

where

$$\underline{\pi}_i^{(k)}[\alpha]=\min\left\{f_{ij}^{(k)}(\pi)\middle|\pi\in\widetilde{\mathrm{Dom}}[\alpha]\right\},$$

and

$$\overline{\pi}_i^{(k)}[\alpha]=\max\left\{f_{ij}^{(k)}(\pi)\middle|\pi\in\widetilde{\mathrm{Dom}}[\alpha]\right\}.$$

CHAPTER 4 SIMPLE ADDITIVE WEIGHTING METHOD

Criteria weights and alternative performance matrix

Alternatives	Criteria				
	c_1	$\cdots$	c_j	$\cdots$	c_n
	w_1	$\cdots$	w_j	$\cdots$	w_n
a_1	x_{11}	$\cdots$	x_{1j}	$\cdots$	x_{1n}
$\vdots$	$\vdots$		$\vdots$		$\vdots$
a_i	x_{i1}	$\cdots$	x_{ij}	$\cdots$	x_{in}
$\vdots$	$\vdots$		$\vdots$		$\vdots$
a_m	x_{m1}	$\cdots$	x_{mj}	$\cdots$	x_{mn}
Aspired value	x_1^*	$\cdots$	x_j^*	$\cdots$	x_n^*
The worst value	x_1^-	$\cdots$	x_j^-	$\cdots$	x_n^-

Data matrix

$$\left[x_{ij}\right]_{m\times n}\xrightarrow{\text{normalization}}\left[r_{ij}\right]_{m\times n}\xrightarrow{\text{weight } w_j}\left[w_j r_{ij}\right]_{m\times n}\longrightarrow$$

$$R_i=\sum_{j=1}^{n}w_j r_{ij},\ i=1,2,\ldots,m$$

Larger is better $r_{ij}=(x_{ij}-x_j^-)/(x_j^*-x_j^-)$
Smaller is better $r_{ij}=(x_j^- - x_{ij})/(x_j^- - x_j^*)$
General $r_{ij}=|x_{ij}-x_j^-|/|x_j^*-x_j^-|$
Fuzzy data matrix

$$\left[\tilde{x}_{ij}\right]_{m\times n}\xrightarrow{\text{normalization}}\left[\tilde{r}_{ij}\right]_{m\times n}\xrightarrow{\text{weight } \tilde{w}_j}\left[\tilde{w}_j\tilde{r}_{ij}\right]_{m\times n}\longrightarrow$$

$$\tilde{R}_i=\sum_{j=1}^{n}\tilde{w}_j\tilde{r}_{ij},\ i=1,2,\ldots,m$$

CHAPTER 5 TOPSIS AND VIKOR

SUMMARY

TOPSIS

TOPSIS (Technique for Order Preference by Similarity to Ideal Solution)

Alternatives	Criteria				
	c_1	$\cdots$	c_i	$\cdots$	c_n
	w_1	$\cdots$	w_i	$\cdots$	w_n
a_1	x_{11}	$\cdots$	x_{1j}	$\cdots$	x_{1n}
$\vdots$	$\vdots$		$\vdots$		$\vdots$
a_i	x_{i1}	$\cdots$	x_{ij}	$\cdots$	x_{in}
$\vdots$	$\vdots$		$\vdots$		$\vdots$
a_m	x_{m1}	$\cdots$	x_{mj}	$\cdots$	x_{mn}
Aspired value	x_1^*	$\cdots$	x_j^*	$\cdots$	x_n^*
The worst value	x_1^-	$\cdots$	x_j^-	$\cdots$	x_n^-

$$\text{Data matrix}\left[x_{ij}\right]_{m\times n}\xrightarrow{\text{normalization}}\left[r_{ij}\right]_{m\times n}\xrightarrow{\text{weight}}\left[w_j r_{ij}\right]_{m\times n}\longrightarrow\left[v_{ij}\right]_{m\times n}$$

(larger is better) $r_{ij}=\left(x_{ij}-x_j^-\right)/\left(x_j^*-x_j^-\right)$

The distance from point v_{ij} to positive ideal point v_j^* and negative ideal point v_j^- for $j = 1, 2,..., n$ is as follows:

$$d_i^+=\left[\sum_{j=1}^{n}\left(v_{ij}-v_j^*\right)^2\right]^{1/2} \text{ or } \sqrt{\sum_{j=1}^{n}\left(v_{ij}-v_j^*\right)^2},$$

$$d_i^-=\left[\sum_{j=1}^{n}\left(v_{ij}-v_j^-\right)^2\right]^{1/2} \text{ or } \sqrt{\sum_{j=1}^{n}\left(v_{ij}-v_j^-\right)^2}.$$

Ranking index for achieving level (large is better)

$$R_i=\frac{d_i^-}{d_i^*+d_i^-} \text{ or } R_i=1-\frac{d_i^*}{d_i^*+d_i^-}.$$

Ranking index for gap to positive ideal point (small is better)

$$R_i = \frac{d_i^*}{d_i^* + d_i^-}.$$

VIKOR

Multicriteria ranking and compromise solution

Alternatives		**$\max_j$ $\min_j$**						
Criteria	weights	a_1	$\cdots$	a_j	$\cdots$	a_m	(or aspired value)	
c_1	w_1	x_{11}	$\cdots$	x_{1j}	$\cdots$	x_{1m}	x_1^*	x_1^-
$\vdots$	$\vdots$	$\vdots$		$\vdots$		$\vdots$	$\vdots$	$\vdots$
c_i	w_i	x_{i1}	$\cdots$	x_{ij}	$\cdots$	x_{im}	x_i^*	x_i^-
$\vdots$	$\vdots$	$\vdots$		$\vdots$		$\vdots$	$\vdots$	$\vdots$
c_n	w_n	x_{n1}	$\cdots$	x_{nj}	$\cdots$	x_{nm}	x_n^*	x_n^-

Note: Data matrix: larger is better.

$$d_j^p \left\{ \sum_{i=1}^{n} \left[w_i \left(\frac{x_i^* - x_{ij}}{x_i^* - x_i^-} \right) \right]^p \right\}^{1/p} \text{ or } d_j^p \left\{ \sum_{i=1}^{n} w_i^p \left(\frac{x_i^* - x_{ij}}{x_i^* - x_i^-} \right)^p \right\}^{1/p},$$

when

$$d_j^{p=1} = S_j = \sum_{i=1}^{n} w_i \left(\frac{x_i^* - x_{ij}}{x_i^* - x_i^-} \right)$$

$$d_j^{p=\infty} = Q_j = \max_i \left\{ \left(\frac{x_i^* - x_{ij}}{x_i^* - x_i^-} \right) \middle| i = 1, 2, \ldots, n \right\}.$$

Ranking (small is better for distance S_j and Q_j)

$$R_j = v\left[\left(S_j - S^*\right) / \left(S^- - S^*\right) \right] + (1 - v)\left[\left(Q_j - Q^*\right) / \left(Q^- - Q^*\right) \right].$$

We can set $v = 1$ or 0.5 or 0, if we let $v = 0.5$ be the majority criteria, where $S^* = \min_j S_j$, $S^- = \max_j S_j$, and $Q^* = \min_j Q_j$, $Q^- = \max_j Q_j$.

We also re-write $R_j = vS_j + (1 - v)\, Q_j$, when $S^* = 0$ and $Q^* = 0$ (i.e., all criteria have been achieved to the aspired level) and $S^- = 1$ and $Q^- = 1$ (i.e., the worst situation).

CHAPTER 6 ELECTRE METHOD

SUMMARY FOR ELECTRE I, II, III, AND IV

ELECTRE I Model

ELECTRE I is a discrete model. The algorithm is to search for a "kernel," which is a non-inferior solution. The condition of the "kernel" is based on the assumption of intransitive ordering of alternatives and the following formula: alternative i is preferred to alternative j ($i > j$; $i,j \in \{1,\ldots, m\}$) if and only if

$$\text{Concord index: } c(i,j) \geq p$$

and

$$\text{Discord index: } d(i,j) \leq q$$

where p and q are determined by decision makers. Concord index $c(i,j)$ and discord index $d(i,j)$ are defined as follows.

$$\text{Concord index: } c(i,j) = \frac{\sum\limits_{k\in\{i_k > j_k | k=1,\ldots,m\}} w_k + \frac{1}{2} \sum\limits_{k\in\{i_k = j_k | k=1,\ldots,m\}} w_k}{\sum\limits_{k=1}^{m} w_k},$$

$$\text{Discord index: } d(i,j) = \max_{k\in\{i_k < j_k | k=1,\ldots,n\}} \left\{ \frac{\left| i_k(f/l) - j_k(\bar{f}/l) \right|}{k(l)} \middle| k = 1,\ldots,n \right\},$$

where:

w_k: kth criterion weight obtained by using AHP or ANP
$\{i_k > j_k | k = 1,\ldots, m\}$: Performance of $i > j$ at kth criterion
$\{i_k = j_k | k = 1,\ldots, m\}$: Performance of alternative i and j is not different ($i = j$) at kth criterion
$\{i_k < j_k | k = 1,\ldots, m\}$: Performance of alternative i is inferior to the performance of alternative j at kth criterion
$i_k(f/l) - j_k(\bar{f}/l)$: Discomfort is caused by going from level $j_k(\bar{f}/l)$ to $i_k(f/l)$ of criterion k
$k(l)$: Total range of scale

We can use the concord index matrix and the discord index matrix to find the kernel of the inferior and non-inferior clusters. This model has some shortcomings: (1) the ranking of alternatives is just partial; (2) the more exact the decision makers' requirements, the more prior alternatives selected; and (3) the decision makers' preference cannot be fully expressed.

ELECTRE II Model

ELECTRE II is a method that refines ELECTRE I. The main difference between the two models is the ranking of the alternatives. ELECTRE I only divides the alternatives into two sets: inferior and non-inferior, while ELECTRE II can put all the alternatives in rank order according to priority. In contrast to ELECTRE I, there are multiple levels of concordance and discordance specified, and these are used to construct a strong relationship, R_S, and a weak relationship, R_w. These two relationships, in turn, are used to obtain the ranking of the alternatives. The strong relationship, R_S, is defined if and only if one (or both) of the following sets of conditions hold:

$$\text{I: } c(i,j) \geq p^*$$

$$d(i,j) \leq q^*$$

$$\sum_{k \in \{i_k > j_k | k=1,\ldots,m\}} w_k \geq \sum_{k \in \{i_k < j_k | k=1,\ldots,m\}} w_k,$$

$$\text{II: } c(i,j) \geq p^{\text{o}}$$

$$d(i,j) \leq q^{\text{o}}$$

$$\sum_{k \in \{i_k > j_k | k=1,\ldots,m\}} w_k \geq \sum_{k \in \{i_k < j_k | k=1,\ldots,m\}} w_k,$$

The weak relationship, R_w, is defined if and only if the following conditions hold:

$$\text{III: } c(i,j) \geq p^-$$

$$d(i,j) < q^*$$

$$\sum_{k \in \{i_k > j_k | k=1,\ldots,m\}} w_k \geq \sum_{k \in \{i_k < j_k | k=1,\ldots,m\}} w_k,$$

where $0 \leq p^- \leq p^\circ \leq p^* \leq 1$ and $0 < q^\circ < q^* < 1$.

If I or II (or both) hold, then alternative *i* strongly outranks alternative *j*. If III holds, then alternative *i* weakly outranks alternative *j*.

The concordance and discordance definitions of ELECTRE II differ from those of ELECTRE I, which are defined as

$$\text{Concord index: } c(i,j) = \frac{\sum_{k \in \{i_k > j_k | k=1,\ldots,m\}} w_k + \sum_{k \in \{i_k = j_k | k=1,\ldots,m\}} w_k}{\sum_{k=1}^{m} w_k},$$

$$\text{Discord index: } d(i,j) = \frac{R_i(y) - R_i(x)}{\max\{R_i(x), S(k)\}}, \quad (x,y) \in X \times X$$

where $X = \{x_i | i = 1,\ldots, m\}$ is the set of alternatives. Also, R_i is a bounded positive function mapping X to E_i, where E_i is an interval scale and $S(l)$ is a parameter related to the type L of scale adapted for the criterion k.

According to the strong relationship, R_S, and the weak relationship, R_w, we can obtain the ranking of the alternatives. The algorithm is described as follows:

1. Strong ranking $V'(x)$
   Notation G_S represents the set of alternatives that satisfy the strong relationship, and notation G_W represents the set of alternatives that satisfy the weak relationship. Let $Y^{(l)}$ be the partial set of G_S.
   a. Let l=0 and $Y^{(0)} = G_S$ exists.
   b. Choose the non-inferior solutions in G_S; let it be the set of D.
   c. Find the alternatives that satisfy the weak relationship in D; let it be the set of U.
   d. Choose the non-inferior solutions in U; let it be the set of B.
   e. The set of the best alternatives $A^{(l)} = (D - U) \cup B$, where $D - U = \{x | x \in D, x \notin U\}$.
   f. For each $x \in A^{(k)}$, the ranking order is $V'(x) = l + 1$.
   g. Let $Y^{(l+1)} = \varnothing$, then stop; otherwise, let $l = l + 1$, and the algorithm goes back to step b.
2. Weak rank $V''(x)$
   a. Reverse the directions of G_S and G_W.
   b. The algorithm is the same as in the strong ranking, and the ranking order $a(x)$ exists.
   c. Adjust the ranking order $a(x)$, and the weak ranking $V''(x)$ exists, $V''(x) = 1 + a_{\max} - a(x)$, $x \in X$, $a_{\max} = \max_{x \in X} a(x)$.
3. Median (final) ranking $V(x)$

   $V(x) = [V'(x) + V''(x)]/2 \quad \forall x \in X.$

ELECTRE III Model

The evaluation procedures of the ELECTRE III model encompass the establishment of a threshold function, disclosure of concordance index and discordance index, confirmation of credibility degree, and the ranking of alternatives. These data may be represented by fuzzy data in case that subjective judgment by evaluators and decision makers, in what following are the further description.

Let $q(g)$ and $p(g)$ represent the indifference threshold and preference threshold, respectively.

If $g(a) \geq g(b)$:

$$g(a) > g(b) + p(g(b)) \Leftrightarrow aPb$$

$$g(b)+q(g(b))<g(a)<g(b)+p(g(b))\Leftrightarrow aQb$$
$$g(b)<g(a)<g(b)+q(g(b))\Leftrightarrow aIb,$$

where P denotes strong preference, Q denotes weak preference, I denotes indifference, and $g(a)$ is the criterion value of alternative a.

The establishment of a threshold function has to satisfy the subsequent constraint equations:

$$g(a)>g(b)\Rightarrow\begin{cases}g(a)+q(g(a))>g(b)+q(g(b))\\ g(a)+p(g(a))>g(b)+p(g(b))\end{cases},$$

for all criteria, $p(g)>q(g)$.

Furthermore, $p_j(g_j(a))$ and $q_j(g_j(a))$ can be calculated according to Roy's formula:

$$p_j(g_j(a))=\alpha_p+\beta_p g_j(a)$$
$$q_j(g_j(a))=\alpha_q+\beta_q g_j(a),$$

where $p_j(g_j(a))$ and $q_j(g_j(a))$ can be solved in such a way that the threshold values are one case of which as follows (Roy et al. 1986):

1. Either constant (β equals zero and α has to be determined) or
2. Proportional to $g_j(a)$ (β has to be determined and α equals zero) or
3. Of a form combining these two (both α and β have to be determined)

A concordance index, $C(a,b)$, is computed for each pair of alternatives:

$$C(a,b)=\frac{\sum_{i=1}^{m}w_j C_i(a,b)}{\sum_{i=1}^{m}w_j},$$

where $C_i(a,b)$ is the outranking degree of alternative a with alternative b under criterion i, and

$$C_i(a,b)=\begin{cases}0, & \text{if } g_i(b)-g_i(a)>p_i(g_i(a))\\ 1, & \text{if } g_i(b)-g_i(a)\le q_i(g_i(a))\end{cases},$$

and $0<c_i(a,b)<1$ when $q_i(g_i(a))<g_i(b)-g_i(a)\le p_i(g_i(a))$.

The veto threshold, $v_j(g_j(a))$, is defined for each criterion j:

$$v_j(g_j(a))=\alpha_v+\beta_v g_j(a).$$

A discordance index, $d(a,b)$, for each criterion is then defined as

$$d_j(a,b)=0 \quad \text{if } g_j(b)-g_j(a)\le p_j(g_j(a)),$$
$$d_j(a,b)=1 \quad \text{if } g_j(b)-g_j(a)> v_j(g_j(a)),$$
$$0<d_j(a,b)<1,$$

when

$$p_j(g_j(a))<g_j(b)-g_j(a)\le v_j(g_j(a)).$$

Finally, the degree of outranking is defined by $S(a,b)$:

$$S(a,b)=\begin{cases} c(a,b) & \text{if } d_j(a,b)\le c(a,b) \quad \forall j\in J \\ c(a,b)\times \prod\limits_{j\in J(a,b)} \dfrac{1-d_j(a,b)}{1-c(a,b)} & \text{otherwise} \end{cases},$$

where $J(a,b)$ is the set of criteria for which $d_j(a,b) > c(a,b)$.

ELECTRE IV Model

Roy and Bouyssou (1983) proposed ELECTRE IV to simplify the procedure of ELECTRE III. The basic difference between ELECTRE III and ELECTRE IV is that ELECTRE IV does not introduce any weight expressing the weights of the criteria, which may be hard to measure in practice. However, this does not mean that the weights of the criteria are assumed to be equal. Therefore, the pseudo-criteria are used as in ELECTRE III.

Five outranking relations are defined in ELECTRE (Roy and Bouyssou 1993):

1. Quasi-dominance
   The couple (b,a) verifies the relation of quasi-dominance if and only if:
   - For every criterion, b is either preferred or indifferent to a.
   - If the number of criterion for which the performance of a is better than the one of b (a staying indifferent to b) is strictly inferior to the number of criteria for which the performance of b is better than the one of a.
2. Canonic-dominance
   The couple (b,a) verifies the relation of canonic-dominance if and only if:
   - For no criterion, a is strictly preferred to b.
   - If the number of criteria for which a is weakly preferred to b is inferior or equal to the number of criteria for which b is strictly preferred to a.
   - If the number of criteria for which the performance of a is better than the one of b is strictly inferior to the number of criteria for which the performance of b is better than the one of a.

3. Pseudo-dominance
   The couple (b,a) verifies the relation of pseudo-dominance if and only if:
   - For no criterion, a is strictly preferred to b.
   - If the number of criteria for which a is weakly preferred to b is inferior or equal to the number of criteria for which b is strictly or weakly preferred to a.
4. Sub-dominance
   The couple (b,a) verifies the relation of sub-dominance if and only if, for no criterion, a is strictly preferred to b.
5. Veto-dominance
   The couple (b,a) verifies the relation of veto-dominance if and only if:
   - Either for no criterion, a is strictly preferred to b.
   - a is strictly preferred to b for only one criterion but this criterion not vetoing the outranking of a by b and, furthermore, b is strictly preferred to a for at least half the criteria.

The partial preorder is performed as in ELECTRE III, but is made simpler by the fact that there are only two outranking levels.

CHAPTER 7 PROMETHEE

PROMETHEE: A new family of outranking methods in multi-criteria analysis (J.P. Brans, B. Mareschal, Ph. Vincke (1984) INFORS, Operational Research'84, Elsevier Science Publishers)

- PROMETHEE (Preference Ranking Organization METHods for Enrichment Evaluations)
- Brans, Mareschal, Vincke (1984) consider a new family of *outranking methods* for solving multicriteria problems

Alternatives	Criteria				
	c_1	⋯	c_h	⋯	c_k
	w_1	⋯	w_h	⋯	w_k
⋮	⋮		⋮		⋮
a	$f_1(a)$	⋯	$f_h(a)$	⋯	$f_k(a)$
⋮	⋮		⋮		⋮
Aspired value	f_1^*	⋯	f_h^*	⋯	f_k^*
The worst value	f_1^-	⋯	f_h^-	⋯	f_k^-

max $\{f_1(a), ..., f_h(a), ..., f_k(a)\}, a \in A$.

Define a preference function $p(A \times A \to [0,1])$ giving the following intensity of preference of the action a over the action b and having the following meaning:

- $p_h(a, b) = 0$ No preference of a over b, indifference between a and b,
- $p_h(a, b) \sim 0$ Weak preference of a over b ($f_h(a) > f_h$(b)),
- $p_h(a, b) \sim 1$ Strong preference of a over b ($f_h(a) >> f_h$(b)),
- $p_h(a, b) = 1$ Strict preference of a over b ($f_h(a) >>> f_h$(b)),

where $a, b \in A$,

$$d_h = f_h(a) - f_h(b),$$

$$p_h = (a, b) = p(d_h),$$

Multicriteria preference index

$$\pi(a, b) = \sum_{h=1}^{k} w_h p_h(a, b),$$

where the weight w_h can be obtained by AHP or ANP, which depend on criteria structure. How can we know the criteria structure? We can use some techniques, such as interpretive structure modeling (ISM), DEMATEL, fuzzy cognitive mapping (FCM), and so on.

For evaluating the actions/alternatives of set A by using the outranking relation:

1. The leaving flow: $\phi^+(a) = \sum_{b \in A} \pi(a, b)$
2. The entering flow: $\phi^-(a) = \sum_{b \in A} \pi(b, a)$
3. The net flow: $\phi(a) = \phi^+(a) - \phi^-(a)$

$$\begin{cases} aP^+b & \text{iff} \quad \phi^+(a) > \phi^+(b); \\ aI^+b & \text{iff} \quad \phi^+(a) = \phi^+(b); \end{cases}$$

$$\begin{cases} aP^-b & \text{iff} \quad \phi^-(a) < \phi^-(b); \\ aI^-b & \text{iff} \quad \phi^-(a) = \phi^-(b), \end{cases}$$

where P and I represent preference and indifference, respectively.

PROMETHEE I

According to Brans et al. (1984, 1985), PROMETHEE I determines the partial preorder (P^I, I^I, R) on the alternatives of A that satisfy the following principle:

$$aP^Ib \text{ (} a \text{ outranks } b\text{), if } \begin{cases} aP^+b & \text{and} \quad aP^-b \\ aP^+b & \text{and} \quad aI^-b\,, \\ aI^+b & \text{and} \quad aP^-b \end{cases}$$

aI^Ib (a is indifferent from b), if aI^+b *and* aI^-b,
aRb (a and b are incomparable), otherwise.

From the above equations, we can obtain a partial order for alternatives, while some others are incomparable (i.e., if aRb cases exist, a and b are incomparable).

PROMETHEE II

Furthermore, PROMETHEE II gives a complete preorder (P^{II}, I^{II}) induced by the net flow and defined by :

$aP^{II}b$ (a outranks b), iff $\phi(a) > \phi(b)$
$aI^{II}b$ (a is indifferent to b), iff $\phi(a) = \phi(b)$

PROMETHEE III

Base on the same reasons as above, PROMETHEE III associates to each action a an interval $[x_a, y_a]$ and defines a complete interval order $(P^{III,} I^{III})$ as follows:

$aP^{III}b$ (a outranks b), iff $x_a > y_b$
$aI^{III}b$ (a is indifferent to b), iff $x_a \le y_b$ and $x_b \le y_a$,

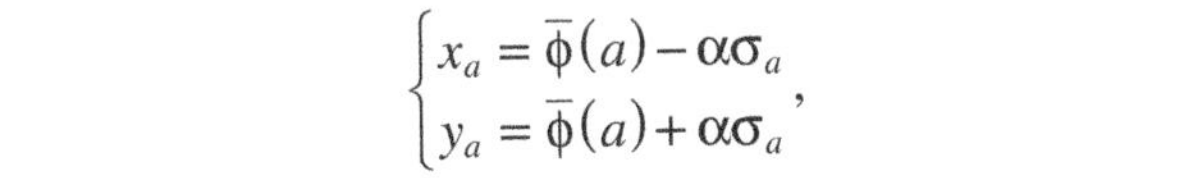

Relations of a, b, c: $aI^{III}b$, $bI^{III}c$, $aP^{III}c$

The interval $[x_a, y_a]$ is given by

$$\begin{cases} x_a = \bar{\phi}(a) - \alpha\sigma_a \\ y_a = \bar{\phi}(a) + \alpha\sigma_a \end{cases},$$

where n is the number of actions (or criteria):

$$\bar{\phi}(a) = \frac{1}{n}\sum_{b \in A}\left(\pi(a,b) - \pi(b,a)\right) = \frac{1}{n}\phi(a),$$

$$\sigma_a^2 = \frac{1}{n}\sum_{b\in A}\left(\pi(a,b) - \pi(b,a) - \bar{\phi}(a)\right)^2,$$

and where $\alpha > 0$ in general.

PROMETHEE IV

Furthermore, PROMETHEE IV extends PROMETHEE II to the case of a continuous set of actions (or alternatives) A; such a set arises when the actions are, for instance, percentages, dimensions of a product, compositions of an alloy, investments, and so on.

The generalized criteria of PROMETHEE IV are defined by extending PROMETHEE II from preference functions $P_h(a, b)$ such that $P_h(a, b) = \wp(d)$, where $d_h = f_h(a) - f_h(b)$ and $h = 1, 2 \ldots, k$. In addition, the leaving flow, the entering flow, and the net flow for continuous set A are defined as follows:

$$\phi^+(a) = \int_A \pi(a,b)\,db,$$

$$\phi^-(a) = \int_A \pi(b,a)\,db,$$

$$\phi(a) = \phi^+(a) - \phi^-(a).$$

In fact, it is not always easy to integrate the preference index $\pi(a, b)$ on set A. Brans et al. (1984) suggested a simplification as follows:

$$\phi^+(a) = \int_A P_h(a,b)\,db,$$

$$\phi^-(a) = \int_A P_h(b,a)\,db,$$

and to deduce

$$\phi(a) = \sum_{h=1}^{k} w_h\left[\phi_h^+(a) - \phi_h^-(a)\right].$$

CHAPTER 8 GRAY RELATION MODEL

Summary

Gray Relation for Evaluation

Alternatives	Criteria				
	c_1	$\cdots$	c_j	$\cdots$	c_n
	w_1	$\cdots$	w_j	$\cdots$	w_n
x_1	$x_1(1)$	$\cdots$	$x_1(j)$	$\cdots$	$x_1(n)$
$\vdots$	$\vdots$		$\vdots$		$\vdots$
x_i	$x_i(1)$	$\cdots$	$x_i(j)$	$\cdots$	$x_i(n)$
$\vdots$	$\vdots$		$\vdots$		$\vdots$
x_m	$x_m(1)$	$\cdots$	$x_m(j)$	$\cdots$	$x_m(j)$
Aspired value x^*	$x^*(1)$	$\cdots$	$x^*(j)$	$\cdots$	$x^*(n)$
The worst value x^-	$x^-(1)$	$\cdots$	$x^-(j)$	$\cdots$	$x^-(n)$

Note: Data matrix: normalization.

1. Coefficients of gray relation for aspired values

$$\gamma\left(x^*(j),x_i(j)\right)=\frac{\min_i\min_j\left|x^*(j)-x_i(j)\right|+\varsigma\max_i\max_j\left|x^*(j)-x_i(j)\right|}{\left|x^*(j)-x_i(j)\right|+\varsigma\max_i\max_j\left|x^*(j)-x_i(j)\right|}$$

Grade (degree) of gray relation (larger is better)

$$\gamma\left(x^*,x_i\right)=\sum_{j=1}^{n}w_j\gamma\left(x^*(j),x_i(j)\right) \tag{A8.1}$$

where the weight w_j can be obtained by AHP or ANP, which depend on criteria structure. How can we know the criteria structure? ISM, DEMATEL, FCM, and so on.

2. Coefficients of gray relation for worst values

$$\gamma\left(x^-(j),x_i(j)\right)=\frac{\min_i\min_j\left|x^-(j)-x_i(j)\right|+\varsigma\max_i\max_j\left|x^-(j)-x_i(j)\right|}{\left|x^-(j)-x_i(j)\right|+\varsigma\max_i\max_j\left|x^-(j)-x_i(j)\right|}$$

Grade (degree) of gray relation (larger is worse, small is better)

$$\gamma(x^-, x_i) = \sum_{j=1}^{n} w_j \gamma(x^-(j), x_i(j)) \tag{A8.2}$$

3. Combine A8.1 and A8.2 for ranking based on the concept of TOPSIS

$$R_i = \frac{\gamma(x^*, x_i)}{\gamma(x^-, x_i)}$$

CHAPTER 9 FUZZY INTEGRAL TECHNIQUES

SUMMARY

Fuzzy Integral (MADM, Evaluation in Basic Concepts)

- Fuzzy measure
  Basic idea: if $A \cap B = \phi$, λ–Fuzzy measure
  $g_\lambda(A \cup B) = g_\lambda(A) + g_\lambda(B) + \lambda g_\lambda(A) g_\lambda(B)$, $-1 < \lambda < \infty$
  If $\lambda > 0$, $g_\lambda(A \cup B) > g_\lambda(A) + g_\lambda(B)$ Multiplicative
  If $\lambda = 0$, $g_\lambda(A \cup B) = g_\lambda(A) + g_\lambda(B)$ Additive
  If $\lambda < 0$, $g_\lambda(A \cup B) < g_\lambda(A) + g_\lambda(B)$ Substitutive
- Fuzzy integral (Sugeno fuzzy integral and Choquet integral)
  - Evaluation items (criteria/attributes): $x_1, x_2, \ldots, x_n$
  - Weights of each item (criterion/attribute): $w_1, w_2, \ldots, w_n$
  - Performance (evaluator): $h_1 = h(x_1)$, $h_2 = h(x_2), \ldots, h_n = h(x_n)$
  - Total performance (total evaluator): $h_T = h_1 w_1 + h_2 w_2 + \cdots + h_n w_n$
    $g(\bullet)$: Fuzzy measure and assume $h(x_1) \geq h(x_2) \geq \cdots \geq h(x_n)$

[Summation]

$$(C)\int h dg = h(x_n) g(H_n) + [h(x_{n-1}) - h(x_n)] g(H_{n-1}) + \cdots + [h(x_1) - h(x_2)] g(H_1).$$

[Example 1] We compare the qualities and capabilities of computer products by using evaluation based on dependence/interrelationship among criteria.

x_1 = easy use and x_2 = functional capabilities

λ–Fuzzy measure

$g_\lambda(\{x_1\}) = 0.5$ $(u(x_1^*, x_2^0) = g_\lambda(\{x_1\})$, when $g_\lambda(\{x_1^*, x_2^*\}) = g_\lambda(\{x_1, x_2\}) = 1)$

$g_\lambda(\{x_1\}) = 0.3$ $(u(x_2^*, x_1^0) = g_\lambda(\{x_2\})$, when $g_\lambda(\{x_1^*, x_2^*\}) = g_\lambda(\{x_1, x_2\}) = 1)$

where x_1^* and x_2^* are to show the best and x_1^0 and x_2^0 are to show the worst individually.

$$g_\lambda(\{x_1, x_2\}) > g_\lambda(\{x_1\}) + g_\lambda(\{x_2\})$$

$$\Rightarrow 1 = 0.5 + 0.3 + 0.5 \times 0.3\lambda \Rightarrow 0.2 = 0.15\lambda \Rightarrow \lambda = 1.333 > 0 \text{ Multiplicative}$$

Scores of Computer Products

Products	x_1(Easy Use)	x_2(Functional Capacities)	Additive Model	Fuzzy Integral
P	90	20	63.75 (1)	55 (2)
Q	60	60	60 (2)	60 (1)

Additive model (expected/average value)

P: $90 \times 0.5/(0.5 + 0.3) + 20 \times 0.3/(0.5 + 0.3) = 56.25 + 7.5 = 63.75$
Q: $60 \times 0.5/(0.5 + 0.3) + 60 \times 0.3/(0.5 + 0.3) = 60 \times (0.5 + 0.3)/(0.5 + 0.3) = 60$

Fuzzy integral

$$\text{P}: (C)\int h_1 dg = 70 \times 0.5 + 20 \times 1 = 55$$

$$\text{Q}: (C)\int h_2 dg = 60 \times 1 = 60$$

[Example 2] One company raises the employee using evaluation items from test and interview.

x_1 = salesman ability and x_2 = information process ability

$$g_\lambda(\{x_1\}) = g_\lambda(\{x_2\}) = 0.9, \quad g_\lambda(\{x_1, x_2\}) = 1$$

(i.e., $u(x_1^*, x_2^0) = u(x_2^*, x_1^0) = 0.9$, when $g_\lambda(\{x_1^*, x_2^*\}) = g_\lambda(\{x_1, x_2\}) = 1$)

$$g_\lambda(\{x_1, x_2\}) < g_\lambda(\{x_1\}) + g_\lambda(\{x_2\})$$

$$\Rightarrow 1 = 0.9 + 0.9 + 0.9 \times 0.9\lambda \Rightarrow -0.8 = 0.81\lambda \Rightarrow \lambda = -0.988 < 0 \text{ Substitutive}$$

Ability Cores of Test and Interview for Raising Employee

Products	x_1 (Salesman)	x_2 (Information Process)	Additive Model	Fuzzy Integral
P	90	20	55 (2)	83 (1)
Q	60	60	60 (1)	60 (3)
R	30	70	50 (3)	66 (2)

Additive Model (Expected Value)

P: 90 × 0.9/(0.9 + 0.9) + 20 × 0.9/(0.9 + 0.9) = 55
Q: 60 × 1 = 60
R: 30 × 0.9/(0.9 + 0.9) + 70 × 0.9/(0.9 + 0.9) = 50

Fuzzy integral

$$\mathrm{P}:(C)\int h_1 dg = 70\times 0.9 + 20\times 1 = 83$$

$$\mathrm{Q}:(C)\int h_2 dg = 60\times 1 = 60$$

$$\mathrm{R}:(C)\int h_3 dg = 40\times 0.9 + 30\times 1 = 66$$

[Example 3] Strategies for evaluation are implemented by using criteria as follows: organization (x_1), management resource (x_2), competition (x_3), and customers' or social needs (x_4).

$$\begin{array}{c} \mathrm{AHP} \\ \{x_1\} \\ \{x_2\} \\ \{x_3\} \\ \{x_4\} \end{array} \begin{array}{c} \begin{array}{cccc} \{x_1\} & \{x_2\} & \{x_3\} & \{x_4\} \end{array} \\ \begin{bmatrix} 1 & 1/3 & 1/3 & 1/2 \\ 3 & 1 & 1/2 & 1/4 \\ 3 & 2 & 1 & 1 \\ 2 & 4 & 1 & 1 \end{bmatrix} \end{array}$$

Initial weights (0.110, 0.179, 0.326, 0.384) can be obtained by using AHP, then we let fuzzy measure weights $\boldsymbol{w} = c(0.110, 0.179, 0.326, 0.384)$, i.e.,

$$w_1\text{: } g_\lambda(\{x_1\}) = 0.110c, \quad w_2\text{: } g_\lambda(\{x_2\}) = 0.179c,$$

$$w_3\text{: } g_\lambda(\{x_3\}) = 0.326c, \quad w_4\text{: } g_\lambda(\{x_4\}) = 0.384c,$$

where

$$g_\lambda(X) = g_\lambda(\{x_1, x_2, x_3, x_4\}) = 1,$$

$$g_\lambda(X) = \sum_i w_i + \lambda \sum_{i,j>i} w_i w_j + \lambda^2 \sum_{i,j,k} w_i w_j w_k + \lambda^3 w_1 w_2 w_3 w_4.$$

Assuming $\lambda = 3$, then $c = 0.569$, we can obtain the fuzzy measure:

g_λ	$\{x_1\}$	$\{x_2\}$	$\{x_3\}$	$\{x_4\}$		
	0.063	0.102	0.186	0.219		
g_λ	$\{x_1,x_2\}$	$\{x_1,x_3\}$	$\{x_1,x_4\}$	$\{x_2,x_3\}$	$\{x_2,x_4\}$	$\{x_3,x_4\}$
	0.184	0.283	0.322	0.344	0.388	0.526
g_λ	$\{x_1,x_2,x_3\}$	$\{x_1,x_2,x_4\}$	$\{x_1,x_3,x_4\}$	$\{x_2,x_3,x_4\}$		
	0.472	0.523	0.688	0.789		
g_λ	$\{x_1,x_2,x_3,x_4\}$					
	1.000					

The performance matrix of operational strategies in four alternative strategies is shown as follows:

Performance Matrix of Operational Strategies

Alternative Strategies	Organization x_1	Management Resource x_2	Competition x_3	Customer or Social Needs x_4
A	60	90	90	50
B	60	50	80	90
C	90	80	60	70
D	80	70	70	80

$\mu_A = 30 \times 0.344 + 10 \times 0.472 + 50 \times 1.000 = 65.0$
$\mu_B = 10 \times 0.219 + 20 \times 0.526 + 10 \times 0.688 + 50 \times 1.000 = 69.5$
$\mu_C = 10 \times 0.063 + 10 \times 0.184 + 10 \times 0.523 + 60 \times 1.000 = 67.7$
$\mu_D = 10 \times 0.322 + 70 \times 1.0 = 73.2$

The concept based on general form of multi-attribute utility function

$$u(x_1, x_2, \ldots, x_n) = \sum_{i=1}^{n} w_i u(x_i) + \lambda \sum_{\substack{i=1,\\ j>i}}^{n} w_i w_j u(x_i) u(x_j)$$

$$+\lambda^2 \sum_{\substack{i=1,\\ j>i\\ l>j}}^{n} w_i w_j w_l u(x_i) u(x_j) u(x_l) + \cdots$$

$$+\lambda^{n-1} w_1 w_2 \ldots w_n u(x_1) u(x_2) \ldots u(x_n)$$

$\Leftrightarrow$

$$g_\lambda(\{x_1, x_2, \ldots, x_n\}) = \sum_{i=1}^{n} g_\lambda(\{x_i\}) + \lambda \sum_{\substack{i=1,\\ j>i}}^{n} g_\lambda(\{x_i\}) g_\lambda(\{x_j\}) + \cdots$$

$$+\lambda^{n-1} g_\lambda(\{x_1\}) g_\lambda(\{x_2\}) \ldots g_\lambda(\{x_n\})$$

where

1. $u(x_1^0, x_2^0, ..., x_n^0) = 0$ and
   $u(x_1^*, x_2^*, ..., x_n^*) = 1 \Leftrightarrow g_\lambda(\{x_1^*, x_2^*, ..., x_n^*\}) = g_\lambda(\{x_1, x_2, ..., x_n\}) = 1$
2. $u(x_i)$ is a conditional utility function of x_i,
   $u(x_i^0) = 0 \Leftrightarrow g_\lambda(\{x_i^0\}) = 0, u(x_i^*) = 1 \Leftrightarrow g_\lambda(\{x_i^*\}) = 1, i = 1, 2, ..., n$
3. $w_i = u(x_i^*, x_{\bar{i}}^0) = g_\lambda(\{x_i\})$
   $\rightarrow$ These concepts for λ–measure by using questionnaire
4. λ is a solution of $1 + \lambda = \Pi_{i=1}^n (1 + \lambda w_i) \Leftrightarrow 1 + \lambda = \Pi_{i=1}^n (1 + \lambda g_\lambda(\{x_i\}))$

Above formation

1. If $\Sigma_{i=1}^n w_i = 1$, in other words, if $\lambda = 0$, then attribute utility function can be written as follows:

$$u(x_1, x_2, ..., x_n) = \sum_{i=1}^{n} w_i u(x_i).$$

2. If $\Sigma_{i=1}^n w_i \neq 1$, in other words, if $\lambda \neq 0$, then multiplicative utility function can be written as follows:

$$u(x_1, x_2, ..., x_n) = \sum_{i=1}^{n} w_i u(x_i) + \lambda \sum_{\substack{i=1,\\ j>i}}^{n} w_i w_j u(x_i) u(x_j)$$

$$+ \lambda^2 \sum_{\substack{i=1,\\ j>i,\\ l>j}}^{n} w_i w_j w_l u(x_i) u(x_j) u(x_l) + \cdots$$

$$+ \lambda^{n-1} w_1 w_2 ... w_n u(x_1) u(x_2) ... u(x_n)$$

$$1 + \lambda u(x_1, x_2, ..., x_n) = \prod_{i=1}^{n} (1 + \lambda w_i u(x_i))$$

$$\text{or } u(x_1, x_2, ..., x_n) = \frac{1}{\lambda}\left[\prod_{i=1}^{n} (1 + \lambda w_i u(x_i)) - 1\right]$$

CHAPTER 10 ROUGH SETS

Basic Concepts of Rough Set Theory (RST)

Information System and Approximations

Information System

$IS = (U, A, V, f)$, where U is the *universe* (a finite set of objects, $U = \{x_1, x_2, \ldots, x_n\}$), A is a finite set of attributes (features, variables), $V = \cup_{a \in A} V_a$, where V_a is the set of values for each attribute a (called the domain of attribute a), and $f: U \times A \rightarrow V$ is a description function such that $f(x,a) \in V_a$ for all $x \in U$ and $a \in A$ is called an information function. Let $B \subseteq A$ and $x,y \in U$.

Indiscernibility Relation

We say x and y are indiscernible by the set of attributes B in S iff $f(x,b) = f(y,b)$ for every $b \in B$. Thus every $B \subseteq A$ generates a binary relation on U, called *B indiscernibility relation*, denoted by I_B. In other words, I_B is an equivalence relation for any B.

Lower and Upper Approximations

The lower approximation of X in S by B, denoted by $\underline{B}X$, and the upper approximation of X in S by B, denoted by $\overline{B}X$, are defined as $\underline{B}X = \{x \in U | I_B[x] \subseteq X\}$ and $\overline{B}X = \{x \in U | I_B[x] \cap X \neq \varnothing\}$.

Boundary

The boundary of X in S by B is defined as $BN_B(X) = \overline{B}X - \underline{B}X$.

Accuracy of Approximation

An accuracy measure of the set X in S by $B \subseteq A$ is defined as $\alpha_B(X) = \text{card}(\underline{B}X)/\text{card}(\overline{B}X)$, where card(•) is the cardinality of a set. Let $F = \{X_1, X_2, \ldots, X_n\}$ be a classification of U, i.e., $X_i \cap X_j = \varnothing$, $\forall i, j \leq n$, $i \neq j$, and $\cup_{i=1}^{n} X_i = U$, X_i are called classes of F. The lower and upper approximations of F by $B \subseteq A$ are defined as $\underline{B}F = \{\underline{B}X_1, \underline{B}X_2, \ldots, \underline{B}X_n\}$ and $\overline{B}F = \{\overline{B}X_1, \overline{B}X_2, \ldots, \overline{B}X_n\}$, respectively. The quality of approximation of classification F by the set B of attributes, or in short, quality of classification F is defined as $\gamma_B(F) = \sum_{i=1}^{n} \text{card}(\underline{B}X_i)/\text{card}(U)$. It expresses the ratio of all B-correctly classified objects to all objects in the system.

Reductions and Core

An important issue in RST is attribute reduction, which is performed in such a way that the reduced set of attributes B, $B \subseteq A$, provides the same quality of classification $\gamma_B(F)$ as the original set of attributes A. The minimal subset $C \subseteq B \subseteq A$, such that $\gamma_B(F) = \gamma_C(F)$, is called the *F-reduct* of B and is denoted by $RED_F(B)$. A *reduct* is a minimal subset of attributes that enables the same classification of elements of the universe as the whole set of attributes. In other words, attributes that do not belong to a reduct are superfluous in terms of classification of elements of the universe. The *core* is the common part of all reducts. For example, $CORE_F(B)$ is called the *F*-core of *B*, if $CORE_F(B) = \cap RED_F(B)$.

Decision Rules

An information table $A = C \cup D$ can be seen as a *decision table*, where *C* and *D* are condition and decision attributes, respectively, and $C \cap D = \varnothing$. The decision attribute *D* induces an indiscernibility relation I_D that is independent of the conditional attributes *C*; objects in the same I_D are grouped together in decision elementary sets (decision classes). The reducts of the condition attribute set will preserve the relevant relationship between condition attributes and decision classes, and this relationship can be expressed by a decision rule. A decision rule in *S* is expressed as $\Phi \rightarrow \Psi$, which is read as if Φ then Ψ (a logical statement). Φ and Ψ are referred to as conditions and decisions of the rule, respectively. In data mining, we usually take into account relevant confirmation measures and apply them within RST to data analysis. They are presented as follows [13]. The *strength* of the decision rule $\Phi \rightarrow \Psi$ in *S* is expressed as $\sigma_s(\Phi, \Psi) = \text{supp}_s(\Phi, \Psi)/\text{card}(U)$ where $\text{supp}_s(\Phi, \Psi) = \text{card}(\|\Phi \wedge \Psi\|_s)$ is called the *support* of the rule $\Phi \rightarrow \Psi$ in *S* and card(*U*) is the cardinality of *U*. With every decision rule $\Phi \rightarrow \Psi$ we associate a *coverage factor/covering ratio* (*CR*) defined as $\text{cov}_s(\Phi, \Psi) = \text{supp}_s(\Phi, \Psi)/\text{card}(\|\Psi\|_s)$.

Approximation of the Dominance Relation for DRSA

A rough set-based rule induction technique can be expressed by a pair of crisp sets called the lower and upper approximations. The dominance-based rough set approach (DRSA) uses a dominance relation instead of an indiscernible relation, which is based on there being at least one conditional attribute and classes being preference-ordered. The approximation is a collection of upward and downward unions of classes. The formula is as follows: Firstly, let $\succeq_a$ be an outranking relation on *U* with respect to criterion $a \in Q$, such that $x \succeq_a y$ means that "*x* is at least as good as *y* with respect to criterion *a*." Suppose that $\succeq_a$ is a complete preorder. Outstanding relation $\succeq_a$ is designed on *U* on the basis of evaluations. For instance, $f(x, a) \geq f(y, a)$ means that the greater *a* is, the more preferred the object, while $f(x, a) \leq f(y, a)$ means that the smaller *a* is, the more preferred the object. Furthermore, let $\boldsymbol{Cl} = \{Cl_t, t \in T\}$, $T = \{1,\ldots,n\}$, be a set of decision classes of *U* such that each $x \in U$ belongs to one and only one class $Cl_t = \boldsymbol{Cl}$. Assume that for all $r, s \in T$, $r \succ s$, or the elements of Cl_r are preferred to the elements of Cl_s. In addition, if $\succeq$ is a comprehensive outranking relation on *U*, then suppose that:

$$\left[x \in Cl_r, y \in Cl_s, r > s\right] \Leftrightarrow x \succ y, \tag{A10.1}$$

where $x \succ y$ means $x \succeq y$ and not $y \succeq x$. Then, given the set of the decision class $\boldsymbol{Cl}$, it is possible to define upward and downward unions of classes, respectively, as the following:

$$Cl_t^{\geq} = \bigcup_{s \geq t} Cl_s,\ Cl_t^{\leq} = \bigcup_{s \leq t} Cl_s,\ t = 1, \ldots, n. \tag{A10.2}$$

For example, $x \in Cl_t^{\geq}$ means that "x belongs at least in class Cl_t," whereas $x \in Cl_t^{\leq}$ means that "x belongs to, at most, class Cl_t."

The indiscernibility relation generated in the universe of discourse is the mathematical basis for the rough set theory. However, in dominance-based approaches, where condition attributes are criteria and classes are preference-ordered, the knowledge is to be approximated using a dominance relation instead of indiscernibility. It is said that object x P-dominates objects y with respect to $P \subseteq C$ if $x \succeq_a y$ for all $a \in P$. Given $P \subseteq C$ and $x \subseteq U$, let $D_p^+(x) = \{y \in U : y \succeq x\}$ represent a set of objects dominating x, called a P-dominating set, and $D_p^-(x) = \{y \in U : x \succeq y\}$ represent a set of objects dominated by x, called a P-dominated set. We can adopt $D_p^+(x)$ and $D_p^-(x)$ to approximate a collection of upward and downward unions of decision classes.

The P-lower approximation of $\underline{P}(Cl_t^{\geq})$ of the unions of class $Cl_t^{\geq}$, $t \in \{2,3,...,n\}$ with respect to $P \subseteq C$ contains all objects x in the universe U, such that objects y that have at least the same evaluations for all of the considered ordered attributes from P also belong to class Cl_t or such that:

$$\underline{P}\left(Cl_t^{\geq}\right) = \left\{x \in U : D_P^+(x) \subseteq Cl_t^{\geq}\right\}. \tag{A10.3}$$

Similarly, the P-upper approximation of $\overline{P}(Cl_t^{\geq})$ is composed of all objects x in the universe U, whose evaluations on the criteria from P are not worse than the evaluations of at least one object y belonging to class Cl_t or such that:

$$\overline{P}\left(Cl_t^{\geq}\right) = \left\{x \in U : D_P^-(x) \cap Cl_t^{\geq} \neq \varnothing\right\} = \bigcup_{x \in Cl_t^{\geq}} D_P^+(x) \text{ for } t = 1, \ldots, n. \tag{A10.4}$$

Analogously, the P-lower and P-upper approximations of $\underline{P}(Cl_t^{\leq})$ and $\underline{P}(Cl_t^{\leq})$, respectively, of the union of class $Cl_t^{\geq}$ for which $t \in \{2,3,\ldots,n\}$, with respect to $P \subseteq C$, are defined as:

$$\underline{P}\left(Cl_t^{\leq}\right) = \left\{x \in U : D_P^-(x) \subseteq Cl_t^{\leq}\right\}, \tag{A10.5}$$

$$\overline{P}\left(Cl_t^{\leq}\right) = \left\{x \in U : D_P^+(x) \cap Cl_t^{\leq} \neq \varnothing\right\} = \bigcup_{x \in Cl_t^{\leq}} D_P^-(x) \text{ for } t = 1, \ldots, n. \tag{A10.6}$$

The P-boundaries (P-doubtable regions) of $Cl_t^{\geq}$ and $Cl_t^{\leq}$ are defined as follows:

$$Bn_P\left(Cl_t^{\geq}\right) = \overline{P}\left(Cl_t^{\geq}\right) - \underline{P}\left(Cl_t^{\geq}\right), \tag{A10.7}$$

$$Bn_P\left(Cl_t^{\leq}\right) = \overline{P}\left(Cl_t^{\leq}\right) - \underline{P}\left(Cl_t^{\leq}\right) \text{ for } t = 1, \ldots, n. \tag{A10.8}$$

With each set $P \subseteq U$, we can estimate the accuracy of the approximation of $Cl_t^{\geq}$ and $Cl_t^{\leq}$ using the following expression:

$$\alpha_P\left(Cl_t^{\geq}\right) = \frac{\left|\underline{P}\left(Cl_t^{\geq}\right)\right|}{\left|\overline{P}\left(Cl_t^{\geq}\right)\right|},\ \alpha_P\left(Cl_t^{\leq}\right) = \frac{\left|\underline{P}\left(Cl_t^{\leq}\right)\right|}{\left|\overline{P}\left(Cl_t^{\leq}\right)\right|}, \tag{A10.9}$$

and the ratio

$$\gamma_P(\boldsymbol{Cl}) = \left|\frac{U - \left(\bigcup_{t\in\{2,\ldots,n\}} Bn_P\left(Cl_t^{\geq}\right)\right)}{U}\right| = \left|\frac{U - \left(\bigcup_{t\in\{1,\ldots,n-1\}} Bn_P\left(Cl_t^{\leq}\right)\right)}{U}\right|. \tag{A10.10}$$

The ratio $\gamma_p(\boldsymbol{Cl})$ is called the quality of approximation of classification $\boldsymbol{Cl}$ by the set of attributes P or, in short, the quality of classification. It indicates the ratio of all of the P-correctly classified objects (all of the nonambiguous objects to all of the objects in the system). Each minimal subset $P \subseteq C$ such that $\gamma_p(\boldsymbol{Cl}) = \gamma_c(\boldsymbol{Cl})$ is called a reduct of C with respect to $\boldsymbol{Cl}$, and is denoted by $RED_{Cl}(P)$. A data table may have more than one *reduct*, and the intersection of all the *reducts* is known as the *core*, denoted by $CORE_{Cl}$.

Extraction of Decision Rules

The end result of the DRSA is a representation of the information contained in the considered information table. The decision table is a deterministic or exact decision rule and can be expressed in a logical manner in the *if* (antecedent) *then* (consequence) type of decision. For a given upward union of classes $Cl_t^{\geq}$, the decision rule included under the hypothesis that all objects belonging to $\overline{P}(Cl_t^{\geq})$ are positive and others are negative suggests an assignment to "at least class Cl_t." Analogously, for a given downward union $Cl_s^{\leq}$, the rule induced under a hypothesis for which all items belonging to $\underline{P}(Cl_s^{\leq})$ are positive and all others are negative suggests an assignment to "at most class Cl_s." There are two types of decision rules:

1. $D_{\geq}$ decision rules ("at least" decision rules):
   If $f(x, a_1) \geq r_{a_1}$ and $f(x, a_2) \geq r_{a_2}$ and … $f(x, a_p) \geq r_{a_p}$, then $x \in Cl_t^{\geq}$.
   These rules are supported only by objects from P-lower approximations of the upward unions of classes $Cl_t^{\geq}$.
2. $D_{\leq}$ decision rules ("at most" decision rules):
   If $f(x, a_1) \leq r_{a_1}$ and $f(x, a_2) \leq r_{a_2}$ and … $f(x, a_p) \leq r_{a_p}$, then $x \in Cl_t^{\leq}$.
   These rules are supported only by objects from P-lower approximations of the upward unions of classes $Cl_t^{\leq}$.

3. $D_{\geq\leq}$ decision rules (ambiguous for "at most" otherwise):
If $f(x, a_1) \geq r_{a_1}$ and $f(x, a_2) \geq r_{a_2}$ and ... $f(x, a_k) \geq r_{a_k}$ and $f(x, a_{k+1}) \geq r_{a_{k+1}}$, and $f(x, a_p) \leq r_{a_p}$ then $x \in Cl_t \cup Cl_{t+1} \cup \cdots \cup Cl_s$.

Causal-and-Effect of Decision Rules based on Flow Network Graph

More precisely, a flow network graph is a directed acyclic finite graph $G = (V,\beta,h)$ where V is a set of nodes, $\beta \subseteq V \times V$ is a set of directed branches, $\beta \rightarrow R^+$ is a flow function, and R^+ is the set of non-negative real numbers. The throughflow of a branch is $(x, y) \in \beta$ and can be defined as $r(x, y)$. The input of a node $x \in V$ is the set $I(x) = \{y \in V|(y, x) \in \beta\}$ and the output of a node $x \in V$ is defined as $O(x) = \{y \in V|(x, y) \in \beta\}$. Based on these concepts, the input and output of a graph G are defined as $I(G) = \{x \in V|\, I(x) \neq \varnothing\}$ and $O(G) = \{x \in V|\, O(x) \neq \varnothing\}$. For every node x in the flow graph, inflow is defined as $h_+(y) = \sum_{x \in I(y)} h(x, y)$ and outflow is defined as $h_-(y) = \sum_{y \in O(x)} h(x, y)$. Similarly, the inflow and outflow of the whole flow graph can be defined as $h_+(G) = \sum_{x \in I(G)} h_-(x)$ and $h_-(G) = \sum_{x \in O(G)} h_+(x)$, respectively. This research assumes that for any node x in a flow graph G, $h_+(x) = h_-(x) = h(x)$.

To measure the strength of every branch (x, y) in a flow graph $(G) = (V, \beta, h)$, this research defines the strength $\rho(x, y) = h(x, y)/r(G)$. Obviously, $0 = \rho(x, y) \leq 1$. The strength of the branch simply expresses the amount of total flow through the branch. Every branch (x, y) of a flow graph G is associated with certainty and coverage coefficients. The certainty and coverage of every branch are defined as $\mathbf{cer}(x, y) = \rho(x, y)/\rho(x)$ and $\mathbf{cov}(x, y) = \rho(x, y)/\rho(y)$, respectively, where $\rho(x, y) = h(x, y)/h(G)$, $\rho(x) = h(x)/h(G)$, and $\rho(y) = h(y)/h(G)$ are normalized throughflow, and $\rho(x) \neq 0$, $\rho(y) \neq 0$, and $0 \leq \rho(x,y) \leq 1$.

CHAPTER 11 STRUCTURAL MODELING

Summary

Interpretive Structure Modeling (ISM)

If the answer is "yes" then $\pi_{ij} = 1$, otherwise $\pi_{ij} = 0$. The general form of the relation matrix can be presented as follows:

$$\boldsymbol{D} = \begin{array}{c} \\ e_1 \\ e_2 \\ \vdots \\ e_n \end{array} \begin{array}{c} \begin{array}{cccc} e_1 & e_2 & \cdots & e_n \end{array} \\ \begin{bmatrix} 0 & \pi_{12} & \cdots & \pi_{1n} \\ \pi_{21} & 0 & \cdots & \pi_{2n} \\ \vdots & \vdots & \ddots & \vdots \\ \pi_{n1} & \pi_{n2} & \cdots & 0 \end{bmatrix} \end{array}$$

$$M = D + I, \qquad M^* = M^k = M^{k+1}, k > 1,$$

$$R(t_i) = \{e_i \mid m^*_{ji} = 1\}, \quad \text{and} \quad A(t_i) = \{e_i \mid m^*_{ji} = 1\},$$

$$R(t_i) \cap A(t_i) = R(t_i).$$

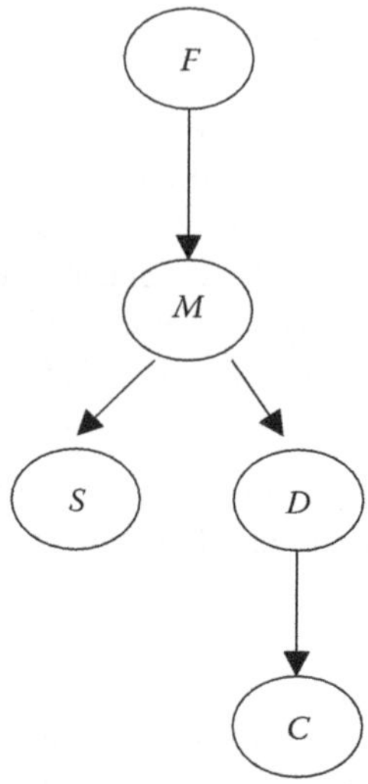

Example

$$\begin{array}{c} \\ F \\ M \\ S \\ D \\ C \end{array}\begin{array}{c} \begin{array}{ccccc} F & M & S & D & C \end{array} \\ \begin{bmatrix} 1 & 1 & 0 & 0 & 0 \\ 0 & 1 & 1 & 1 & 0 \\ 0 & 0 & 1 & 0 & 0 \\ 0 & 0 & 0 & 1 & 1 \\ 0 & 0 & 0 & 0 & 1 \end{bmatrix} \end{array}$$

$$\begin{array}{c} \\ F \\ M \\ S \\ D \\ C \end{array}\begin{array}{c} \begin{array}{ccccc} F & M & S & D & C \end{array} \\ \begin{bmatrix} 1 & 1 & 0 & 0 & 0 \\ 0 & 1 & 1 & 1 & 0 \\ 0 & 0 & 1 & 0 & 0 \\ 0 & 0 & 0 & 1 & 1 \\ 0 & 0 & 0 & 0 & 1 \end{bmatrix} \end{array} \otimes \begin{array}{c} \begin{array}{ccccc} F & M & S & D & C \end{array} \\ \begin{bmatrix} 1 & 1 & 0 & 0 & 0 \\ 0 & 1 & 1 & 1 & 0 \\ 0 & 0 & 1 & 0 & 0 \\ 0 & 0 & 0 & 1 & 1 \\ 0 & 0 & 0 & 0 & 1 \end{bmatrix} \end{array} \Rightarrow \begin{array}{c} \\ F \\ M \\ S \\ D \\ C \end{array}\begin{array}{c} \begin{array}{ccccc} F & M & S & D & C \end{array} \\ \begin{bmatrix} 1 & 1 & 1^* & 1^* & 0 \\ 0 & 1 & 1 & 1 & 1^* \\ 0 & 0 & 1 & 0 & 0 \\ 0 & 0 & 0 & 1 & 1 \\ 0 & 0 & 0 & 0 & 1 \end{bmatrix} \end{array}$$

$$
\begin{array}{c} \\ F \\ M \\ S \\ D \\ C \end{array}
\begin{array}{c} \begin{array}{ccccc} F & M & S & D & C \end{array} \\ \begin{bmatrix} 1 & 1 & 1^* & 1^* & 0 \\ 0 & 1 & 1 & 1 & 1^* \\ 0 & 0 & 1 & 0 & 0 \\ 0 & 0 & 0 & 1 & 1 \\ 0 & 0 & 0 & 0 & 1 \end{bmatrix} \end{array}
\otimes
\begin{array}{c} \begin{array}{ccccc} F & M & S & D & C \end{array} \\ \begin{bmatrix} 1 & 1 & 0 & 0 & 0 \\ 0 & 1 & 1 & 1 & 0 \\ 0 & 0 & 1 & 0 & 0 \\ 0 & 0 & 0 & 1 & 1 \\ 0 & 0 & 0 & 0 & 1 \end{bmatrix} \end{array}
\Rightarrow
\begin{array}{c} \\ F \\ M \\ S \\ D \\ C \end{array}
\begin{array}{c} \begin{array}{ccccc} F & M & S & D & C \end{array} \\ \begin{bmatrix} 1 & 1 & 1^* & 1^* & 1^{**} \\ 0 & 1 & 1 & 1 & 1^* \\ 0 & 0 & 1 & 0 & 0 \\ 0 & 0 & 0 & 1 & 1 \\ 0 & 0 & 0 & 0 & 1 \end{bmatrix} \end{array}
$$

DEMATEL

Finding the Weights by DEMATEL-based ANP (DANP)

After DEMATEL confirming the influential relation of criteria, DANP is then used to obtain their most accurate weights. The ANP presented by Saaty (1996) to decrease the limitations associated with the analytic hierarchy process (AHP) creates a solution to determining nonlinear and complex network relationships. Therefore, the research applies the strength of ANP onto DEMATEL to solving the dependence and feedback problems associated with the interrelation between the criteria. The DANP is processed as follows (Kuan et al., 2011; Hsu et al., 2011):

First, develop an unweighted supermatrix. Normalize each level with a total degree of influence from the total influence matrix $\boldsymbol{T}$ of DEMATEL as shown in Equation A11.1.

$$
\boldsymbol{T}_c = \begin{array}{c} \\ D_1 \begin{array}{c} c_{11} \\ c_{12} \\ \vdots \\ c_{1m_1} \end{array} \\ \vdots \\ D_i \begin{array}{c} c_{i1} \\ c_{i2} \\ \vdots \\ c_{im_i} \end{array} \\ \vdots \\ D_n \begin{array}{c} c_{n1} \\ c_{n2} \\ \vdots \\ c_{nm_n} \end{array} \end{array}
\begin{array}{c} \begin{array}{ccccc} D_1 & & D_j & & D_n \\ c_{11}\ldots c_{1m_1} & \cdots & c_{j1}\ldots c_{jm_j} & \cdots & c_{n1}\ldots c_{nm_n} \end{array} \\
\begin{bmatrix} \boldsymbol{T}_c^{11} & \cdots & \boldsymbol{T}_c^{1j} & \cdots & \boldsymbol{T}_c^{1n} \\ \vdots & & \vdots & & \vdots \\ \boldsymbol{T}_c^{i1} & \cdots & \boldsymbol{T}_c^{ij} & \cdots & \boldsymbol{T}_c^{in} \\ \vdots & & \vdots & & \vdots \\ \boldsymbol{T}_c^{n1} & \cdots & \boldsymbol{T}_c^{nj} & \cdots & \boldsymbol{T}_c^{nn} \end{bmatrix} \end{array}. \tag{A11.1}
$$

Next, normalize $\boldsymbol{T}_c$ with a total degree of effect; $\boldsymbol{T}_c^{\alpha}$ can be obtained as shown in Equation A11.2.

$$
\boldsymbol{T}_C^{\alpha} = \begin{array}{c} \\ D_1 \begin{array}{c} c_{11} \\ c_{12} \\ \vdots \\ c_{1m_1} \end{array} \\ \vdots \\ D_i \begin{array}{c} c_{i1} \\ c_{i2} \\ \vdots \\ c_{im_i} \end{array} \\ \vdots \\ D_n \begin{array}{c} c_{n1} \\ c_{n2} \\ \vdots \\ c_{nm_n} \end{array} \end{array}
\begin{array}{c} \begin{array}{ccccc} D_1 & & D_j & & D_n \\ c_{11}\ldots c_{1m_1} & \cdots & c_{j1}\ldots c_{jm_j} & \cdots & c_{n1}\ldots c_{nm_n} \end{array} \\
\begin{bmatrix} \boldsymbol{T}_c^{\alpha 11} & \cdots & \boldsymbol{T}_c^{\alpha 1j} & \cdots & \boldsymbol{T}_c^{\alpha 1n} \\ \vdots & & \vdots & & \vdots \\ \boldsymbol{T}_c^{\alpha i1} & \cdots & \boldsymbol{T}_c^{\alpha ij} & \cdots & \boldsymbol{T}_c^{\alpha in} \\ \vdots & & \vdots & & \vdots \\ \boldsymbol{T}_c^{\alpha n1} & \cdots & \boldsymbol{T}_c^{\alpha nj} & \cdots & \boldsymbol{T}_c^{\alpha nn} \end{bmatrix} \end{array}. \tag{A11.2}
$$

Then, normalize $T_c^{\alpha 11}$ using Equations A11.3 and A11.4. Repeating this, $T_c^{\alpha nn}$ shall be obtained:

$$d_i^{11} = \sum_{j=1}^{m_1} t_{cij}^{11},\ i = 1,2,\ldots,m_1, \tag{A11.3}$$

$$T_c^{\alpha 11} = \begin{bmatrix} t_{c11}^{11}/d_1^{11} & \cdots & t_{c1j}^{11}/d_1^{11} & \cdots & t_{c1m_1}^{11}/d_1^{11} \\ \vdots & & \vdots & & \vdots \\ t_{ci1}^{11}/d_i^{11} & \cdots & t_{cij}^{11}/d_i^{11} & \cdots & t_{cim_1}^{11}/d_i^{11} \\ \vdots & & \vdots & & \vdots \\ t_{cm_11}^{11}/d_{m_1}^{11} & \cdots & t_{cm_1j}^{11}/d_{m_1}^{11} & \cdots & t_{cm_1m_1}^{11}/d_{m_1}^{11} \end{bmatrix} = \begin{bmatrix} t_{c11}^{\alpha 11} & \cdots & t_{c1j}^{\alpha 11} & \cdots & t_{c1m_1}^{\alpha 11} \\ \vdots & & \vdots & & \vdots \\ t_{ci1}^{\alpha 11} & \cdots & t_{cij}^{\alpha 11} & \cdots & t_{cim_1}^{\alpha 11} \\ \vdots & & \vdots & & \vdots \\ t_{cm_11}^{\alpha 11} & \cdots & t_{cm_1j}^{\alpha 11} & \cdots & t_{cm_1m_1}^{\alpha 11} \end{bmatrix}. \tag{A11.4}$$

Till then, the total effect matrix is normalized into a supermatrix according to the dependent relationship in the group and the unweighted supermatrix thus can be obtained as shown in Equation A11.5.

$$W = \left(T_c^{\alpha}\right)' = \begin{matrix} & \\ D_1 & \begin{matrix} c_{11} \\ c_{12} \\ \vdots \\ c_{1m_1} \end{matrix} \\ \vdots & \\ D_j & \begin{matrix} c_{j1} \\ c_{j2} \\ \vdots \\ c_{jm_j} \end{matrix} \\ \vdots & \\ D_n & \begin{matrix} c_{n1} \\ c_{n2} \\ \vdots \\ c_{nm_n} \end{matrix} \end{matrix} \overset{\begin{matrix} D_1 & & D_i & & D_n \\ c_{11}\ldots c_{1m_1} & \cdots & c_{i1}\ldots c_{im_i} & \cdots & c_{n1}\ldots c_{nm_n} \end{matrix}}{\begin{bmatrix} W^{11} & \cdots & W^{i1} & \cdots & W^{n1} \\ \vdots & & \vdots & & \vdots \\ W^{1j} & \cdots & W^{ij} & \cdots & W^{nj} \\ \vdots & & \vdots & & \vdots \\ W^{1n} & \cdots & W^{in} & \cdots & W^{nn} \end{bmatrix}}. \tag{A11.5}$$

At the same time, matrices W^{11} and W^{12} are obtained by Equation A11.6. If a space is blank or 0 in the matrix, it shows the group or criterion is independent. In the same way, the matrix W^{nn} is obtained.

$$W^{11} = \left(T^{11}\right)' = \begin{matrix} c_{11} \\ \vdots \\ c_{1j} \\ \vdots \\ c_{1m_1} \end{matrix} \overset{\begin{matrix} c_{11} & \cdots & c_{1i} & \cdots & c_{1m_1} \end{matrix}}{\begin{bmatrix} t_{c11}^{\alpha 11} & \cdots & t_{ci1}^{\alpha 11} & \cdots & t_{cm_11}^{\alpha 11} \\ \vdots & & \vdots & & \vdots \\ t_{c1j}^{\alpha 11} & \cdots & t_{cij}^{\alpha 11} & \cdots & t_{cm_1j}^{\alpha 11} \\ \vdots & & \vdots & & \vdots \\ t_{c1m_1}^{\alpha 11} & \cdots & t_{cim_1}^{\alpha 11} & \cdots & t_{cm_1m_1}^{\alpha 11} \end{bmatrix}} \tag{A11.6}$$

Third, obtain the weighted supermatrix. Derive the matrix of the total effect of dimensions T_D using Equation A11.7. Then normalize T_D to obtain T_D^{α}, as shown in Equation A11.8.

$$d_i = \sum_{j=1}^{n} t_D^{ij},\ i = 1, 2, \ldots, n$$

$$T_D = \begin{bmatrix} t_D^{11} & \cdots & t_D^{1j} & \cdots & t_D^{1n} \\ \vdots & & \vdots & & \vdots \\ t_D^{i1} & \cdots & t_D^{ij} & \cdots & t_D^{in} \\ \vdots & & \vdots & & \vdots \\ t_D^{n1} & \cdots & t_D^{nj} & \cdots & t_D^{nn} \end{bmatrix} \quad \text{(A11.7)}$$

$$T_D^{\alpha} = \begin{bmatrix} t_D^{11}/d_1 & \cdots & t_D^{1j}/d_1 & \cdots & t_D^{1n}/d_1 \\ \vdots & & \vdots & & \vdots \\ t_D^{i1}/d_2 & \cdots & t_D^{ij}/d_2 & \cdots & t_D^{in}/d_2 \\ \vdots & & \vdots & & \vdots \\ t_D^{n1}/d_n & \cdots & t_D^{nj}/d_n & \cdots & t_D^{nn}/d_n \end{bmatrix} = \begin{bmatrix} t_D^{\alpha 11} & \cdots & t_D^{\alpha 1j} & \cdots & t_D^{\alpha 1n} \\ \vdots & & \vdots & & \vdots \\ t_D^{\alpha i1} & \cdots & t_D^{\alpha ij} & \cdots & t_D^{\alpha in} \\ \vdots & & \vdots & & \vdots \\ t_D^{\alpha n1} & \cdots & t_D^{\alpha nj} & \cdots & t_D^{\alpha nn} \end{bmatrix} \quad \text{(A11.8)}$$

Then, transform the normalized T_D^{α} into the unweighted supermatrix W to obtain the weighted supermatrix W^{α}, as shown in Equation A11.9.

$$W^{\alpha} = T_D^{\alpha} W = \begin{bmatrix} t_D^{\alpha 11} \times W^{11} & \cdots & t_D^{\alpha i1} \times W^{i1} & \cdots & t_D^{\alpha n1} \times W^{n1} \\ \vdots & & \vdots & & \vdots \\ t_D^{\alpha 1j} \times W^{1j} & \cdots & t_D^{\alpha ij} \times W^{ij} & \cdots & t_D^{\alpha nj} \times W^{nj} \\ \vdots & & \vdots & & \vdots \\ t_D^{\alpha 1n} \times W^{1n} & \cdots & t_D^{\alpha in} \times W^{in} & \cdots & t_D^{\alpha nn} \times W^{nn} \end{bmatrix} \quad \text{(A11.9)}$$

Finally, obtain the limit supermatrix. Let the weighted supermatrix W^{α} multiply itself multiple times to obtain the limit supermatrix. Then, the DANP weights of each criterion can be obtained by $\lim_{z \to \infty} (W^{\alpha})^z$, where z represents any number for power.

CHAPTER 14

Appendix 14.1 Practical Data Given to the Experts

The data given to the experts are summarized in Tables A14.1 through A14.5.

TABLE A14.1

Comparison of the Functions of Diesel and Alternative-fuel Buses

Items	Diesel Bus	EVs	HEVs	Methanol/Ethanol Bus	Natural Gas Bus	Fuel-cell Bus
Route	NSR[a]	Flat	Flat	NSR[a]	NSR[a]	Flat
Depot	Small	Large (REP[b])	Large (REP[b])	Small	Large (REP[b])	Small
Passengers	60–80 (S, M, L)	20–60 (S, M, L)	20–60 (S, M, L)	60–80 (L)	60–80 (L)	60–80 (L)
Max. speed (km/hr)	100–120	45–80	60–80	100–120	80	70–80
Cruising dist. (km)	400–500	60–220	90–400	200–250	200–300	300–350
Gradeability	<18	<16	<16	<18	<18	<16
Recharge time	10 min	slow: 8–10 hr fast: 30 min	slow: 8–10 hr fast: 30 min	10 min	slow: 6–8 hr fast: 5–20 min	10 min methanol system

Source: Institute of Transportation, 2000.

[a] NSR, no special restriction.

[b] REP, recharge equipment provided.

TABLE A14.2
Energy Efficiency of Diesel and Alternative-fuel Buses

Bus	Energy Efficiency	Fuel Heating Value[a]	Comp of Energy Efficiency[b]
Diesel bus	1.5 km/L (1.5–1.6)	8800 kcal/L	1.0
Pure electric bus	1.6 km/kWh (1.6–2.4)	860 kcal/kWh	10.9
Hybrid electric bus	2.31 km/L	8800 kcal/L	1.5
CNG bus	1.27 km/m^3 (1.27–1.45)	8900 kcal/m[c]	0.8
Methanol bus	0.6 km/L (0.6–0.7)	4200 kcal/L	0.8
Fuel-cell bus	2.79 km/L (diesel equivalent)	8800 kcal/L	1.9

Source: Institute of Transportation, 2000.

[a] The energy efficiency of the hybrid electric and fuel-cell buses is represented by diesel equivalent, so their heating values are represented by the heating value of diesel.

[b] Comparison of energy efficiency = (alternative-fuel bus energy efficiency/fuel heating value)/(diesel bus energy efficiency/diesel heating value).

TABLE A14.3
Exhaust Emission Characteristics of Alternative-fuel Bus

Bus	PM	NO_X	HC	CO
Pure electric	0.00	0.00	0.00	0.00
Hybrid electric	0.23	8.64	–	–
Natural gas	0.02	7.25	9.87	0.73
Methanol	0.07	4.28	1.31	5.25
Ethanol	0.35	11.06	7.59	24.66
Fuel-cell	0.00	0.03	0.32	6.23
Diesel	1.26	15.66	1.30	10.23

Source: Institute of Transportation, 2000.

TABLE A14.4
The Emission Characteristics of Carbon Dioxide of Alternative-fuel Buses

Bus Type	CO_2(kg/km)	AFV/Diesel	(Diesel-AFV)/Diesel
Diesel bus	1.7	–	–
Pure electric bus	0.3	0.18	82%
Hybrid electric bus	1.1	0.64	36%
Natural gas bus	1.4	0.82	18%
Methanol bus	1.8	1.06	–6%
Fuel-cell bus	0.2	0.12	88%

Source: Institute of Transportation, 2000.

Note: The major pollutions of pure electric buses are generated from plants.

TABLE A14.5

Cost of Diesel and Alternative-fuel Buses (1000 NT$)

Cost Items		Diesel Bus	Pure Electric	Hybrid Electric	Natural Gas	Methanol Bus	Fuel-cell Bus
Attainment cost	Purchase cost	90,000	300,000	360,000	300,000	120,000	600,000
	Recharge equipment cost	10,000	40,000	40,000	120,000	24,000	24,000
	Total cost	100,000	340,000	400,000	420,000	144,000	624,000
Operation cost	Fuel cost	12,000	1,875	5,880	9,450	12,495	6,450
	Management cost	2,000	2,000	2,000	2,000	2,000	2,000
	Total cost	14,000	3,875	7,880	11,450	14,495	46,600
Maintenance cost	Vehicle maintenance cost	11,400	18,495	22,200	7,440	9,840	30,720
	Recharge equipment cost				2,970	4,860	
	Total cost	11,400	18,495	22,200	10,410	14,700	30,720
Lifecycle cost		298,901	521,950	643,328	598,011	373,127	1,239,482

Source: Institute of Transportation, 2000.

Note: Management cost is assumed to be 2000 thousand/yr.

Bibliography

Abdulaal, M., and L.J. LeBlanc. (1979). Continuous equilibrium network design models. *Transportation Research (Part B)* 13 (1): 19–32.

Al-Harbi, K.M. (2001). Application of the AHP in project management. *International Journal of Project Management* 19 (1): 19–27.

Al Khalil, M.I. (2002). Selecting the appropriate project delivery method using AHP. *International Journal of Project Management* 20 (6): 469–74.

Al-Najjar, B., and I. Alsyouf. (2003). Selecting the most efficient maintenance approach using fuzzy multiple criteria decision making. *International Journal of Production Economics Volume* 84 (1): 85–100.

Altrock, C.V., and B. Krause. (1994). Multi-criteria decision-making in German automotive industry using fuzzy logic. *Fuzzy Sets and Systems* 63 (3): 375–80.

Andreou, A.S., N.H. Mateou, and G.A. Zombanakis. (2005). Soft computing for crisis management and political decision making: The use of genetically evolved fuzzy cognitive maps. *Soft Computing* 9 (3): 194–210.

Armstrong, R.D., W.D. Cook, and L.M. Seiford. (1982). Priority ranking and consensus formation: The case of ties. *Management Science* 28 (6): 638–45.

Axelrod, R. (1976). *Structure of decision, the cognitive maps of political elite*. London: Princeton University Press.

Azibi, R., and D. Vanderpooten. (2002). Construction of rule-based assignment models. *European Journal of Operational Research* 138 (2): 274–93.

Baas, S.M., and H. Kwakernaak. (1997). Rating and ranking of multiple aspect alternative using fuzzy sets. *Automatica* 13 (1): 47–58.

Beliakov, G. (2002). Monotone approximation of aggregation operators using least squares splines. *International Journal of Uncertainty, Fuzziness and Knowledge-Based Systems* 10 (6): 659–76.

Bell, D.E., H. Raiffa, and A. Tversky. (1988). *Decision making: Descriptive, normative, and prescriptive interactions*. Cambridge: Cambridge University Press.

Bellman, R.E., and L.A. Zadeh. (1970). Decision-making in a fuzzy environment. *Management Science* 17 (4): 141–64.

Belton, V., and T.J. Stewart. (2002). *Multiple criteria decision analysis: An integrated approach*. Boston: Kluwer Academic Publishers.

Benayoun, R., B. Roy, and N. Sussman. (1966). Manual de Reference du Program ELECTRE. Note de Synthese et Formation, Direction Scientifique SEMA, No. 25, Paris, France.

Benayoun, R., B. Roy, and N. Sussman. (1966). Manual de reference du programme electre. Note De Synthese et Formaton, No. 25, Direction Scientifique SEMA, Paris, France.

Bernoulli, D. (1738). Specimen Theoriae Novae de Mensura Sortis. *Commentarri Academiae Scientiarum Imperialis Petropolitanae* 5:175–92.

Bernstein, P. (1996). *Against the Gods: The Remarkable story of Risk*. John Wiley & Sons: New York.

Bertier, P., G. Gagey, J. de Montgolfier, and B. Roy. (1972). Choix de traces autoroutiers en milieu suburbain: Faut-il vraiment endommager des sites urbains et/ou forestiers, lesquels? Communication aux Journees de Formation "Transports et Agglomerations," Nice.

Beynon, M.J., and M.J. Peel. (2001). Variable precision rough set theory and data discretisation: An application to corporate failure prediction. *Omega* 29 (6): 561–76.

Bialas, W., and M. Karwan. (1984). Two-level linear programming. *Management Science* 30 (8): 1004–20.

Bouyssou, D., and J.C. Vansnick. (1986). Noncompensatory and generalized noncompensatory preference structures. *Theory and Decision* 21 (3): 251–66.

Brans, J.P., B. Mareschal, and Ph. Vincke. (1984a). PROMETHEE: A new family of outranking methods in MCDM, Operational Research, IFORS'84, North Holland, 477–90.

Brans, J.P., B. Mareschal, and Ph. Vincke. (1984b). PROMETHEE: A new family of outranking methods in multi-criteria analysis. In *Operational Research*, ed. J.P. Brans, 408–21. North-Holland: Elsevier Science Publishers B.V.

Brans, J.P., and Ph. Vincke. (1985). A preference ranking organization method (The PROMETHEE method for MCDM). *Management Science* 31 (6): 647–56.

Brodrick, C.J., T.E. Lipman, M. Farshchi, N.P. Lutsey, H.A. Dwyer, D. Sperling, S.W. Gouse III, D.B. Harris, and F.G. King Jr. (2002). Evaluation of fuel cell auxiliary power units for heavy-duty diesel trucks. *Transportation Research (Part D)* 7 (4): 303–15.

Buckley, J.J. (1985a). Ranking alternatives using fuzzy numbers. *Fuzzy Sets and Systems* 15 (1): 21–31.

Buckley, J.J. (1985b). Fuzzy hierarchical analysis. *Fuzzy Sets and Systems* 17 (3): 233–47.

Buckley, J.J. and E. Eslami. (2002). Fuzzy Markov chains: Uncertain probabilities. *Mathware and Soft Computing* 9 (1): 33–41.

Büyüközkan, G., T. Ertay, and C. Kahraman. (2004). Determining the importance weights for the design requirements in the house of quality using the fuzzy analytic network approach. *International Journal of Intelligent Systems* 19 (5): 443–61.

Candler, W., and R. Townsley. (1982). A linear two-level programming problem. *Computer and Operations Research* 9 (1): 59–76.

Chan, C.C. (1998). A rough set approach to attribute generalization in data mining. *Journal of Information Sciences* 107 (1–4): 169–76.

Chan, C.C., and G.H. Tzeng. (2008a). Computing approximations of dominance-based rough sets by bit-vector encodings, RSCTC 2008, Springer-Verlag in the Lecture Notes on Artificial Intelligence (LNAI) series.

Chan, C.C., and G.H. Tzeng. (2008b). Dominance-based rough sets using indexed blocks as granules. *Fundamentals of Knowledge Technology* 94 (2).

Chang, P.L., and Y.C. Chen. (1994). A fuzzy multi-criteria decision making method for technology transfer strategy selection in biotechnology. *Fuzzy Sets and Systems* 63 (1): 131–39.

Chang, P.T., and E.S. Lee. (1995). The estimation of normalized fuzzy weights. *Computers & Mathematics with Applications* 29 (5): 21–42.

Chen, C.H. (1978). *The methodologies for architecture design*. Taipei: Dong-Da (in Chinese).

Chen, T.Y. (1998). Static and dynamic analyses of importance-assessing for evaluation criteria and applications on transportation, PhD Dissertation. Institute of Traffic and Transportation, College of Management, National Chiao Tung University, Taiwan.

Chen, Y.C. (2002). An application of fuzzy set theory to external performance evaluation of distribution centers in logistics. *Soft Computing* 6 (1): 64–70.

Chen, M., and A.S. Alfa. (1991). A network design algorithm using a stochastic incremental traffic assignment approach. *Transportation Science* 25 (3): 215–24.

Chen, T.Y., C.C. Chang, and G.H. Tzeng. (2001a). Applying fuzzy measures to establish priority-setting procedures for the pavement management system. *PanPacific Management Review* 4 (1): 23–33.

Chen, T.Y., H.L. Chang, and G.H. Tzeng. (2001b). Using a weight-assessing model to identify route choice criteria and information effects. *Transportation Research (Part A)* 35 (3): 197–224.

Chen, T.Y., H.L. Chang, and G.H. Tzeng. (2002). Using fuzzy measure and habitual domains to analyze the public attitude and apply to the gas taxi policy. *European Journal of Operational Research* 137 (2): 145–61.

Chen, S.H., and C.H. Hsieh. (1998). Graded mean integration representation of generalized fuzzy number. In *Proceeding of 1998 Sixth Conference on Fuzzy Theory and Its Application, (CD-ROM)*, filename: 031.wdl, 1–6. Taiwan: Chinese Fuzzy Systems Association.

Chen, H.C., Y.C. Hu, J.Z. Shyu, and G.H. Tzeng. (2005a). Comparing possibility grey forecasting with neural network-based fuzzy regression by an empirical study. *Journal of Grey System* 18 (2): 93–106.

Chen, H.C., Y.C. Hu, J.Z. Shyu, and G.H. Tzeng. (2005b). Comparison analysis of possibility grey forecasting and fuzzy regression to the stock-market price in Taiwan. *Journal of Information Management* 12 (1): 195–214.

Chen, S.J., and C.L. Hwang. (1992). *Fuzzy multiple attribute decision making-methods and applications*. New York: Springer-Verlag.

Chen, S.J., and C.L. Hwang. (1993). F*uzzy multiple attribute decision making, methods and applications*. Lecture Notes in Economics and Mathematical Systems, vol. 375. New York: Springer-Verlag.

Chen, P.T., Z.Y. Lee, H.C. Yu, and G.H. Tzeng. (2003). Analysis of the theoretical basis and practical applicability of a college teacher's achievement evalution model: The case study of a National University in Hsinchu. *Bulletin of Educational Research* 49 (4): 191–218 (Chinese, Taiwan).

Chen, Y.C., H.P. Lien, C.H. Liu, James, J.H. Lion, G.H. Tzeng, and L.S. Yang. (2011). Fuzzy MCDM approach for selecting the best environment-watershed plan. *Applied Soft Computing* 11 (1): 265–275.

Chen, Y.W., and G.H. Tzeng. (1999). A fuzzy multi-objective model for reconstructing post-earthquake road-network by genetic Algorithm. *International Journal of Fuzzy Systems* 1 (2): 85–95.

Chen, Y.W., and G.H Tzeng. (2000a). Fuzzy multi-objective approach to the supply chain model. *International Journal of Fuzzy Systems* 1 (3): 220–27.

Chen, T.Y., and G.H. Tzeng. (2000b). Reducing the amount of information necessary to compute fuzzy measures used in multiattribute decision making. *Journal of Management* 17 (3): 483–514 (Chinese, Taiwan).

Chen, Y.W., and G.H. Tzeng. (2001). Using fuzzy integral for evaluating subjectively perceived travel costs in a traffic assignment model. *European Journal of Operational Research* 130 (3): 653–64.

Chen, T.Y., and G.H. Tzeng. (2003). Exploring the public attitude using fuzzy integrals. *PanPacific Management Review* 6 (1): 45–58.

Chen, M.F. and G.H. Tzeng. (2004). Combining grey relation and TOPSIS concepts for selecting an expatriate host country. *Mathematical and Computer Modelling* 40 (13): 1473–90.

Chen, M.F., G.H. Tzeng, and C.G. Ding. (2008). Combing fuzzy AHP with MDS in identifying the preference similarity of alternatives. *Applied Soft Computing* 8 (1): 110–17.

Chen, Y.W., G.H. Tzeng, and P.J. Lou. (1997). Fuzzy multi-objectives facility location planning: A case study of C.K.S. International Airport in Taiwan. *The Chinese Public Administration Review* 6 (2): 17–42 (Chinese, Taiwan).

Chen, M.F., G.H. Tzeng, and T.I. Tang. (2005). Fuzzy MCDM approach for evaluation of expatriate assignments. *International Journal of Information Technology and Decision Making* 4 (2): 277–96.

Chen, T.Y., G.H. Tzeng, and J.C. Wang. (1999). Dynamic analysis of importance-assessing for evaluation criteria using habitual domain. *Journal of Management* 16 (2): 255–83 (Chinese, Taiwan).

Chen, T.Y., and J.C. Wang. (2001). Identification of λ-fuzzy measures using sampling design and genetic algorithms. *Fuzzy Sets and Systems* 123 (3): 321–41.

Chen, T.Y., J.C. Wang, and G.H. Tzeng. (2000). Identification of general fuzzy measures by genetic algorithms based on partial information. *IEEE Transactions on Systems, Man, and Cybernetics Part B: Cybernetics* 30B (4): 517–28.

Cheng, Y.C. (1993). The theory and characteristics of school-based management. *International Journal of Educational Management* 7 (6): 6–17.

Cheng, C.H., and Y. Lin. (2002). Evaluating the best main battle tank using fuzzy decision theory with linguistic criteria evaluation. *European Journal of Operational Research* 142 (1): 174–86.

Cheng, C.C., J.Z. Shyu, and G.H. Tzeng. (2004). The decision/evaluation of civil aeronautics administration public construction plan. *Civil Aviation Journal Quarterly* 6 (1): 37–63 (Chinese, Taiwan).

Cheng, C.C., L.M. Tsai, J.Z. Shyu, and G.H. Tzeng. (2004). Fuzzy heuristic algorithm method for transport route-choice of low-radiation waste junk. *Chung Hua Journal of Management* 5 (1): 41–56 (Chinese, Taiwan).

Cheng, H.J., and G.H. Tzeng. (2003). Multiobjective planning for relief distribution systems. *Transportation Planning Journal* 32 (3): 561–80.

Cheng, J.Z., Y.W. Yu, M.J. Tsai, and G.H. Tzeng. (2004). Setting a business strategy to weather the telecommunications industry downturn by using fuzzy MCDM. *International Journal of Services Technology and Management* 5 (4): 346–61.

Cheung, S.O., T.I. Lam, M.Y. Leung, and Y.W. Wan. (2001). An analytical hierarchy process based procurement selection method. *Construction Management and Economics* 19 (1): 427–37.

Cheung, F.K.T., J.L.F. Kuen, and M. Skitmore. (2002). Multi-criteria evaluation model for the selection of architecture consultants. *Construction Management and Economics* 20 (7): 569–80.

Chiang, J.H. (1999). Choquet fuzzy integral-based hierarchical networks for decision analysis. *IEEE Transactions on Fuzzy Systems* 7 (1): 63–71.

Chiang, C.I., J.M. Li, G.H. Tzeng, and P.L. Yu. (2000). A revised minimum spanning table method for optimal expansion of competence sets. *Research and Practice in Multiple Criteria Decision Making* 487:238–47.

Chiang, C.I., and G.H. Tzeng. (2000a). A multiple objective programming approach to data envelopment analysis. In *New frontiers of decision making for the information technology era,* eds. Y. Shi and M. Zeleny, 270–85. London: World Science Publishing Company.

Chiang, C.I., and G.H. Tzeng. (2000b). A new efficiency measure for DEA: Efficiency achievement measure established on fuzzy multiple objectives programming. *Journal of Management* 17 (2): 369–88 (Chinese, Taiwain).

Chiou, H.K., and G.H. Tzeng. (2001). Fuzzy hierarchical evaluation with grey relation model of green engineering for industry. *International Journal of Fuzzy System* 3 (3): 466–75.

Chiou, H.K., and G.H. Tzeng. (2002). Fuzzy multiple-criteria decision-making approach for industrial green engineering. *Environmental Management* 30 (6): 816–30.

Chiou, H.K., and G.H. Tzeng. (2003). An extended approach of multicriteria optimization for MODM problems. In *Multi-objective programming and goal-programming: Theory and applications*, eds. T. Tanino, T. Tanaka, and M. Inuiguchi, 111–16. Berlin: Springer.

Chiou, H.K., G.H. Tzeng, and D.C. Cheng. (2005). Evaluating sustainable fishing development strategies using a fuzzy MCDM approach. *OMEGA-International Journal of Management Science* 33 (3): 223–34.

Chiu, Y.C., B. Chen, J.Z. Shyu, and G.H. Tzeng. (2006a). An evaluation model of new product launch strateg. *Technovation* 26 (11): 1244–52.

Chiu, Y.J., H.C. Chen, J.Z. Shyu, and G.H. Tzeng. (2006b). Marketing strategy based on customer behavior for the LCD-TV. *International Journal and Decision Making* 7 (2/3): 143–65.

Chiu, Y.C., J.Z. Shyu, and G.H. Tzeng. (2004). Fuzzy MCDM for evaluating the E-commerce strategy. *International Journal of Computer Applications in Technology* 19 (1): 12–22.

Chiu, Y.C., and G.H. Tzeng. (1999). The market acceptance of electric motorcycles in Taiwan: Experience through a stated preference analysis. *Transportation Research (Part D)* 4 (2): 127–46.

Choi, Y.L. (1984). Land use transport optimization study. *Hong Kong Engineer* Nov.:47–57.
Choi, Y.L. (1985). The LUTO model and its applications in Hong Kong. *Planning & Development* 1: 21–31.
Choi, Y.L. (1986). LUTO model for strategic urban development planning. *Asian Geographer* 5 (1/2): 155–76.
Choquet, G. (1953). Theory of capacities. *Annales de L'Institut Fourier* 5 (1): 131–295.
Chou, Ch.-Ch. (2003). The canonical representation of multiplication operation on triangular fuzzy numbers, *Computers and Mathematics with Applications* 45 (10–11): 1601–10.
Chou, T.Y., S.T. Chou, and G.H. Tzeng. (2006). Evaluating IT/IS investments: A fuzzy multi-criteria decision model approach. *European Journal of Operational Research* 173 (3): 1026–46.
Chu, M.T., J. Shyu, and G.H. Tzeng. (2007a). Comparison among three analytical methods for knowledge communities group-decision analysis. *Expert Systems with Applications* 33 (4): 1011–24.
Chu, M.T., J.Z. Shyu, and G.H. Tzeng. (2007b). Using non-additive fuzzy integral to assess performances of organization transformation via communities of practice. *IEEE Transactions on Engineering Management* 54 (2): 327–39.
Chu, P.Y., G.H. Tzeng, and M.J. Teng. (2004). A multivariate analysis of the relationship among market share, growth and profitability: The case of the science-based industrial park. *Yat-Sen Management Review* 12 (3): 507–34.
Churchman, C.W., and R.L. Ackoff. (1954). An approximate measure of value. *Journal of Operations Research Society of America* 2 (1): 172–87.
Cogger, K.O., and P.L. Yu. (1985). Eigen weight vectors and least distance approximation for revealed preference in pairwise weight ratios. *Journal of Optimization Theory and Applications* 46 (4): 483–91.
Combarro, E.F., and P. Miranda. (2006). Identification of fuzzy measures from sample data with genetic algorithms. *Computers & Operations Research* 33 (10): 3046–66.
Cook, W.D., and L.M. Seiford. (1978). Priority ranking and consensus formation. *Management Science* 24 (16): 1721–32.
Csutora, R., and J. Buckley. (2001). Fuzzy hierarchical analysis: The Lambda-Max method. *Fuzzy Sets and Systems* 120 (2): 181–95.
Current, J.R., and H. Min. (1986). Multiobjective design of transportation networks: Taxonomy and annotation. *European Journal of Operational Research* 26 (2): 187–201.
Current, J.R., C.S. Revelle, and J.L. Cohon. (1987). The median shortest path problem: A multiobjective approach to analyze cost vs. accessibility in the design of transportation betworks. *Transportation Science* 21 (3): 188–97.
D'Ambra, J., and R.E. Rice. (2001). Emerging factors in user evaluation of the world wide web. *Information & Management* 38 (6): 373–84.
David, J.L. (1989). Synthesis of research on school-based management. *Educational Leadership* 46 (8): 45–53.
Delgado, M., J.L. Verdegay, and M.A. Vila. (1992). Linguistic decision-making models. *International Journal of Intelligent System* 7 (5): 479–92.
DeLucchi, M A. (1989). Hydrogen vehicles: An evaluation of fuel storage, performance, safety, environmental impacts, and cost. *International Journal of Hydrogen Energy* 14 (2): 81–130.
DeLucchi, M.A., and T.E. Lipman. (2001). An analysis of the retail and lifecycle cost of battery-powered electric vehicles. *Transportation Research (Part D)* 6 (6): 371–404.
DeLucchi, M.A., J.J. Murphy, and D.R. McCubbin. (2002). The health and visibility cost of air Pollution: A comparison of estimation methods. *Journal of Environmental Management* 64 (2): 139–52.
Deng, J. (1982). Control problem of grey systems. *Systems and Control Letters* 1 (5): 288–94.

Deng, J. (1985). *Grey system fundamental method.* Huazhoug University of Science and Technology, Wuhan, China (in Chinese).

Deng, J. (1988). *Grey system book.* Windsor: Sci-Tech Information Services.

Deng, J. (1989). Introduction to grey system theory. *The Journal of Grey System* 1 (1): 1–24.

DeRoche, E.F. (1987). *An administratior's guide for evaluation programs and personnel: An effective schools approach* (2nd ed.). Newton Massachusetts: Allyn and Bacon.

Dimitras, A.I., R. Slowinski, R. Susmaga, and C. Zopounidis. (1999). Business failure prediction using rough sets. *European Journal of Operational Research* 114 (2): 263–80.

Doumpos, M., and C. Zopounidis. (2002). *Multicriteria decision aid classification methods.* Boston: Kluwer Academic Publishers.

Dubois, D., and H. Prade. (1980). *Fuzzy sets and systems.* New York: Academic Press.

Dubois, D., and H. Prade. (1987). *Fuzzy numbers: An overview, analysis of fuzzy information*, Vol. I, 3–39 (J. Bezdek, ed.). Boca Raton, FL: CRC Press.

Dubois, D., and H. Prade. (1988). *Possibility theory. An approach to computerized processing of uncertainty.* New York: Plenum Press.

Dubois, D., and H. Prade. (1992). Putting rough sets and fuzzy sets together. In *Intelligent decision support: Handbook of applications and advances of the rough sets theory*, ed. R. Slowinski, 203–32. Dordrecht: Kluwer Academic Publishers.

Dubois, D., E. Kerre, R. Mesiar, and H. Prade. (2000). Fuzzy interval analysis. In *Fundamentals of fuzzy sets*, eds. D. Dubois and H. Prade, The handbooks of fuzzy sets series, 483–581. Boston, Mass: Kluwer.

Dubois, D., H. Prade, and R.R. Yager. (1993). *Readings in fuzzy sets for intelligent systems.* San Mateo, CA: Morgan & Kaufmann.

Energy Committee of the Ministry of Economic Affairs. (1989). *Energy master plan for Taiwan area (1989–2000)*, Taipei, Taiwan.

Ertay, T., G. Büyüközkan, C. Kahraman, and D. Ruan. (2005). Quality function deployment implementation based on analytic network process with linguistic data: An application in automotive industry. *Journal of Intelligent and Fuzzy Systems* 16 (3): 221–32.

Fan, T.F., D.R. Liu, and G.H. Tzeng. (2005). Arrow Decision Logic. In *Rough sets, fuzzy sets, data mining, and guanular computing*, eds. Ivo Duntsch and YiYu Yao, 3641 (Part I), 651–59.

Fan, T.F., D.R. Liu, and G.H. Tzeng. (2006). Arrow decison logic for relational information systems. *Transactions on Rough Sets V* 4100:240–62.

Fan, T.F., D.R. Liu, and G.H. Tzeng. (2007). Rough set-based logics for multicriteria decision analysis. *European Journal of Operational Research* 182 (1): 340–55.

Fidler, B., and G. Bowles. (1989). *Effective local management of school.* London: Longman.

Fishburn, P.C. (1970). *Utility theory for decision-making.* New York: Wiley.

Fitz-Gibbon, C.T. (1994). Indicator systems for school and teacher evaluation: Fire-fighting it is! Paper presented at the Annual National Evaluation Institute of the Center for Research on Educational Accountability and Teacher Evaluation (3rd, Gatlinburg, TN, July 10–15, 1994) (ERIC Document Reproduction Service NO. ED 379340).

Fong, S.W., and S.K.Y. Choi. (2000). Final contractor selection using the analytical hierarchy process. *Construction Management and Economics* 18 (5): 547–57.

Fontela, E., and A. Gabus. (1974). DEMATEL, Innovative methods, Report no. 2, Structural analysis of the world problematique. Battelle Geneva Research Institute.

Fontela, E., and A. Gabus. (1976). The DEMATEL observer. Battelle Institute, Geneva Research Center.

Fortuny-Amat, J., and B. McCarl. (1981). A representation and economic interpretation of a two-level programming problem. *Journal of Operational Research and Society* 32 (9): 783–92.

Freimer, M., and P.L. Yu. (1976). Some new results on compromise solutions for group decision problems. *Management Science* 22 (6): 688–93.

Friesz, T.L., and P.T. Harker. (1983). Multicriteria spatial price equilibrium network design: Theory and computational results. *Transportation Research (Part B)* 17 (5): 411–26.

Funk, K., and A. Rabl. (1999). Electric versus conventional vehicles: Social costs and benefits in france. *Transportation Research (Part D)* 4 (6): 397–411.

Gabus, A., and E. Fontela. (1972). *World problems, an invitation to further thought within the framework of DEMATEL.* Battelle Geneva Research Centre. Geneva, Switzerland.

Gabus, A., and E. Fontela. (1973). *Perceptions of the world problematique: Communication procedure, communicating with those bearing collective responsibility* (DEMATEL Report No. 1), Battelle Geneva Research Centre, Geneva, Switzerland.

Geoffrion, A.M. (1987). An introduction to structured modeling. *Management Science* 33 (5): 547–88.

Goh, C., and R. Law. (2003). Incorporating the rough sets theory into travel demand analysis. *Tourism Management* 24 (5): 511–17.

Gorman, W.M. (1968a). Conditions for additive separability. *Econometrica* 36 (3/4): 605–9.

Gorman, W.M. (1968b). The structure of utility function. *Review of Economic Studies* 35 (4): 367–90.

Grabisch, M. (1995a). A new algorithm for identifying fuzzy measures and its application to pattern recognition, Int. Joint Conf. of the 4th IEEE International Conference on Fuzzy Systems and the 2nd International Fuzzy Engineering Symposium, March 1995, Yokohama, Japan, 145–50.

Grabisch, M. (1995b). Fuzzy integral in multicriteria decision making. *Fuzzy Sets and Systems* 69 (3): 279–98.

Grabisch, M. (1996). The application of fuzzy integrals in multicriteria decision making. *European Journal of Operational Research* 89 (3): 445–56.

Grabisch, M. (1997). k-Order additive discrete fuzzy measures and their representation, integral in multicriteria decision making. *Fuzzy Sets and Systems* 92 (2): 167–89.

Greco, S., B. Matarazzo, and R. Slowinski. (2001). Rough sets theory for multicriteria decision analysis. *European Journal of Operational Research* 129 (1): 1–47.

Griffith, P., and G. Gleason. (1996). Six years of battery-electric bus operation at the Santa Barbara metropolitan transit district. Proceedings of 13th Electric Vehicle Conference.

Harary, F., R. Norman, and D. Cartwright. (1965). *Structural models: An introduction to the theory of directed graphs.* New York: Wiley.

Hardie, B.G.S., E.J. Johnson, and P.S. Fader. (1993). Modeling loss aversion and reference dependence effects on brand choice. *Marketing Science* 12 (4): 378–94.

Harding, G.G. (1999). Electric vehicles in the next millennium. *Journal of Power Sources* 78 (2): 193–98.

Hassan, Y., and E. Tazaki. (2001). Rule extraction based on rough set theory combined with genetic programming and its application to medical data analysis. 7th Australian and New Zealand Intelligent Information System Conference, Perth, Australia.

Hastak, M. (1998). Advanced automation or conventional construction process. *Automation in Construction* 7 (4): 299–314.

Hennig-Thurau, T., and A. Klee. (1997). The impact of customer satisfaction and relationship quality on customer retention: A critical reassessment and model development. *Psychology & Marketing* 14 (8): 737–64.

Herrera, F., E. Herrera-Viedma, and J.L. Verdegay. (1996). A model of consensus in group decision making under linguistic assessments. *Fuzzy Sets and Systems* 78 (1): 73–87.

Hess, P., and J. Siciliano. (1996). *Management: Responsibility for performance.* New York: McGraw-Hill.

Highhouse, S., and M.A. Johnson. (1996). Gain/loss asymmetry and riskless choice: Loss aversion in choices among job finalists. *Organizational Behavior and Human Decision Processes* 68 (3): 225–33.

Hillier, F.S. (2001). *Evaluation and decision models: A critical perspective.* Boston: Kluwer Academic Publishers.

Holland, J.M. (1975). *Adaptation in natural and artificial systems.* Ann Arbor: The University of Michigan Press.

Hori, S., and Y. Shimizu. (1999). Designing methods of human interface for supervisory control systems. *Control Engineering Practice* 7 (11): 1413–19.

Hougaard, J.L., and H. Keiding. (1996). Representation of preferences on fuzzy measures by a fuzzy integral. *Mathematical Social Sciences* 31 (1): 1–17.

Hsieh, T.Y., S.T. Lu, and G.H. Tzeng. (2004). Fuzzy MCDM approach for planning and design tenders selection in public office buildings. *International Journal of Project Management* 22 (7): 573–84.

Hsu, C.Y., K.T. Chen, and G.H. Tzeng. (2007). FMCDM with fuzzy DEMATEL approach for customers' choice behavior model. *International Journal of Fuzzy Systems* 9 (4): 236–46.

Hsu, C.S., Z.Y. Lee, C.Y. Hung, C. Shih, H.C. Yu, and G.H. Tzeng. (2004). Key factors in performance appraisal for R&D organizations: The case of the industrial technology research institute in Taiwan. *Journal of Biomedical Fuzzy Systems* 10 (1/2): 19–29.

Hsu, C.S., Z.Y. Lee, C.S. Shi, C.Y. Hung, and G.H. Tzeng. (2003). The clustering and performance efficiency of ITRI's R&D units. *Journal of Technology Management* 8 (1): 33–60 (Chinese, Taiwan).

Hsu, C.S., Z.Y. Lee, C.S. Shi, C.Y. Hung, H.C. Yu, and G.H. Tzeng. (2003). Application of the DEA approach to evaluate management performance of R&D organizations in the industrial technology research institute of Taiwan. *Management Review* 22 (2): 25–53 (Chinese, TSSCI).

Hsu, C.F., and G.H. Tzeng. (2002). Evaluation analysis for Taiwan agriculture productivity: An application of DEA method. *Journal of the Land Bank of Taiwan* 39 (2): 139–57 (Chinese).

Hsu, Y.G., G.H. Tzeng, and J.Z. Shyu. (2003). Fuzzy multiple criteria selection of government-sponsored frontier technology R&D projects. *R&D Management* 33 (5): 539–51.

Hsu, Y.G., J.Z. Shyu, and G.H. Tzeng. (2005). Policy tools on the formation of new biotechnology firms in Taiwan. *Technovation* 25 (3): 281–92.

Hu, Y.C., R.S. Chen, Y.T. Hsu, and G.H. Tzeng. (2002). Grey self-organizing feature maps. *Neurocomputing* 48 (4): 863–77.

Hu, Y.C., R.S. Chen, and G.H. Tzeng. (2002). Generating learning sequences for decision makers through data mining and competence set expansion. *IEEE Transactions on Systems, Man and Cybernetics (Part B)* 32 (5): 679–86.

Hu, Y.C., R.S. Chen, and G.H. Tzeng. (2003a). An effective learning algorithm for discovering fuzzy sequential patterns. *International Journal Uncertainty, Fuzziness and Knowledge-Based Systems* 11 (2): 173–93.

Hu, Y.C., R.S. Chen, and G.H. Tzeng. (2003b). Discovering fuzzy association rules using fuzzy partition methods. *Knowledge-Based Systems* 16 (3): 137–47.

Hu, Y.C., R.S. Chen, and G.H. Tzeng. (2003c). Finding fuzzy classification rules using data mining techniques. *Pattern Recognition Letters* 24 (1–3): 509–19.

Hu, Y.C., R.S. Chen, G.H. Tzeng, and Y.J. Chiu. (2003). Acquisition of compound skills and learning costs for expanding competence sets. *Computers and Mathematics with Applications* 46 (5–6): 831–48.

Hu, Y.C., R.S. Chen, G.H. Tzeng, and J.H. Shieh. (2003). A fuzzy data mining algorithm for finding sequential patterns. *International Journal of Uncertainty, Fuzziness and Knowledge-Based Systems* 11 (2): 173–93.

Hu, Y.C., Y.J. Chiu, C.M. Chen, and G.H. Tzeng. (2003). Competence set expansion for obtaining scheduling plans in intelligent transportation security systems. In *Multi-objective programming and goal-programming: Theory and applications*, eds. T. Tanino, T. Tanaka, and M. Inuiguchi, 347–52. Berlin: Springer.

Hu, Y.C., Y.J. Chiu, J.S. Hu, G.H. Tzeng, and Y.S. Huang. (2002). Acquisitions of learning costs for expanding competence sets using grey relations and neural networks. *Journal of The Chinese Grey System Association* 5 (2): 75–82 (Chinese).

Hu, Y.C., Y.J. Chiu, and G.H. Tzeng. (2003). Grey theory and competence sets for multiple criteria project scheduling. *Chung Yuan Management Review* 1 (2): 257–73 (Chinese, Taiwan).

Hu, Y.C., J.S. Hu, R.S. Chen, and G.H. Tzeng. (2004). Assessing weights of product attributes from fuzzy knowledge in a dynamic environment. *European Journal of Operational Research* 154 (1): 125–43.

Hu, Y.C., and G.H. Tzeng. (2002). Mining fuzzy association rule for classification problems. *Computers & Industrial Engineering* 43 (4): 735–50.

Hu, Y.C., and G.H. Tzeng. (2003). Elicitation of classification rules by fuzzy data mining. *Engineering Applications of Artificial Intelligence* 16 (7–8): 709–16.

Hu, Y.C., G.H. Tzeng, and C.M. Chen. (2004). Deriving two stage learning sequence from knowledge in fuzzy sequential pattern mining. *Information Sciences (An International Journal)* 159 (1–2): 69–86.

Huang, J.J., C.S. Ong, and G.H. Tzeng. (2004). Using rough set theory for detecting the interaction terms in a generalized logit model. Rough Sets and Current Trends in Computing, LNAI 3066, 624–29.

Huang, J.J., C.S. Ong, and G.H. Tzeng. (2005a). A novel hybrid model for portfolio selection. *Applied Mathematics and Computation* 169 (2): 1195–210.

Huang, J.J., C.S. Ong, and G.H. Tzeng. (2005b). Building credit scoring models using genetic programming. *Expert Systems With Applications* 29 (1): 41–47.

Huang, J.J., C.S. Ong, and G.H. Tzeng. (2005c). Model identification of ARIMA family using genetic algorithms. *Applied Mathematics and Computation* 164 (3): 885–912.

Huang, J.J., G.H. Tzeng, and C.S. Ong. (2005d). Multidimensional data in multidimensional scaling using the analytic network process. *Pattern Recognition Letters* 26 (6): 755–67.

Huang, J.J., C.S. Ong, and G.H. Tzeng. (2006a). Fuzzy principal component regression (FPCR) for fuzzy input and output data. *International Journal of Uncertainty and Knowledge-based Systems* 14 (1): 87–100.

Huang, J.J., C.S. Ong, and G.H. Tzeng. (2006b). Optimal fuzzy multi-criteria expansion of competence sets using multi-objectives evolutionary algorithms. *Expert Systems with Applications* 30 (4): 739–45.

Huang, C.Y., J.Z. Shyu, and G.H. Tzeng. (2007). Reconfiguring the innovation policy portfolios for Taiwan's SIP mall industry. *Technovation* 27 (12): 744–65.

Huang, J.J., and G.H. Tzeng. (2007a). Marketing segmentation using support vector clustering. *Expert Systems With Applications* 32 (2): 313–17.

Huang, C.Y., and G.H. Tzeng. (2007b). Post-merger high technology R&D human resources optimization through the de novo perspective. In *Advances in multiple criteria decision making and human systems management: Knowledge and wisdom*, eds. Y. Shi, D.L. Olson, and A. Stam, 47–64. IOS Press.

Huang, C.Y., and G.H. Tzeng. (2008). Multiple generation product life cycle predictions using a novel two-stage fuzzy piecewise regression analysis method. *Technological Forecasting and Social Change* 75 (1): 12–31.

Huang, C.Y., G.H. Tzeng, C.C. Chan, and H.C. Wu. (2008). Semiconductor market fluctuation indicators and rules derivations by using the rough set theory. *International Journal of Innovative Computing, Information and Control* (Minor revised, possible publication).

Huang, J.J., G.H. Tzeng, and C.S. Ong. (2005). Multidimensional data in multidimensional scaling using the analytic network process. *Pattern Recognition Letters* 26 (6): 755–67.

Huang, J.J., G.H. Tzeng, and C.S. Ong. (2006a). A novel algorithm for dynamic factor analysis. *Applied Mathematics and Computation* 175 (2): 1288–97.

Huang, J.J., G.H. Tzeng, and C.S. Ong. (2006b). A novel algorithm for uncertain portfolio selection. *Applied Mathematics and Computation* 173 (1): 350–59.

Huang, J.J., G.H. Tzeng, and C.S. Ong. (2006c). Choosing best alliance partners and allocating optimal alliance resources using the fuzzy multi-objective dummy programming model. *Journal of Operational Research Society* 57 (10): 1216–23.

Huang, J.J., G.H. Tzeng, and C.S. Ong. (2006d). Interval multidimensional scaling for group decision using rough set concept. *Expert Systems with Applications* 31 (3): 525–30.

Huang, J.J., G.H. Tzeng, and C.S. Ong. (2006e). Two-stage genetic programming (2SGP) for the credit scoring model. *Applied Mathematics and Computation* 174 (2): 1039–53.

Hugonnard, J., and B. Roy. (1982). Le plan d'extension du metro en banlieue parisienne, uncas type d'application de l'analyse multicritere. *Les Cahiers Scientifiques de la Revue Transports* 6:77–108.

Hwang, M.J., C.I. Chiang, I.C. Chiu, and G.H. Tzeng. (2001). Multi-stages optimal expansion of competence sets in fuzzy environment. *International Journal of Fuzzy Sets* 3 (3): 486–92.

Hwang, M.J., G.H. Tzeng, and Y.H. Liu. (1996). A dynamic model for optimal expansion of multistage competence sets. *Journal of Management* 14 (1): 115–33.

Hwang, C.L., and K. Yoon. (1981). Multiple attribute decision making, methods and applications. Lecture Notes in Economics and Mathematical Systems, vol.186. Now York: Springer-Verlag.

Inuiguchi, M. (2004). Generalizations of rough sets: From crisp to fuzzy cases. rough sets and current trends in computing. In *Lecture Notes in Artificial Intelligence (LNAI)*, eds. S. Tsumoto, R. Slowinski, J. Komoroski, J.W. Grzymala-Busse, 3066(1), 26–37. Berlin: Springer-Verla.

Ishii, K., and M. Sugeno. (1985). A model of human evaluation process using fuzzy measure. *International Journal of Man-Machine Studies* 22 (1): 19–38.

Jackson, A.G., S.R. Leclair, M.C. Ohmer, W.W. Ziarko, and H. Al-kamhwi. (1996). Rough sets applied to materials data. *ACTA Material* 44 (11): 4475–84.

James, J.H., L.Y. Liou, and G.H. Tzeng. (2008). Building an effective safety management system for airlines. *Journal of Air Transport Management* 14 (1): 20–26.

Johnsson, B., and M. Ahman. (2002). A comparison of technologies for carbon-neutral passenger transport. *Transportation Research* (*Part D*) 7 (3): 175–96.

Kahraman, C., T. Ertay, and G. Büyüközkan. (2006). A fuzzy optimization model for QFD planning process using analytic network approach. *European Journal of Operational Research* 171 (22): 390–411.

Kahneman, D., J.L. Knetsch, and R.H. Thaler. (1991). Anomalies: The endowment effect, loss aversion, and status quo bias. *Journal of Economic Perspectives* 5 (1): 193–206.

Kahneman, D., and A. Tversky. (1979). Prospect theory: An analysis of decision under risk. *Econometrica* 47 (2): 263–91.

Kahneman, D., and A. Tversky. (1984). Choices, values, and frames. *American Psychologist* 39 (4): 341–50.

Kahneman, D., and A. Tversky. (2000). *Choices, values, and frames*. Cambridge: Cambridge University Press.

Kaiser, H.F. (1958). The varimax criteria for analysis rotation in factor analysis. *Psychometric* 23 (2): 187–200.

Karsak, E.E., S. Sozer, and S.E. Alptekin. (2002). Product planning in quality function deployment using a combined analytic network process and goal programming approach. *Computers & industrial engineering* 44 (1): 171–90.

Kaufmann, A., and M.M. Gupta. (1985). *Introduction to fuzzy arithmetic: Theory and applications*. New York: Van Nostrand Reinhold.

Kaufmann, A., and M.M. Gupta. (1988). *Fuzzy mathematical models in engineering and management science*. Amsterdam: North-Holland.

Kazimi, C. (1997). Evaluating the environmental impact of alternative-fuel vehicles. *Journal of Environmental Economics and Management* 33 (2): 163–85.

Keeney, R.L., and H. Raiffa. (1976). *Decision with multiple objectives: Preferences and value tradeoffs*. New York: John Wiley and Sons.

Kempton, W., and T. Kubo. (2000). Electric-drive vehicles for peak power in Japan. *Energy Policy* 28 (1): 9–18.

Kim, D.J., D.L. Ferrin, and H.R. Rao. (2003). A study of the effect of consumer trust on consumer expectations and satisfaction: The Korean experience. In *ACM International Conference Proceeding Series: Proceedings of the 5th International Conference on Electronic commerce*, 310–15. Pittsburgh: ACM Press.

Kleindorfer, P.R., H.C. Kunreuther, and P.J.H. Schoemaker. (1993). *Decision sciences: An integrative perspective*. Cambridge: Cambridge University Press.

Klir, G.J., and T.A. Folger. (1988). *Fuzzy sets, uncertainty, and information*. Englewood Cliffs, NJ: Prentice Hall.

Klir, G.J., and Y. Pan. (1997). Fuzzy arithmetic with requisite constraints. *Fuzzy Sets and Systems* 91 (2): 165–75.

Klir, G.J., and Y. Pan. (1998). Constrained fuzzy arithmetic: Basic questions and some answer. *Soft Computing* 2 (2): 100–8.

Kosko, B. (1988). Hidden patterns in combined and adaptive knowledge networks. *International Journal of Approximate Reasoning* 2 (4): 377–93.

Koza, J. (1992). *Genetic programming: On the programming of computers by means of natural selection*. MA: MIT Press.

Kristense, K., A. Martensen, and L. Gronholdt. (1999). Measuring the impact of buying behavior on customer satisfaction. *Total Quality Management* 10 (4): 602–14.

Krusinska, E., R. Slowinski, and J. Stefanowski. (1992). Discriminant versus rough sets approach to vague data analysis. *Applied Stochastic Models and Data Analysis* 8 (1): 43–56.

Kuhn, H.W., and A.W. Tucker. (1951). Nonlinear programming. In *Proceedings of the Second Berkeley Symposium on Mathematical Statistics and Probability*, 481–91. Berkeley: University of California Press.

Kuo, M.S., G.H. Tzeng, and W.C. Huang. (2007). Group decision making based on concepts of ideal and anti-ideal points in fuzzy environment. *Mathematical and Computer Modeling* 45 (3–4): 324–39.

Laarhoven, P.J.M., and W. Pedrycz. (1983). A fuzzy extension of saaty's priority theory. *Fuzzy Sets and Systems* 11 (3): 229–41.

Lebesgue, H. (1966). *Measure and the integral*. San Francisco: Holden-Day.

LeBlanc, L.J, and D.E. Boyce. (1986). A bilevel programming algorithm for exact solution of the network design problem with user-optimal flows. *Transportation Research (Part B)* 20 (3): 259–65.

LeBlanc, L.J. (1975). An algorithm for the discrete network design problem. *Transportation Science* 9 (3): 183–99.

LeBlanc, L.J., E. Morlok, and W. Pierskalla. (1975). An efficient approach to solving road network equilibrium traffic assignment problems. *Transportation Research (Part B)* 18:309–18.

Lee, J.W., and S.H. Kim. (2000). Using analytic network process and goal programming for interdependent information system project selection. *Computers & Operations Research* 27 (4): 367–82.

Lee, K.M., and H. Leekwang. (1995). Identification of λ-fuzzy measure by genetic algorithms. *Fuzzy Sets and Systems* 75 (3): 301–9.

Lee, C., L.C. Liu, and G.H. Tzeng. (2000). Multi-criteria evaluation of the schooling efficiency for promoting outstanding junior colleges to technological institutes. *Journal of The Chinese Grey System Association* 3 (2): 95–113 (Chinese, Taiwan).

Lee, C., L.C. Liu, and G.H. Tzeng. (2001). Hierarchical fuzzy integral evaluation approach for vocational education performance: Case of junior colleges in Taiwan. *International Journal of Fuzzy Systems* 3 (3): 476–85.

Lee, E.S., and R.J. Li. (1993). Fuzzy multiple objective programming and compromise programming with Pareto optimum. *Fuzzy Sets and Systems* 53 (2): 275–88.

Lee, W.S., G.H. Tzeng C.Y. Hsu, and J.M. Huang. (2008). Combined MCDM techniques for exploring stock selection based on Gordon model. *Expert Systems With Applications* 43 (4).

Lee, C.F., G.H. Tzeng, and S.Y. Wang. (2005a). A fuzzy set approach to generalize CRR model: An empirical analysis of S&P 500 index option. *Review of Quantitative Finance and Accounting* 25 (3): 255–75 (Springer).

Lee, C.F., G.H. Tzeng, and S.Y. Wang. (2005b). A new application of fuzzy set theory to the black-scholes option pricing model. *Expert Systems With Applications* 29 (2): 330–42.

Leszczyński, K., P. Penczek, and W. Grochulski. (1985). Sugeno's fuzzy measure and fuzzy clustering. *Fuzzy Sets and Systems* 15 (2): 147–58.

Li, H.L. (1982). *The use of hierarchical multiobjective programming in urban transportation network design*, PhD Dissertation. University of Pennsylvania.

Li, H.L. (1988). Expert system for evaluating multicriteria transportation networks. *Microcomputer in Civil Engineering* 3 (3): 259–65 (in Chinese).

Li, R.J. (1999). Fuzzy method in group decision making. *Computers and Mathematics with Applications* 38 (1): 91–101.

Li, R., and Z.O. Wang. (2004). Mining classification rules using rough sets and neural networks. *European Journal of Operational Research* 157 (2): 439–48.

Li, R.J., and E.S. Lee. (1990). Multi-criteria *de Novo* programming with fuzzy parameters. *Computers & Mathematics with Applications* 19 (5): 13–20.

Lin, C.M., J.J. Huang, M. Gen, and G.H. Tzeng. (2006). Recurrent neural network for dynamic portfolio selection, applied mathematics and computation. *Applied Mathematics and Computation* 175 (2): 1139–46.

Liou, J.J.H., and G.H. Tzeng. (2007). A non-additive model for evaluating airline service quality. *Journal of Air Transport Management* 13 (3): 131–38.

Liou, J.J.H., G.H. Tzeng, and H.C. Chang. (2007). Airline safety measurement using a hybrid model. *Journal of Air Transport Management* 13 (4): 243–49.

Liou, J.J.H., L. Yen, G.H. Tzeng. (2008). Building an effective safety management system for airlines. *Journal of Air Transport Management* 14 (1): 20–26.

Liu, L.C., C. Lee, and G.H. Tzeng. (2001). Hierarchical fuzzy integral evaluation approach for vocational education performance: Case of junior colleges in Taiwan. *International Journal of Fuzzy Sets* 3 (3): 476–85.

Liu, L.C., C. Lee, and G.H. Tzeng. (2003a). DEA for evaluating the current-period and cross-period efficiency of Taiwan's upgraded technical institutes. In *Multi-objective programming and goal-programming: Theory and applications*, eds. T. Tanino, T. Tanaka, and M. Inuiguchi, 353–36. Berlin: Springer.

Liu, L.C., C. Lee, and G.H. Tzeng. (2003b). Using DEA of REM and EAM for efficiency assessment of technology institutes upgraded from junior colleges: The case in Taiwan. In *Multi-objective programming and goal-programming: Theory and applications*, eds. T. Tanino, T. Tanaka, and M. Inuiguchi, 361–66. Berlin: Springer.

Liu, L.C., C. Lee, and G.H. Tzeng. (2004). DEA approach of the current-period and cross-period efficiency for evaluating the vocational education. *International Journal of Information Technology and Decision Making* 3 (2): 353–74.

Liu, B.C., G.H. Tzeng, and C.T. Hsieh. (1992). Energy planning and environment—quality management—a decision support system approach. *Energy Economics* 14 (4): 302–7.

Liu, Y.H., G.H. Tzeng, and D.H. Park. (2004). Set covering problem and the reliability of the covers. *International Journal of Reliability and Applications* 5 (4): 145–54.

Lu, S.T., T.Y. Hsieh, and G.H. Tzeng. (2004). Fuzzy MCDM approach for planning and design tenders selection in public office buildings. *International Journal of Project Management* 22 (7): 574–84.

Maggetto, G., and J. Van Mierlo. (2001). Electric vehicles, hybrid electric vehicles and fuel cell electric vehicles: State of the art and perspectives. *Annales de Chimie Science des Materiaux* 26 (4): 9–26.

Mahdi, I.M., M.J. Riley, S.M. Fereig, and A.P. Alex. (2002). A multi-criteria approach to contractor selection engineering. *Construction and Architectural Management* 9 (1): 29–37.

Mandal, A., and S.G. Deshmukh. (1994). Vendor selection using interpretive structural modelling (ISM). *International Journal of Operations & Production Management* 14 (6): 52–59.

Mares, M. (1994). *Computation over fuzzy quantities*. Boca Raton, FL: CRC Press.

Marglin, S. (1967). *Public investment criteria*, p. 103. Cambridge, Massachusetts: MIT Press.

Marichal, J.L., and M. Roubens. (2000). Determination of weights of interacting criteria from a reference set. *European Journal of Operational Research* 124 (3): 641–50.

Martinson, F.K. (1993). Fuzzy vs. Minmax weighted multiobjective linear programming illustrative comparisons. *Decision Sciences* 24 (5): 809–24.

Matheny, M.S., P.A. Erickson, C. Niezrecki, and V.P. Roan. (2002). Interior and exterior noise emitted by a fuel cell transit bus. *Journal of Sound and Vibration* 251 (5): 937–43.

McIntyre, C., and M.K. Parfitt. (1998). Decision support system for residential land development site selection process. *Journal of Architectural Engineering ASCE* 4 (4): 125–31.

McNicol, B.D., D.A.J. Rand, and K.R. Williams. (2001). Fuel cells for road transportation purpose—yes or no? *Journal of Power Sources* 100 (1): 47–59.

Meade, L.M., and A. Presley. (2002). R&D project selection using the analytic network process. *IEEE Transactions on Engineering Management* 49:59–66.

Mi, J., W. Wu, and W. Zhang. (2004). Approaches to knowledge reduction based on variable precision rough set model. *Information Sciences* 159 (3/4): 255–72.

Mikhailov, L. (2000). A fuzzy programming method for deriving priorities in the analytic hierarchy process. *Journal of Operational Research Society* 51 (3): 341–49.

Mikhailov, L. (2003). Deriving priorities form fuzzy pairwise comparision judgements. *Fuzzy Sets and Systems* 134 (3): 365–85.

Mikhailov, L., and M.G. Singh. (2003). Fuzzy analytic network process and its application to the development of decision support systems. *IEEE Transactions on Systems, Man, and Cybernetics* 33 (1): 33–41.

Miranda, P., M. Grabisch, and P. Gil. (2002). p Symmetric fuzzy measures, international journal of uncertainty. *Fuzziness and Knowledge-Based Systems* 10 (supplement). 105–23.

Mohanty, R.P., R. Agarwal, A.K. Choudhury, and M.K. Tiwari. (2005). A fuzzy ANP-based approach to R&D project selection: A case study. *International Journal of Production Research* 43 (24): 5199–216.

Momoh, J.A., and J. Zhu. (2003). Optimal generation scheduling based on AHP/ANP. *IEEE Transactions on Systems, Man, and Cybernetics—Part B: Cybernetics* 33 (3): 531–35.

Mon, D.L., C.H. Cheng, and J.C. Lin. (1994). Evaluating weapon system using fuzzy analytic hierarchy process based on entropy weight. *Fuzzy Sets and Systems* 62 (2): 127–34.

Mon, O.L., G.H. Tzeng, and H.C. Lu. (1995). Grey decision making in weapon system evaluation. *Journal of Chung Cheng Institute of Technology* 24 (1): 73–84.

Mori, T., and T. Murofushi. (1989). An analysis of evaluation model using fuzzy measure and the choquet integral. *Proceedings of the 5th Fuzzy Systems Symposium*, 207–12 (Japanese).

Mori, N., H. Tanaka, and K. Inoue. (2004). Rough Sets and KANSEI: Knowledge Acquisition and Reasoning from KANSEI Data. Kaibundo: Japan, Tokyo.

Morita, K. (2003). Automotive power source in 21st century. *Journal of Society of Automotive Engineers of Japan* 24 (1): 3–7.

Morris, V.C., and P. Young. (1976). *Philosophy and the American school: An introduction to philosophy of education*. Boston: Houghton Mifflin.

Moscarola, J., and B. Roy. (1977). Procedure automatique d'examen de dossiers fondee surune segmentation trichotomique en presence de criteres multiples. *RAIRO Recherche Operationnelle* 11 (2): 145–73.

Moseley, P.T. (1999). High-rate, valve-regulated lead-acid batteries: Suitable for hybrid electric vehicles? *Journal of Power Sources* 84 (2): 237–42.

Murchland, J. (1970). Braess' paradox of traffic flow. *Transportation Research (Part B)* 16 (1): 45–55.

Murofushi, T., and M. Sugeno. (1989). An interpretation of fuzzy measure and the Choquet integral as an integral with respect to a fuzzy measure. *Fuzzy Sets and Systems* 29 (2): 201–27.

Murofushi, T., and M. Sugeno. (1991). A theory of fuzzy measures representations, the Choquet integral, and null sets. *Journal of Mathematical Analysis and Applications* 159 (2): 532–49.

Murofushi, T., and M. Sugeno. (1993). Some quantities represented by the Choquet integral. *Fuzzy Sets and Systems* 56 (2): 229–35.

Nesbitt, K., and D. Sperling. (1998). Myths regarding alternative fuel vehicle demand by light-duty vehicle fleets. *Transportation Research (Part D)* 3(4): 259–69.

Ohuchi, A., and I. Kaji. (1989). Correction procedures for flexible interpretive structural modeling, IEEE transactions on systems. *Man and Cybernetics* 19 (1): 85–94.

Ohuchi, A., S. Kase, and I. Kaji. (1988). MINDS: A flexible interpretive structural modeling system, proceedings of the 1988. *IEEE International Conference on Systems, Man, and Cybernetics* 2: 1326–29.

Ong, C.S., J.J. Huang, and G.H. Tzeng. (2005). Motivation and resource allocation for strategic alliance through de novo perspective. *Mathematical and Computer Modeling* 41(6–7): 711–21.

Opricovic, S. (1998). *Multicriteria optimization of civil engineering systems*. Faculty of Pennsylvania, Belgrade.

Opricovic, S., and G.H. Tzeng. (2002). Multicriteria planning of post-earthquake sustainable reconstruction. *Computer-Aided Civil and Infrastructure Engineering* 17 (3): 211–20.

Opricovic, S., and G.H. Tzeng. (2003a). Comparing DEA and MCDM method. In *multi-objective programming and goal-programming: Theory and applications*, eds. T. Tanino, T. Tanaka, and M. Inuiguchi, 227–32. Berlin: Springer.

Opricovic, S., and G.H. Tzeng. (2003b). Defuzzification within a fuzzy multicriteria decision model. *International Journal of Uncertainty, Fuzziness and Knowledge-based Systems* 11 (5): 635–52.

Opricovic, S., and G.H. Tzeng. (2003c). Fuzzy multicriteria model for post-earthquake land-use planning. *Natural Hazards Review* 4 (2): 59–64.

Opricovic, S., and G.H. Tzeng. (2003d). Multicriteria expansion of a competence set using genetic algorithm. In *Multi-objective programming and goal-programming: Theory and applications*, eds. T. Tanino, T. Tanaka, and M. Inuiguchi, 221–26. Berlin: Springer.

Opricovic, S., and G.H. Tzeng. (2004). Compromise solution by MCDM methods: A comparative analysis of VIKOR and TOPSIS. *European Journal of Operational Research* 156 (2): 445–55.

Opricovic, S., and G.H. Tzeng. (2007). Extended VIKOR method in comparison with outranking methods. *European Journal of Operational Research* 178 (2): 514–29.

Ou Yang, Y.P., J.D. Leu, and G.H. Tzeng. (2009). A VIKOR-based multiple criteria decision method for improving information security risk. *International Journal or Information Technology and Decision Making* 8 (2): 267–87.

Ou Yang, Y.P., H.M. Shieh, J.D. Leu, and G.H. Tzeng. (2008). A novel hybrid MCDM model combined with DEMATEL and ANP with applications. *International Journal of Operations Research* 5 (3): 1–9.

Ou Yang, Y.P., H.M. Shieh, G.H. Tzeng, L. Yen, and C.C. Chan. (2008). Business aviation decision-making using rough sets. RSCTC 2008, Lecture Notes on Artificial Intelligence (LNAI) series, Springer-Verlag.

Pagageorgiou, E.I., and P.P. Groumpos. (2005). A new hybrid method using evolutionary algorithms to train fuzzy cognitive maps. *Applied Soft Computing* 5 (4): 409–31.

Pawlak, Z. (1982). Rough sets. *International Journal of Computer and Information Science* 11 (5): 341–56.

Pawlak, Z. (1984). Rough classification. *International Journal of Man-Machine Studies* 20 (5): 469–83.

Pawlak, Z. (1997). Vagueness: A rough set view. In *Structures in logic and computer science,* eds. J. Mycielski and A. Salomaa, 106–17. New York: Springer-Verlag.

Pawlak, Z. (2002a). Rough set theory and its applications. *Journal of Telecommunications and Information Technology* 3 (1): 7–10.

Pawlak, Z. (2002b). Rough sets, decision algorithms and Bayes' theorem. *European Journal of Operational Research* 136 (1): 181–89.

Pawlak, Z. (2004). Decision networks. In *Rough sets and current trends in computing*, eds. S. Tsumoto, R. Slowinski, J. Komoroski, and J.W. Grzymala-Busse. Lecture Notes in Artificial Intelligence (LNAI), 3066 (1): 1–7.

Pawlak, Z., and R. Słowinski. (1994). Decision analysis using rough sets. *International Transactions in Operational Research* 1 (1): 107–14.

Pawlak, Z., and R. Słowinski. (1994). Rough set approach to multi-attribute decision analysis. *European Journal of Operational Research* 72:443–59.

Poh, K.L., and B.W. Ang. (1999). Transportation fuels and policy for Singapore: An AHP planning approach. *Computers and Industrial Engineering* 37 (4): 507–25.

Polkowski, L. (2004). Toward rough set foundations, mereological approach. In *Rough sets and current trends in computing*, eds. S. Tsumoto, R. Slowinski, J. Komoroski, and J. W. Grzymala-Busse. Lecture Notes in Artificial Intelligence (LNAI), 3066 (1): 8–25.

Quafafou, M. (2000). α-RST: A generalization of rough set theory. *Information Sciences* 124 (4): 301–16.

Quattrone, G.A., and A. Tversky. (1988). Contrasting rational and psychological analyses of political choice. *American Political Science Review* 82 (3): 719–36.

Ralescu, D.A. and G. Adams, (1980). Fuzzy integral. *Journal of Mathematical Analysis and Applications* 75 (2): 562–70.

Rechenberg, I. (1973). *Evolutionsstrategie: Optimierung Technischer Systeme und Prinzipien der Biologischen Evolution*. Stuttgart: Frommann-Holzboog.

Reynolds, D., and P. Cuttance. (1992). *School effectiveness: Research, policy and practice*. London: Villiers House.

Rong-Jun Li, R.J., and E.S. Lee. (1990) Fuzzy approaches to multicriteria de Novo programs. *Journal of Mathematical Analysis and Applications* 153 (1): 97–111.

Roy, B. (1968). Classement et Choix en Presence de Points de vue Multiples (la methode Electre). *Revue Francaise d'Informatique et de Recherche Operationnelle* 8 (1): 57–75.

Roy, B. (1971). Problems and methods with multiple objective functions. *Mathematical Programming* 1 (1): 239–66.

Roy, B. (1976). From optimization to multicriteria decision aid: Three main operational attitudes. In *Multiple Criteria Decision Making,* volume 130 of *Lecture Notes in Economics and Mathematical Systems,* eds. H. Thiriez and S. Zionts, 1–34. Springer-Verlag, Berlin.

Roy, B. (1977). Partial preference analysis and decision-aid: The fuzzy outranking relation concept. In *Conflicting Objectives in Decisions*, eds. D.E. Bell, R.L. Keeney, and M. Raiffa, 40–75, New York: Wiley.

Roy, B. (1978). ELECTRE III: Un algorithme de classements fondé sur une représentation floue des préférences en présence de critéres multiples. *Cahiers du Centre d'Etudes de Recherche Opérationnelle* 20: 3–24.

Roy, B. (1980). Selektieren, Sortieren und Ordnen mit Hilfe yon Pr/ivalenzrelationen: Neue Ans/itzeauf dem Gebiet der Entscheidungshilfe f'tir Multikriteria-Probleme. Deutsche Bearbeitung **und** Erweiterung yon Heinz-Michael Winkels und Arno Jaeger. *ZfbF* 32:465–97.

Roy, B. (1981). A multicriteria analysis for trichotomic segmentation problems. In *Multiple Criteria Analysis: Operational Methods,* eds. P. Nijkamp and J. Spronk, 245–57. Farnborough: Gower Press.

Roy, B. (1989). *The outranking approach and the foundations of ELECTRE methods.* University of Paris-Dauphine, Document Du Lamsade.

Roy, B. (1990). The outranking approach and the foundation of the ELECTRE method. In *Readings in multiple criteria decision aid*, ed. Carios A. Bana e Costa, 155–83. Berlin: Spring-Verlag.

Roy, B. (1991). The outranking approach and the foundations of ELECTRE methods. *Theory and Decision* 31 (1): 49–73.

Roy, B., and P. Bertier. (1971). La methode ELECTRE II: Une methode de classement en presence de critteres multiples. SEMA (Metra International), Direction Scientifique, Note de Travail No. 142, p. 25, Paris.

Roy, B., and P. Bertier. (1973). La methode ELECTRE II—Une application au media planning. In *OR'72*, ed. M. Ross, 291–302. Amsterdam: North-Holland Publishing Company.

Roy, B., and D. Bouyssou. (1983). Comparasion, sur un cas précis, de deux modéles concurrents d'aide á la décision. *Université de Paris-Dauphine, Document du LAMSADE* no. 22.

Roy, B., and D. Bouyssou. (1993). Aide multicritère à la décision: Méthodes et cas. *Economica*, Collection Gestion, Paris.

Roy, B., and P. Vincke. (1981). Multicriteria analysis: Survey and new directions. *European Journal of Operational Research* 8:207–18.

Roy, B., and P. Vincke. (1981). Multicriteria analysis: Survey and new directions. *European Journal of Operational Research* 8 (3): 207–18.

Roy, B., M. Present, and D. Silhol. (1986). A programming method for determining which Paris metro stations should be renovated. *European Journal of Operational Research* 24 (2): 318–34.

Saaty, T.L. (1972). *An eigenvalue allocation model for prioritization and planning*. Energy management and policy center, University of Pennsylvania.

Saaty, T.L. (1977). A scaling method for priorities in hierarchical structures. *Journal of Mathematical Psychology* 15 (3): 234–81.

Saaty, T.L. (1980). *The analytic hierarchy process.* New York: McGraw-Hill.

Saaty, T.L. (1992). A Natural way to make momenents decision. *Journal of Science and Industrial Research* 61: 561–71.

Saaty, T.L. (1996). Decision making with dependence and feedback, The analytic network process. RWS Publications, 4922 Ellsworth Ave., Pittsburgh, PA 15213, p.481.

Saaty, T.L. (1999). Fundamentals of the analytic network process. *International Symposium on the Analytic Hierarchy Process*, Kobe.

Saaty, T.L. (2003). Decision-making with the AHP: Why is the principal eigenvector necessary. *European Journal of Operational Research* 145 (1): 85–91.

Saaty, T.L., and K.P. Keams. (1985). *Analytical planning.* New York, NY: Pergamon Press.

Saaty, T.L., and L.G. Vargas. (1982). *The logic of priorities.* Boston, MA: Kluwer.

Saaty, T.L., and L.G. Vargas. (1998). Diagnosis with dependent symptoms: Bayes theorem and the analytic hierarchy process. *Operations Research* 46 (4): 491–502.

Sakawa, M. (1983). Interactive fuzzy decision making for multiobjective linear programming and its application. In *Proceedings of IFAC Symposium on Fuzzy Information, Knowledge Representation and Decision Analysis*, 295–300. New York: Pergamon Press.

Sakawa, M. (1984a). Interactive fuzzy decision making for multiobjective nonlinear programming problems. In *Interactive decision analysis*, eds. M. Grauer and A.P. Wierzbicki, 105–12. Berlin: Springer-Verlag.

Sakawa, M. (1984b). Interactive fuzzy goal programming for multiobjective nonlinear programming problems and its application to water quality management. *Control and Cybernetics* 13 (3): 217–28.

Salomon, V.A.P., and J.A.B. Montevechi. (2001). A compilation of comparisons on the analytic hierarchy process and others multiple criteria decision making methods: Some cases developed in Brazil. ISAHP, Berne, Switzerland.

Sarkis, J. (2003). A strategic decision framework for green supply chain management. *Journal of Cleaner Production* 11(4): 397–409.

Sekimori, T., K. Hayashi, and Y. Kawashima. (1996). Development of Toyota "coaster hybrid EV". *Proceedings of 13th Electric Vehicle Conference.*

Sekitani, K., and I. Takahashi. (2001). A unified model and analysis for AHP and ANP. *Journal of the Operations Research Society of Japan* 44 (1): 67–89.

Sharma, H.D., A.D. Gupta, and Sushil. (1995). The objectives of waste management in India: A futures inquiry. *Technological Forecasting and Social Change* 48 (3): 285–309.

Shee, D.Y., T. Tang, and G.H. Tzeng. (2000). Modeling the supply-demand interaction in electronic commerce: A bi-level programming approach. *Journal of Electronic Commerce Research* 1 (2): 79–93.

Shee, D.Y., and G.H. Tzeng. (2002). The key dimensions of criteria for the evaluation of ISPs: An exploratory study. *The Journal of Computer Information System* 42 (4): 112–21.

Shee, D.Y., G.H. Tzeng, and T. Tang. (2003). AHP, fuzzy measure and fuzzy integral approaches for the appraisal of information service providers in Taiwan. *Journal of Global Information Technology Management* 6 (1): 8–30.

Shi, K. (1990). *Grey relation theory and its applications.* ISUMA 90, USA.

Shi, Y., and Y. H. Liu. (1993). Fuzzy potential solutions in multi-criteria and multi-constraint level linear programming problems. *Fuzzy Sets and Systems* 60 (2): 163–79.

Shing, T.P., G.H. Tzeng, and Y.L. Yeh. (2001). Building an accessing efficiency model of criminal investigation and crime prevention based on the index of subjectivity and objectivity: An application of data envelopment analysis (DEA). *Police Science Quarterly* 31 (6): 175–98. (Chinese, Taiwan).

Shu, J.Z., G.H. Tzeng, and R.S. Chen. (2004). ANP MCDM approach for site selection of Biotechnology Park. *National Policy Quarterly* 3(4). 185–200 (Chinese, Taiwan).

Shuai, J.J., G.H. Tzeng, and H.L. Li. (2004). The Multi-source fabless-foundry partnership selection model. *International Journal of Manufacturing Technology and Management* 6 (1/2): 137–154.

Shyng, J.Y., G.H. Tzeng, and F.K. Wang, (2007). Rough set theory in analyzing the attributes of combination values for insurance market. *Expert Systems with Applications* 32 (1): 56–64.

Simon, H.A. (1977). *The new science of management decision.* New Jersey: Prentice Hall PTR.

Skowron, A., and J.W. Grzymala-Busse. (1993). From the rough set theory to the evidence theory. In *Advances in the Dempster–Shafer theory of evidence*, eds. M. Fedrizzi, J. Kacprzyk, and R.R. Yager, 295–305. New York: Wiley.

Spencer, L.M., and S.M. Spencer. (1993). *Competence at work: Model for superior performance.* New York: John Wiley & Sons Inc.

Sperling, D. (1995). *Future drive—electric vehicles and sustainable transportation.* Washington D.C., Island Press.

Sperling, D., W. Setiawan, and D. Hungerford. (1995). The target market for methanol fuel. *Transportation Research (Part A)* 29 (1): 33–45.

Steenbrink, P.A. (1974). *Optimization of transportation network.* New York: Wiley.

Stylios, C.D., and P.P. Groumpos. (2004). Modeling complex systems using fuzzy cognitive maps. *IEEE Transactions on Systems, Man, and Cybernetics—Part A: Systems and Humans* 34 (1): 155–62.

Sugeno, M. (1974). *Theory of fuzzy integrals and its applications*, PhD Dissertation. Tokyo Institute of Technology, Tokyo, Japan.

Sugeno, M. (1977). Fuzzy measures and fuzzy integrals: A survey. In *Fuzzy automata and decision processes*, eds. M.M. Gupta, G.N. Saridis, and B.R. Gaines, 89–102. North-Holland, Amsterdam and New York.

Sugeno, M., K. Fujimoto, and T. Murofushi. (1995). A hierarchical decomposition of choquet integral model. *International Journal of Uncertainty, Fuzziness and Knowledge-Based Systems* 3 (1): 213–22.

Sugeno, M., and S.H. Kwon. (1995). A clusterwise regression-type model for subjective evaluation. *Journal of Japan Society for Fuzzy Theory and Systems* 7 (2): 291–310.

Sugeno, M., and T. Terano. (1977). A model of learning based on fuzzy information. *Kybernetes*, 16 (3): 157–66.

Suwansirikul, C., T.L. Friesz, and R.L. Tobin. (1987). Equilibrium decomposed optimization: A Heuristic for the continuous equilibrium network design problem. *Science* 21:254–63.

Swiniarski, R.W., and A. Skowron. (2003). Rough set methods in feature selection and recognition. *Pattern Recognition Letters* 24 (6): 833–49.

Tamura, M., H. Nagata, and K. Akazawa. (2002). Extraction and systems analysis of factors that prevent safety and security by structural models. In *41st SICE annual conference*, Osaka, Japan.

Tanaka, K., and M. Sugeno. (1991). A study on subjective evaluation of color printing images. *International Journal of Approximate Reasoning* 5 (3): 213–22.

Tang, T.I., D.Y. Shee, and G.H. Tzeng. (2002). An MCDM framework for assessing ISPs: The fuzzy synthesis decisions of additive and non-additive measurements. *Journal of Information Management* 8 (2): 175–92.

Tang, T.I., and G.H. Tzeng. (1998). Fuzzy MCDM model for pricing strategies in the internet environment. *Journal of Commercial Modernization* 1 (1): 19–34.

Tang, M.T., G.H. Tzeng, and S.W. Wang. (1999). A hierarchy fuzzy MCDM method for studying electronic marketing strategies in the information service industry. *Journal of International Information Management* 8 (1): 1–22.

Tay, E.H., and L. Shen. (2002). A modified chi2 algorithm for discretization. *IEEE Transactions on Knowledge and Data Engineering* 14 (3): 666–70.

Teng, J.Y., and G.H. Tzeng. (1994). Multicriteria evaluation for strategies of improving and controlling air quality in the super city: A case study of Taipie city. *Journal of Environmental Management* 40 (3): 213–29.

Teng, J.Y., and G.H. Tzeng. (1996a). A multiobjective programming approach for selecting non-independent transportation investment alternatives. *Transportation Research (Part B)* 30B (4): 291–307.

Teng, J.Y., and G.H. Tzeng. (1996b). Fuzzy multicriteria ranking of urban transportation investment alternatives. *Transportation Planning and Technology* 20 (1): 15–31.

Ting, S.C., and G.H. Tzeng. (2003a). Ship scheduling and cost analysis for route planning in liner shipping. *Maritime Economics & Logistics* 5 (4): 378–92.

Ting, S.C., and G.H. Tzeng. (2003b). Ship scheduling and service network integration for liner shipping companies and strategic alliances. *Journal of Eastern Asia Society for Transportation Studies* 5:765–77.

Ting, S.C., and G.H. Tzeng. (2004). An optimal containership slot allocation for liner shipping revenue management. *Maritime Policy and Management* 31 (3): 199–211.

Tsai, H.C., C.M. Chen, and G.H. Tzeng. (2006). The comparative productivity efficiency for global telecoms. *International Journal of Production Economics* 103 (2): 509–26.

Tsaur, S.H., and G.H. Tzeng. (1996). Multiattribute decision making analysis for customer preference of tourist hotels. *Journal of Travel and Tourism Marketing* 4 (4): 55–69.

Tsaur, S.H., G.H. Tzeng, and T.Y. Chang. (1997). Travel agency organization buying behavior- an application of logit model. *Journal of Management and Systems* 4 (2): 127–46.

Tsaur, S.H., G.H. Tzeng, and G.C. Wang. (1997). Evaluating tourist risks from fuzzy perspectives. *Annals of Tourism Research* 24 (4): 796–812.

Tsaut, S.H., G.H. Tzeng, and C.I. Chiang. (1996). The comparison of four kinds of prediction methods: Traditional econometic regression, fuzzy linear regression, GMDH, and artificial neural network-an example of the demand forecast of the Taiwanese tourists to Hong Kong. *Journal of Chinese Statistical Association* 34 (2): 132–61.

Tseng, Y.H., P. Durbin, and G.H. Tzeng. (2001). Using a fuzzy piecewise regression to predict the nonlinear time-series of turbulent flows with automatic change-point detection. *Flow, Turbulence and Combustion (An International Journal)* 67 (2): 81–106.

Tseng, F.M., and G.H. Tzeng. (1999a). Forecast seasonal time series by comparing five kinds of hybrid grey models. *Journal of The Chinese Fuzzy Systems Association* 5 (2): 45–55.

Tseng, F.M., and G.H. Tzeng. (1999b). The comparison of four kinds of prediction methods: ARIMA, fuzzy time series, fuzzy regression time series and grey forecasting: An example of the production value forecast of mechanical industry in Taiwan. *Journal of The Chinese Grey System Association* 2 (2): 83–98 (Chinese, Taiwan).

Tseng, F.M., and G.H. Tzeng. (2002). A fuzzy seasonal ARIMA model for forecasting. *Fuzzy Sets and Systems* 126 (3): 367–76.

Tseng, F.M., G.H. Tzeng, and H.C. Yu. (1999). Fuzzy seasonal time series for forecasting the production value of mechanical industry in Taiwan. *Technological Forecasting and Social Change (An International Journal)* 60 (3): 263–73.

Tseng, F.M., G.H. Tzeng, and H.C. Yu. (2000). Combining neural network with seasonal time series ARIMA model to forecast the total production value of Taiwan mechanical industry. *Technological Forecasting and Social Change* 65:1–17.

Tseng, F.M., G.H. Tzeng, B.J.C. Yuan, and H.C.D. Yu. (2001). Fuzzy ARIMA model for forecasting the foreign exchange market. *Fuzzy sets and Systems* 118 (1): 9–19.

Tseng, F.M., H.C. Yu, and G.H. Tzeng. (2001). Applied hybrid grey model to forecast seasonal time series. *Technological Forecasting and Social Change (An International Journal)* 67 (2): 291–302.

Tseng, F.M., H.C. Yu, and G.H. Tzeng. (2002). Combing neural network with seasonal time series ARIMA model. *Technological Forecasting and Social Change* 69 (1): 71–87.

Tzeng, G.H. (2003). Multiple objective decision making in past, present, and future, multiobjective programming and goal-programming: Theory and applications, eds. T. Tanino, T. Tanaka, and M. Inuiguchi, 65–76. Springer.

Tzeng, G.H., J.R. Chang, J.D. Lin, and C.T. Hung. (2002). Non-additive grey relation model for the evaluation of flexible pavement condition. *International Journal of Fuzzy Systems* 4 (2): 715–24.

Tzeng, G.H., C.Y. Chang, and M.C. Lo. (2005a). MADM approach for effecting information quality of knowledge management. *International Journal of Information Systems for Logistics and Management* 1 (1): 55–67.

Tzeng, G.H., C.Y. Chang, and M.C. Lo. (2005b). The simulation and forecast model for human resources of semiconductor wafer fab operation. *International Journal of Industrial Engineering and Management Systems* 4 (1): 47–53.

Tzeng, G.H., S.L. Chang, J.C. Wang, M.J. Hwang, G.C. Yu, and M.C. Juang. (1998). Application of fuzzy multiobjective programming to economic-energy-environment model. *Journal of Management* 15 (4): 683–707 (Chinese, Taiwan).

Tzeng, G.H., and C.H. Chen. (1993). Multiobjective decision making for traffic assignment. *IEEE Transaction on Engineering Management* 40 (2): 180–87.

Tzeng, G.H., and J.J. Chen. (1997). Developing Taipei automobile driving cycles for emissions. *Energy & Environment* 8 (3): 227–38.

Tzeng, G.H., and J.J. Chen. (1998). Developing Taipei motorcycle driving cycle for emissions and fuel economy. *Transportation Research Part D (An International Journal)* 3D (1): 19–27.

Tzeng, G.H., and Y.W. Chen. (1998). Implementating an effective schedule for reconstructing post-earthquake road-network based on asymetric traffic assignment: An application of genetic algorithm. *International Journals of Operations and Quantitative Management* 4 (3): 229–46.

Tzeng, G.H., and Y.W. Chen. (1999). The optimal location of airport fire stations: A fuzzy multi-objective programming through revised genetic algorithm. *Transportation Planning and Technology* 23 (1): 37–55.

Tzeng, G.H., Y.W. Chen, and C.Y. Lin. (2000). Fuzzy multi-objective reconstruction plan for post-earthquake road-network by genetic algorithm. *Research and Practice in Multiple Criteria Decision Making* 487: 238–47.

Tzeng, G.H., J.J. Chen, and J.D. Teng. (1997). Evaluation and selection of suitable battery for electrics motorcycle in Taiwan-application of fuzzy multiple attribute decision making. *Journal of the Chinese Institute of Industrial Engineers* 14 (3): 319–31.

Tzeng, G.H., T.Y. Chen, and J.C. Wang. (1998). A weight-assessing method with habitual domains. *European Journal of Operational Research* 110 (2): 342–67.

Tzeng, G.H., J.J. Chen, and Y.K. Yen. (1996). The strategic model of multicriteria decision making for managing the quality of the environment in metropolitan Taipei. *Asian Journal of Environmental Management* 4 (1): 41–52.

Tzeng, G.H., H.J. Cheng, and T.D. Huang. (2007). Multi-objective optimal planning for designing relief delivery systems. *Transportation Research Part E: Logistics and Transportation Review* 43 (6): 673–86.

Tzeng, G.H., and C.I. Chiang. (1998). Applying possibility regression to grey model. *Journal of the Chinese Grey System Association* 1 (1): 19–31.

Tzeng, G.H., C.I. Chiang, and M.J. Hwang. (1996). Multiobjective programming approach to the allocation of air pollution monitoring station. *Journal of the Chinese Institute of Environment Engineering* 6 (1): 99–105.

Tzeng, G.H., C.H. Chiang, and C.W. Li. (2007). Evaluating intertwined effects in E-learning programs: A novel hybrid MCDM model based on factor analysis and DEMATEL. *Expert Systems with Applications* 32 (4): 1028–44.

Tzeng, G.H., Y.C. Chiou, and S.K. Sheu. (1997). A comparison of genetic and stepwise algorithms for optimal sites of mainline barrier-type toll station. *Journal of the Chinese Institute of Civil and Hydraulic Engineering* 9 (1): 171–78.

Tzeng, G.H., Y.C. Chiu, and J.J. Chen. (1997). Integrated model of fuzzy stated preference. *Journal of The Chinese Fuzzy Systems Association* 2 (3): 39–51.

Tzeng, G.H., H.J. Cho, Y.J. Jou, and C.I. Chiang. (1997). Estimation of dynamic O-D flows from dynamic traffic counts. *Transportation Planning Journal* 26 (4): 615–38 (Chinese, Taiwan).

Tzeng, G.H., G.M. Feng, and C.C. Kang. (2001). The fuzzy set theory and DEA model for forecasting production efficiency: Case study for Taipei city bus company. *Journal of Advance Computational Intelligences* 5 (3): 128–38.

Tzeng, G.H., and Y.C. Hu. (1996). The section of bus system operation and service performance indicators-application of grey relation analysis. *Journal of The Chinese Fuzzy Systems Association* 2 (1): 73–82.

Tzeng, G.H., and W.C. Huang. (1997). The spatial and temporal bi-criteria parallel savings based heuristic algorithm for vehicle routing problem with time windows. *Transportation Planning and Technology* 20 (2): 163–81.

Tzeng, G.H., M.J. Hwang, and Y.H. Liu. (1997). Dynamic optimal expansion for fuzzy competence sets of multistage-multi-objective planning. *Pan Pacific Management Review* 1 (1): 55–70.

Tzeng, G.H., M.J. Hwang, and W.C. Yeh. (1998). Multiobjective planning for integrated land-use and outsideallied transportation systems in recreation area. *Journal of Management* 15 (1): 133–161 (Chinese, Taiwan).

Tzeng, G.H., W. Jen, and K.C. Hu. (2002). Fuzzy factor analysis for selecting service quality factors: A case of the service quality of city bus service. *International Journal of Fuzzy Systems* 4 (4): 911–21.

Tzeng, G.H., and J.S. Kuo. (1996). Fuzzy multiobjective-double sampling plans with genetic algorithms based on Bayesian model. *Journal of The Chinese Fuzzy Systems Association* 2 (2): 57–74.

Tzeng, G.H., and M.Y. Lee. (2001). Intellectual capital in the information industry. In *Made by Taiwan: Booming in the information technology era*, eds. C.-Y. Chang and P.-L. Yu, 298–344. Singapore: World Science Publishing Company.

Tzeng, G.H., and J.J. Lin. (1997). Application of TOPSIS multiobjective programming approach to fire station location in Tanhai New Town. *Journal of City and Planning* 24 (1): 81–98 (in Chinese, Taiwan).

Tzeng, G.H., and C.W. Lin. (2000). Evaluation on new substitute fuel mode of bus vehicle for suitable urban public transportation. *Transportation Planning Journal* 29 (3): 665–92 (in Chinese, Taiwan).

Tzeng, G.H., C.W. Lin, and S. Opricovic. (2005). Multi-criteria analysis of alternative-fuel buses for public transportation. *Energy Policy* 33 (11): 1373–83.

Tzeng, G.H., Y.P. Ou Yang, C.T. Lin, and C.B. Chen. (2005). Hierarchical MADM with fuzzy integral for evaluating enterprise intranet web sites. *Information Sciences* 169 (3–4), 409–26.

Tzeng, G.H., J.Y. Shiah, and C.I. Chiang. (1998). Application of de novo programming to land-use plans of Theory of Chiao Tung University. *Journal of City and Planning* 25 (1): 93–105.

Tzeng, G.H., T.A. Shiau, and C.Y. Lin. (1992). Application of multicriteria decision making to the evaluation of a new energy system development in Taiwan. *Energy* 17 (10): 983–92.

Tzeng, G.H., T.A. Shiau, and J.Y. Teng. (1994). Multiobjective decision making approach to energy supply mix decisions in Taiwan. *Energy Sources* 16 (3): 301–16.

Tzeng, G.H., H.M. Shieh, and T.A. Shiau. (1989). Route choice behavior in transportation: An application of the Multiattribute utility theorem. *Transportation Planning and Technology* 13:289–301.

Tzeng, G.H., T.I. Tang, Y.M. Hung, and M.L. Chang. (2006). Multiple objective planning for a production and distribution model of the supply chain: Case of a bicycle manufacturer. *Journal of Scientific & Industrial Research* 65 (4): 309–20.

Tzeng, G.H., D. Teodorovic, and M.J. Hwang. (1996). Fuzzy bicriteria multi-index transportation problems for coal allocation planning of Taipower. *European Journal of Operational Research* 95 (1): 62–72.

Tzeng, G.H., and J.Y. Teng. (1998). Transportation investment project selection using fuzzy multiobjective programming. *Fuzzy Sets and Systems* 96 (3): 259–80.

Tzeng, G.H., M.H. Teng, J.J. Chen, and S. Opricovic. (2002). Multicriteria selection for a restaurant location in Taipei. *International Journal of Hospitality Management* 21 (2): 171–87.

Tzeng, G.H., and S.H. Tsaur. (1993). Application of multicriteria decision making to old vehicle elimination in Taiwan. *Energy and Environment* 4 (2): 268–83.

Tzeng, G.H., and S.H. Tsaur. (1994). The multiple criteria evaluation of grey relation model. *The Journal of Grey System* 6 (2): 87–108.

Tzeng, G.H., and S.H. Tsaur. (1997). Application of multiple criteria decision making for network improvement. *Journal of Advanced Transportation* 31 (1): 49–74.

Tzeng, G.H., S.H. Tsaur, Y.D. Laiw, and O. Serafim. (2002). Multicriteria analysis of environmental quality in Taipei: Public preferences and improvement strategies. *Journal of Environmental Management* 65 (2): 109–20.

Tzeng, G.H., J.C. Wang, and M.J. Hwang. (1996). Using genetic algorithms and the template path concept to solve the traveling salesman problem. *Transportation Planning Journal* 25 (3): 493–516.

Tzeng, G.H., H.F. Wang, U.P. Wen, and P.L. Yu., eds. (1994). *Multiple criteria decision making*. Amsterdam: Springer-Verlag.

van Laarhoven, P.J., and W. Pedrycz. (1983). A fuzzy extension of Saaty's priority method. *Fuzzy Sets and Systems* 11:229–41.

von Altrock, C. (1996). *Fuzzy logic and neurofuzzy applications in business and finance*. Upper Saddle River, NJ: Prentice Hall.

von Neumann, J., and O. Morgenstern. (1947). *Theory of games and economic behavior*, 2nd ed. Princeton, New Jessie: Princeton University Press.

Waczak, B., and D.L. Massart. (1999). Tutorial rough sets theory. *Chemometrics and Intelligent Laboratory Systems* 47 (1): 1–16.

Wagenknecht, M., and K. Hartmann. (1983). On fuzzy rank-ordering in polyoptimization. *Fuzzy Sets and Systems* 11:243–51.

Wakabayashi, T., L. Itoh, and A. Ohuchi. (1995). A method for constructing system models by fuzzy flexible interpretive structural modeling fuzzy systems. In *International Joint Conference of the 4th IEEE International Conference on Fuzzy Systems and The Second International Fuzzy Engineering Symposium, Proceedings of 1995 IEEE International Conference* on, 2, 913–18.

Walczak, B., and D.L. Massart. (1999). Tutorial rough sets theory. *Chemometrics and Intelligent Laboratory Systems* 47 (1): 1–16.

Wang, J.C., T.Y. Chen, and H.M. Shen. (2001). Using fuzzy densities to determine the λ-value for λ-fuzzy measures. *Proceedings of the Ninth National Conference on Fuzzy Theory and its Applications* (Taiwan), 54–9.

Wang, Z., and G.J. Klir. (1992). *Fuzzy measure theory*. New York: Plenum Press.

Wang, B., J.Z. Shyu, and G.H. Tzeng. (2007). Appraisal model for admitting new tenants to the incubation center at ITRT. *International Journal of Innovative Computing Information and Control* 3 (1): 119–30.

Warfield, J.N. (1973). On arranging elements of a hierarchy in graphic form. *IEEE Transactions on Systems, Man, and Cybernetics* SMC-3 (2): 121–32.

Warfield, J.N. (1974a). Developing interconnection matrices in structural modeling. *IEEE Transactions on Systems, Man, and Cybernetics* 4 (1): 81–7.

Warfield, J.N. (1974b). Developing subsystem matrices in structural modeling. *IEEE Transactions on Systems, Man, and Cybernetics* SMC-4 (1): 74–80.

Warfield, J.N. (1974c). Toward interpretation of complex structural models. *IEEE Transactions on Systems, Man, and Cybernetics* 4 (5): 405–17.

Warfield, J.N. (1976). *Societal systems, planning, policy and complexity*. New York: John Wiley & Sons.

Warfield, J.N. (1977). Crossing theory and hierarchy mapping. *IEEE Transactions on Systems, Man, and Cybernetics* SMC-7 (7): 502–23.

Werners, B. (1987). Interactive multiple objective programming subject to flexible constraints. *European Journal of Operational Research* 31 (2): 342–49.

White, P.A. (1989). An overview of school-based management: What does the research say? *NASSP Bulletin* 73 (518): 1–8.

Winebrake, J.J., and B.P. Creswick. (2003). The future of hydrogen fueling systems for transportation: An application of perspective-based scenario analysis using the analytic hierarchy process. *Technological Forecasting and Social Change* 70 (2): 359–84.

Winkels, H.-M. (1983). tdber einige Weiterentwicklungen der ELECTRE-Verfahren. In Operations Research Proceedings, Papers of the 1 lth Annual Meeting of DGOR 1982, eds. W. Biihler et al., 401–11. Springer, Berlin.

Witlox, F., and H. Tindemans. (2004). The application of rough sets analysis in activity-based modeling, Opportunities and Constraints. *Expert Systems with Application* 27 (2): 171–80.

Wu, W.W. (2007). Choosing knowledge management strategies by using a combined ANP and DEMATEL approach. *Expert Systems with Applications* (Forthcoming).

Wu, D.S., D.S. Chen, and C.G. Zhao. (1990). *Plane design of architecture*. Taipei: Scientific & Technical Publishing House (Chinese).

Wu, I.C., D.R. Liu, and G.H. Tzeng. (2002). Application of AHP method and grey relation model to the evaluation of portal sites performance. *Journal of the Chinese Grey System Association* 5 (1): 41–54 (Chinese).

Wu, W.W., Y.T. Lee, and G.H. Tzeng. (2005). Simplifying the manager competency model by using the rough set approach. *Rough Sets, Fuzzy Sets, Data Mining, and Guanular Computing* 3642 (Part II): 484–95.

Wu, K.H., and G.H. Tzeng. (2002). The application of possibility grey forecasting to stock-market price in Taiwan. *Journal of the Chinese Grey System Association* 5 (1): 7–16.

Wu, C.H., and G.H. Tzeng. (2008). A novel hybrid genetic algorithm for kernel function and parameter optimization in support vector regression. *Expert Systems with Applications* 42 (2).

Wu, C.H., G.H. Tzeng, Y.J. Goo, and W.C. Fang. (2007). A real-valued genetic algorithm to optimize the parameters of support vector machine for predicting bankruptcy. *Expert Systems with Applications* 32 (2): 397–408.

Yang, C.A., G.L. Fu, and G.H. Tzeng. (2005). Creating a win-win in the telecommunications industry: The relationship between MVNOs and MNOs in Taiwan. *Canadian Journal of Administration Sciences* 22 (4): 316–28.

Yang, C., G.L. Fu, and G.H. Tzeng. (2007). A multicriteria analysis of the strategies to open Taiwan's mobile virtual network operators services. *International Journal of Information Technology and Decision Making*. 6 (1): 85–112.

Yang, J.L., H.N. Chiu, and G.H. Tzeng. (2008). Vendor selection by integrated fuzzy MCDM techniques with independent and interdependent relationships. *Information Sciences* 178 (21): 4166–83.

Yang, Y.H., Z.Y. Lee, C.Y. Hung, and G.H. Tzeng. (2003). Fuzzy MCDM approach for analyzing the best mobilization timing of Taiwan. *Journal of Defense Management* 24 (1): 27–44 (Chinese, Taiwan).

Yen, K.C., J.J. Lin, and G.H. Tzeng. (2005). Applying eco-planning and fuzzy AHP on the evaluation of the development potentiality of land use in the waterfront environment. *Journal of Architecture and Planning* 6 (1): 21–42 (Taiwan).

Yen, K.C., G.H. Tzeng, and J.J. Lin. (2005). Performance standards control rules for sustainable management of coastal wetland by using grey relation. *Journal of City and Planning* 32 (4): 421–41.

Yu, P.L. (1973). A class of solutions for group decision problems. *Management Science* 19 (8): 936–46.

Yu, P.L. (1990). *Forming winning strategies: An integrated theory of habitual domains*. Heidelberg, Germany: Springer.

Yu, R.C., and G.H. Tzeng. (2006). A soft computing method for multi-criteria decision making with dependence and feedback. *Applied Mathematics and Computation* 180 (1): 63–75.

Yu, J.R., and G.H. Tzeng. (2008). A fuzzy multiple objective programming in interval piecewise regression model. *International Journal Uncertainty, Fuzziness and Knowledge-Based Systems* (Forthcoming).

Yu, J.R., G.H., Tzeng, C.I. Chiang, and H.J. Sheu. (2007). Raw material supplier ratings in the semiconductor manufacturing industry through fuzzy multiple objectives programming to DEA. *International Journal of Operations and Quantitative Management* 13 (4): 101–11.

Yu, J.R., G.H. Tzeng, and H.L. Li. (1999). A general piecewise necessity regression analysis based on linear programming. *Fuzzy Sets and Systems* 105 (3): 429–36.

Yu, J.R., G.H. Tzeng, and H.L. Li. (2001). General fuzzy piecewise regression analysis with automatic change-point detection. *Fuzzy Sets and Systems* 119 (2): 247–57.

Yu, J.R., G.H. Tzeng, and H.L. Li. (2005). Interval piecewise regression model with automatic change-point detection by quadratic programming. *International Journal of Uncertainty, Fuzziness and Knowledge-Based Systems* 13 (3): 347–61.

Yu, J.R., Y.C. Tzeng, G.H. Tzeng, T.Y. Yu, and H.J. Sheu. (2004). A fuzzy multiple objective programming approach to DEA with imprecise data. *International Journal of Uncertainty, Fuzziness and Knowledge-Based Systems* 12 (5): 591–600.

Yu, P. L., and L. Seiford. (1981). *Multistage decision problems with multicriteria* edited by P. Nijkamp and J. Spronk. London, Britain: Gower Press.

Yuan, B., G.H. Tzeng, T.H. Kang, and C.C. Cheng. (2005). MADM approach for selecting the consignment performance with cooperating the information service computer room of De-Lin Institute of Technology. *Journal of Information Management* 12 (1): 131–48.

Yuan, B.J.C., C.P. Wang, and G.H. Tzeng. (2005). An emerging approach for strategy evaluation in fuel cell development. *International Journal of Technology Management* 32 (3–4): 302–38.

Zadeh, L. (1965). Fuzzy sets. *Information and Control* 8 (3): 338–53.

Zadeh, L.A. (1975a). The concept of a linguistic variable and its application to approximate reasoning-I. *Information Science* 8 (3): 199–249.

Zadeh, L.A. (1975b). The concept of a linguistic variable and its application to approximate reasoning-II. *Information Science* 8 (4): 301–57.

Zadeh, L.A. (1978). Fuzzy sets as a basis for a theory of possibility. *Fuzzy Sets Systems* 1 (1): 3–28.

Zanakis, S.H., A. Solomon, N. Wishart, and S. Dublish. (1998). Multi-attribute decision making: A simulation comparison of select methods. *European Journal of Operational Research* 107 (3): 507–29.

Zeleny, M. (1972). Linear Multiobjective Programming. The University of Rochester, Graduate School of Management, PhD Thesis (Advisor P. L. Yu).

Zeleny, M. (1982). *Multiple criteria decision making*. New York: McGraw-Hill.

Zhai, L.Y., L.P. Khoo, and S.C. Fok. (2002). Feature extraction using rough set theory and genetic algorithms an application for the simplification of product quality evaluation. *Computers & Industrial Engineering* 43 (4): 661–76.

Zhou, H., and D. Sperling, (2001). Traffic emission pollution sampling and analysis on urban streets with high-rising buildings. *Transportation Research (Part D)* 6 (4): 269–81.

Zimmermann, H.J. (1978). Fuzzy programming and linear programming with several objective functions. *Fuzzy Sets and Systems* 1 (1): 45–55.

Index

G

H

I

L

M

N

P

R

S

T

V

For Product Safety Concerns and Information please contact our
EU representative GPSR@taylorandfrancis.com Taylor & Francis
Verlag GmbH, Kaufingerstraße 24, 80331 München, Germany